Lecture Notes in Computer Science 14617

Founding Editors

Gerhard Goos
Juris Hartmanis

Editorial Board Members

Elisa Bertino, *Purdue University, West Lafayette, IN, USA*
Wen Gao, *Peking University, Beijing, China*
Bernhard Steffen, *TU Dortmund University, Dortmund, Germany*
Moti Yung, *Columbia University, New York, NY, USA*

The series Lecture Notes in Computer Science (LNCS), including its subseries Lecture Notes in Artificial Intelligence (LNAI) and Lecture Notes in Bioinformatics (LNBI), has established itself as a medium for the publication of new developments in computer science and information technology research, teaching, and education.

LNCS enjoys close cooperation with the computer science R & D community, the series counts many renowned academics among its volume editors and paper authors, and collaborates with prestigious societies. Its mission is to serve this international community by providing an invaluable service, mainly focused on the publication of conference and workshop proceedings and postproceedings. LNCS commenced publication in 1973.

Barbara König · Henning Urbat
Editors

Coalgebraic Methods in Computer Science

17th IFIP WG 1.3 International Workshop, CMCS 2024
Colocated with ETAPS 2024
Luxembourg City, Luxembourg, April 6–7, 2024
Proceedings

Springer

Editors
Barbara König
University of Duisburg-Essen
Duisburg, Germany

Henning Urbat
FAU Erlangen-Nürnberg
Erlangen, Germany

ISSN 0302-9743 ISSN 1611-3349 (electronic)
Lecture Notes in Computer Science
ISBN 978-3-031-66437-3 ISBN 978-3-031-66438-0 (eBook)
https://doi.org/10.1007/978-3-031-66438-0

This Springer imprint is published by the registered company Springer Nature Switzerland AG
The registered company address is: Gewerbestrasse 11, 6330 Cham, Switzerland

Preface

The 17th International Workshop on Coalgebraic Methods in Computer Science (CMCS) was held on April 6–7, 2024 in Luxembourg City, Luxembourg, as a satellite event of the Joint Conference on Theory and Practice of Software, ETAPS 2024.

The aim of the workshop is to bring together researchers with a common interest in the theory of coalgebras, their logics, and their applications. Coalgebras allow for a uniform treatment of a large variety of state-based dynamical systems, such as transition systems, automata (including weighted and probabilistic variants), Markov chains, and game-based systems. Over the last two decades, coalgebra has developed into a field of its own interest, presenting a deep mathematical foundation, a growing field of applications, and interactions with various other fields such as reactive and interactive system theory, object-oriented and concurrent programming, formal system specification, modal and description logics, artificial intelligence, dynamical systems, control systems, category theory, algebra, analysis, etc.

Previous workshops have been organised in Lisbon (1998), Amsterdam (1999), Berlin (2000), Genoa (2001), Grenoble (2002), Warsaw (2003), Barcelona (2004), Vienna (2006), Budapest (2008), Paphos (2010), London (2012), Grenoble (2014), Eindhoven (2016), Thessaloniki (2018), Dublin (2020, held online because of the COVID pandemic), and Munich (2022). Since 2004, CMCS has been a biannual workshop, alternating with the International Conference on Algebra and Coalgebra in Computer Science (CALCO), which, in odd-numbered years, has been formed by the union of CMCS with the International Workshop on Algebraic Development Techniques (WADT).

The CMCS 2024 program featured a keynote talk by Shin-ya Katsumata (Kyoto Sangyo University) and invited talks by Henning Basold (Leiden University) and Dana Fisman (Ben-Gurion University). In addition, the program included a special session on coalgebras in type theory and proof assistants featuring invited tutorials by Paige Randall North (Utrecht University) and Niccoló Veltri (Tallinn University of Technology). These contributions are not included in the proceedings.

This volume contains the revised regular contributions (10 accepted papers out of 15 submissions). In addition to submissions of full-length regular papers for inclusion in these post-proceedings, the workshop also solicited short 2-page submissions for presentation of work in progress or work published elsewhere that could be of interest to the CMCS community. These short contributions are available from the workshop web page.

The submissions and the reviewing process were handled using Easychair. The reviewing of both types of submissions was carried out single-blind. Each regular submission received three extensive reviews, and each short submission received two short reviews that focused on relevance.

Several people contributed to the success of CMCS 2024. We would like to thank Alexandra Silva as head of the Steering Committee of CMCS, the general chair of ETAPS

Peter Ryan, the event organizer Magali Martin and the workshop chairs Maxime Cordy and Renzo Gaston Degiovanni.

We wish to thank all the authors who submitted to CMCS 2024, and the Program Committee members and external reviewers for their thorough reviewing and help in improving the papers that were accepted for CMCS 2024.

May 2024

Barbara König
Henning Urbat

Organization

Program Committee Chairs

Barbara König	University of Duisburg, Germany
Henning Urbat	FAU Erlangen-Nürnberg, Germany

Steering Committee

Corina Cîrstea	University of Southampton, UK
Helle Hvid Hansen	University of Groningen, The Netherlands
Ichiro Hasuo	National Institute of Informatics, Japan
Bart Jacobs	Radboud University Nijmegen, The Netherlands
Stefan Milius	FAU Erlangen-Nürnberg, Germany
Daniela Petrişan	IRIF, Université Paris-Cité, France
Jurriaan Rot	Radboud University Nijmegen, The Netherlands
Lutz Schröder	FAU Erlangen-Nürnberg, Germany
Alexandra Silva (Chair)	Cornell University, USA
Fabio Zanasi	University College London, UK

Program Committee

Adriana Balan	University Politehnica of Bucharest, Romania
Harsh Beohar	University of Sheffield, UK
Marta Bilkova	Czech Academy of Sciences, Czech Republic
Fredrik Dahlqvist	Queen Mary University London, UK
Jérémy Dubut	National Institute of Advanced Industrial Science and Technology, Japan
Sebastian Enqvist	Lund University, Sweden
Richard Garner	Macquarie University, Australia
Ichiro Hasuo	National Institute of Informatics, Japan
Tobias Kappé	Open University of the Netherlands and University of Amsterdam, The Netherlands
Barbara König (Co-chair)	Universität Duisburg-Essen, Germany
Marina Lenisa	University of Udine, Italy
Larry Moss	Indiana University Bloomington, USA
Fredrik Nordvall Forsberg	University of Strathclyde, UK

Jurriaan Rot	Radboud University Nijmegen, The Netherlands
Matteo Sammartino	Royal Holloway University of London, UK
Pawel Sobocinsky	Tallinn University of Technology, Estonia
David Spivak	Topos Institute, USA
Sam Staton	University of Oxford, UK
Henning Urbat (Co-chair)	FAU Erlangen-Nürnberg, Germany
Tarmo Uustalu	Reykjavik University, Iceland, and Tallinn University of Technology, Estonia
Thorsten Wißmann	FAU Erlangen-Nürnberg, Germany

Additional Reviewers

Atkey, Robert
Basold, Henning
Cruchten, Mike
Hausmann, Daniel
Honsell, Furio
Kori, Mayuko
Marti, Johannes
Roman, Mario

Publicity Chair

Thorsten Wißmann	FAU Erlangen-Nürnberg, Germany

Sponsoring Institutions

IFIP WG 1.3

Contents

Coalgebraic CTL: Fixpoint Characterization and Polynomial-Time Model Checking

Ryota Kojima[1(✉)], Corina Cîrstea[2], Koko Muroya[1], and Ichiro Hasuo[3,4]

[1] RIMS, Kyoto University, Kyoto, Japan
{kojima,kmuroya}@kurims.kyoto-u.ac.jp
[2] University of Southampton, Southampton, UK
cc2@ecs.soton.ac.uk
[3] National Institute of Informatics, Tokyo, Japan
hasuo@nii.ac.jp
[4] The Graduate University for Advanced Studies (SOKENDAI), Tokyo, Japan

Abstract. We introduce a path-based coalgebraic temporal logic, Coalgebraic CTL (CCTL), as a categorical abstraction of standard Computation Tree Logic (CTL). Our logic can be used to formalize properties of systems modeled as coalgebras with branching. We present the syntax and path-based semantics of CCTL, and show how to encode this logic into a coalgebraic fixpoint logic with a step-wise semantics. Our main result shows that this encoding is semantics-preserving. We also present a polynomial-time model-checking algorithm for CCTL, inspired by the standard model-checking algorithm for CTL but described in categorical terms. A key contribution of our paper is to identify the categorical essence of the standard encoding of CTL into the modal mu-calculus. This categorical perspective also explains the absence of a similar encoding of PCTL (Probabilistic CTL) into the probabilistic mu-calculus.

1 Introduction

1.1 Path-Based Temporal Logics and Categorical Generalization

Temporal logics provide specification-description languages in formal verification on transition systems. Among such logics, CTL* and its fragment CTL [12,13] are well-known because of their descriptive power. They are *path-based* temporal logics: they refer not just to immediate successors of the current state but also to states reachable along (infinite) computation paths. Such path-based formulas can express eventual and permanent behaviors of transition systems, such as liveness and safety properties [1,8].

CTL, even though it is a simple fragment, inherits much of the expressive power from CTL*. CTL can express liveness and safety, and its formulas are known to characterize bisimilarity equivalence on transition systems [28].

Published by Springer Nature Switzerland AG 2024
B. König and H. Urbat (Eds.): CMCS 2024, LNCS 14617, pp. 1–22, 2024.
https://doi.org/10.1007/978-3-031-66438-0_1

Table 1. Fixpoint characterization in classical CTL on a Kripke frame $c\colon X \to \mathcal{P}^+X$ and in our generalization CCTL on a TF-coalgebra $c\colon X \to TFX$.

	path-based semantics	step-wise semantics
classical	$\mathrm{CTL} \hookrightarrow \mathrm{CTL}^* \xrightarrow{[14]} \mathbf{2}^X$	$\mathrm{CTL} \xrightarrow{[13]} \mathbf{L}\mu \xrightarrow{[26]} \mathbf{2}^X$
coalgebraic (ours)	$\mathsf{CCTL} \hookrightarrow \mathsf{SFml} \xrightarrow{[\![\_]\!]_{\mathsf{SFml}}} \mathbf{2}^X$	$\mathsf{CCTL} \xrightarrow{\iota^{-1}} \mu^{\mathsf{CCTL}} \xrightarrow{[\![\_]\!]_{\mu^{\mathsf{CCTL}}}} \mathbf{2}^X$

The restriction to CTL gives us computational efficiency in model checking, an advantage over CTL*. It is well-known that, by implementing a naive fixpoint algorithm, verifying a CTL formula on a state takes at most polynomial time [8], in contrast to the known exponential time complexity bound for CTL*.

The technical core behind this efficiency of CTL is an encoding into a fixpoint modal logic, namely the mu-calculus $\mathbf{L}\mu$ [26]. This encoding can then be used to induce another, *step-wise*, semantics of CTL formulas (Table 1, top-right), in contrast to the path-based semantics (Table 1, top-left). The so-called *fixpoint characterization* [13, Lemma 2.6] states that the encoding is semantics-preserving. The fixpoint characterization enables the verification of path-based specifications expressed in CTL by step-wise, iterative calculation on system states and substantially reduces the complexity of the verification.

The fixpoint characterization seems to define what CTL is, as an optimal solution for the trade-off between descriptive power (inherent to the path-based logics) and efficiency in verification (implemented by step-wise iterations).

Nevertheless, the fixpoint characterization does not come for free among known variants of CTL, instantiated over various systems with different branching types. In quantitative variants of CTL [4,30], the fixpoint characterization results hold under some restrictions on its parameters. In contrast, the well-known probabilistic variant of CTL, called PCTL [1,18], does not have a known encoding into a natural probabilistic fixpoint logic, like the probabilistic mu-calculus [7].

We aim to establish a *generic* notion of CTL by which we can uniformly classify known variants of CTL and clarify why the original CTL (with some variants) validates the fixpoint characterization and PCTL does not seem to. To this end, we appeal to *coalgebraic logics* [29,32], a meta-theory of logics on generic systems modeled as coalgebras.

As a coalgebraic generalization of CTL*, the coalgebraic path-based logic $\mu\mathcal{L}$ is proposed in [5]. The original non-deterministic transition systems, which provide the semantic domain for CTL*, are generalized to TF-coalgebras with their branching type and transition type specified by a monad T and a functor F, respectively. The notion of computation path in CTL* is replaced by its categorical abstraction, maximal execution map. As shown in [5], this framework encompasses both classical CTL* and an extension of PCTL, by instantiating the branching type by the non-empty powerset monad $\mathcal{P}^+$ and the Giry monad $\mathcal{G}_1$.

1.2 Contributions: Coalgebraic CTL

We follow [5] and introduce our coalgebraic generalization of CTL, dubbed CCTL. As a fragment of $\mu\mathcal{L}$, our CCTL has the genericity of branching and transition type T, F, and sets of liftings Σ, Λ of these type functors. Furthermore, CCTL has novel syntactic parameters of *μ-schemes* and *ν-schemes*, which restrict the allowed form of the least and greatest fixpoints. We describe the path-based semantics $[\![\_]\!]_{\mathsf{SFml}}$ of CCTL inherited from $\mu\mathcal{L}$ (Table 1, bottom left) on a categorical semantic domain, which we call *BT situation*.

Our theoretical highlight is a coalgebraic version of the fixpoint characterization (Theorem 4.6). We present a bijective and semantics-preserving encoding of CCTL into a restriction μ^{CCTL} of the coalgebraic mu-calculus [7,19,38], yielding the step-wise semantics of CCTL (Table 1, bottom right).

Sufficient semantic conditions (Assumption 4.7) for the fixpoint characterization are identified in purely categorical terms. They classify the non-deterministic and probabilistic situations: while classical CTL enjoys all of them, PCTL violates some. The violation explains the absence of the fixpoint characterization for PCTL, in categorical terms.

As significant by-products of our fixpoint characterization, we discovered a coalgebraic abstraction of the *expansion law* [1], which tells how to expand path-based formulas step by step concretely (Proposition 4.9). The coalgebraic expansion law is obtained under weaker assumptions than the fixpoint characterization, and induces a *partial* fixpoint characterization (Proposition 4.10). Remarkably, these results also apply to a qualitative fragment of PCTL.

Our fixpoint characterization (Theorem 4.6) leads to a polynomial-time model-checking algorithm $\mathsf{MC}^{\mathsf{CCTL}}_{\mathcal{S}}$ of CCTL, which is parametrized by a BT situation $\mathcal{S}$. With an additional finiteness condition on $\mathcal{S}$, we obtain termination and correctness of the coalgebraic algorithm $\mathsf{MC}^{\mathsf{CCTL}}_{\mathcal{S}}$. We further conclude the polynomial-time complexity bound of $\mathsf{MC}^{\mathsf{CCTL}}_{\mathcal{S}}$, which recovers the quadratic bound of the known CTL model checking with fixpoints [8] when precisely instantiated.

This paper is the first step towards a uniform investigation into efficient and expressive coalgebraic path-based logics. It paves the way to classify known examples, like the quantitative CTL [4,30], and unknown ones, like a "monotone neighborhood" version of CTL induced from neighborhood frames [16].

The paper is organized as follows. Section 2 recalls necessary categorical notions. Section 3 defines our semantic domain dubbed *BT situation* and introduces CCTL as a fragment of $\mu\mathcal{L}$ [5]. Section 4 defines a fragment μ^{CCTL} of the coalgebraic mu-calculus, and provides an encoding of CCTL formulas into μ^{CCTL} formulas. Our main result, the fixpoint characterization (Theorem 4.6), shows this encoding is semantic-preserving. Section 5 formulates a polynomial-time model-checking algorithm for CCTL.

2 Preliminaries

We use $\mathbb{C}$ for a category with finite products and countable coproducts. As examples, we will use the category **Set** of sets and functions and the category **SB** of standard Borel spaces and measurable functions [11,34].

Let T be a monad, and F be an endofunctor, on $\mathbb{C}$. We formulate a targeted system as a TF-coalgebra, i.e., a map $c\colon X \to TFX$.

2.1 Functors and Monads

We recall some basic properties of functors and monads. See [24] for details.

A functor on the category $\mathbb{C}$ is a *(simple) polynomial functor* [24, Def. 2.2.1] if it is constructed by the following BNF: $F ::= \mathrm{Id} \mid C \mid \coprod_{b\in B} F_b \mid F_1 \times F_2$ where C is an arbitrary object and B is a countable set. A major example is an *arity functor* [24] $F = \coprod_{\alpha\in A} X^{|\alpha|}$ for some set A with an arity map $|\_|$.[1] For simplicity, we assume any polynomial functor hereafter is an arity functor.

A *commutative* monad [24, Def. 5.2.9] has a strength $\mathsf{st}_{X,Y}\colon X \times FY \to F(X\times Y)$, a swapped strength $\mathsf{st}'_{X,Y}\colon TX\times Y \to T(X\times Y)$ defined by $TX\times Y \cong Y \times TX \xrightarrow{\mathsf{st}_{Y,X}} T(Y \times X) \cong T(X \times Y)$, and the double strength $\mathsf{dst} = \mu_{X\times Y} \circ T\mathsf{st}'_{X,Y} \circ \mathsf{st}_{TX,Y} = \mu_{X\times Y} \circ T\mathsf{st}_{X,Y} \circ \mathsf{st}'_{X,TY}\colon T(TX \times Y) \to T^2(X \times Y)$. More generally, we can also define an *n-ary* strength map $\mathsf{dst}_n\colon TA_1 \times\cdots\times TA_n \to T(A_1 \times\cdots\times A_n)$ likewise.[2]

Example 2.1 (commutative monads).

1. (non-determinism) The powerset monad $\mathcal{P}$ on **Set** is commutative and its strength is given by $(x, S) \mapsto \{(x, s) \mid s \in S\}$. Its double strength is given by $(T, S) \mapsto T \times S$, where $\times$ is the set product.
2. (reliability) The sub-Giry monad $\mathcal{G}$ on **SB** is defined as follows. The object part of $\mathcal{G}$ maps a standard Borel space (X, Σ_X) to $(\mathcal{M}_X, \Sigma_{\mathcal{M}_X})$ where $\mathcal{M}_X$ is the set of sub-probability measures on X and $\Sigma_{\mathcal{M}_X}$ is the Borel set generated from $\{\rho \in \mathcal{M}_X \mid \rho(S) \subset [0,1] \text{ is measurable w.r.t. } ([0,1], \Sigma_{[0,1]})\}$. The sub-Giry monad $\mathcal{G}$ maps a measurable map $f\colon (X,\Sigma_X) \to (Y,\Sigma_Y)$ to $\mathcal{G}f\colon (\mathcal{M}_X, \Sigma_{\mathcal{M}_X}) \to (\mathcal{M}_Y, \Sigma_{\mathcal{M}_Y})$; $(\mathcal{G}f)(\rho) = \lambda S.\, \rho(f^{-1}(S))$.
   Furthermore, $\mathcal{G}$ is indeed a commutative monad. The unit $\eta\colon (X,\Sigma_X) \to (\mathcal{M}_X, \Sigma_{\mathcal{M}_X})$ of $\mathcal{G}$ maps each element x to the Dirac distribution δ_x, and the multiplication $\mu\colon (\mathcal{M}_{\mathcal{M}_X}, \Sigma_{\mathcal{M}_{\mathcal{M}_X}}) \to (\mathcal{M}_X, \Sigma_{\mathcal{M}_X})$ maps $\Phi \in \mathcal{M}_{\mathcal{M}_X}$ to the measure defined by the integration $\int_{\rho\in\mathcal{M}_X} \Phi(\rho)\, d\rho$. The strength of $\mathcal{G}$ is
   $$X \times \mathcal{G}Y \ni (x,\rho) \longmapsto \delta_x \times \rho \in \mathcal{G}(X \times Y)$$
   where $\delta_x \times \rho$ is the product of measures [11] and the double strength is
   $$\mathcal{G}X \times \mathcal{G}Y \ni (\rho_1, \rho_2) \longmapsto \rho_1 \times \rho_2 \in \mathcal{G}(X \times Y).$$

[1] In a category which has enough copowers [27], like **Set** and **SB**, any polynomial functor can be represented as an arity functor, and vice versa.

[2] Such an n-ary strength is defined uniquely by virtue of commutativity.

A commutative monad is *affine* [21, Def. 4.1], if the unit $\eta_1 \colon \mathbf{1} \to T\mathbf{1}$ is an isomorphism. Affine-ness is a categorical generalization of *serial-ness* or *left-totality* [8].

Example 2.2. If the category $\mathbb{C}$ has pullbacks, every monad T has the largest affine submonad T^{a}, called the *affine part* of T [21, Def. 4.5], given by the pullback of $TX \xrightarrow{T!_X} T\mathbf{1} \xleftarrow{\eta_1} \mathbf{1}$. The affine part of a commutative monad is also commutative. The affine part of $\mathcal{P}$ is the non-empty powerset monad $\mathcal{P}^+$, and the affine part of the sub-Giry monad $\mathcal{G} \colon \mathbf{SB} \to \mathbf{SB}$ is the Giry monad $\mathcal{G}_1$ [15], which is defined by restricting sub-probability measures in $\mathcal{G}$ to probability measures.

2.2 Predicate Liftings

The concept of *predicate lifting* [32] was originally defined on **2**-valued predicates and used in interpreting modalities in coalgebraic modal logics. Here we generalize it to any (complete lattice-like) object Ω in $\mathbb{C}$.

Definition 2.3 (logical value object). An object $\Omega \in \mathbb{C}$ is called a *logical value object* if its representation $\Omega^{(-)} \colon \mathbb{C} \to \mathbf{Set}^{\mathrm{op}}$ restricts to the category of complete lattices and $\{\bot, \top, \vee, \wedge\}$-preserving functions.

If Ω is a logical value object, any n-ary boolean operator b induces a monotone natural transformation $\left(\Omega^{(-)}\right)^n \Rightarrow \Omega^{(-)}$.[3] By the Yoneda lemma, we then obtain an n-ary map $\gamma_b \colon \Omega^n \to \Omega$ corresponding to the operator b. Especially, we have $\gamma_\top, \gamma_\bot \colon \mathbf{1} \to \Omega$ and $\gamma_\vee, \gamma_\wedge \colon \Omega^2 \to \Omega$.

With these maps induced from boolean operators, we can treat the object $\Omega \in \mathbb{C}$ as if it were a complete lattice. Hereafter, we will identify a boolean operator b with the induced map γ_b and use the letter Γ for the set of all boolean operators.

Definition 2.4 (predicate lifting). Let $G \colon \mathbb{C} \to \mathbb{C}$ be an endofunctor, and $\Omega \in \mathbb{C}$ be a logical value object.

1. A *(predicate) lifting* or *modality* of G w.r.t. Ω is a natural transformation $\{\lambda_Y \colon \Omega^Y \to \Omega^{GY}\}_{Y \in \mathbb{C}}$ which is monotone w.r.t. the lattice structures on Ω^Y and Ω^{GY}.
2. We write $\mathrm{ev}_\lambda \colon G\Omega \to \Omega$ for the correspondent of a lifting λ via the Yoneda lemma[4]: this is to say, $\mathrm{ev}_\lambda = \lambda_\Omega(\mathrm{id}_\Omega)$ and $\lambda_Y(p) = \mathrm{ev}_\lambda \circ Gp$ for $p \in \Omega^Y$.

Henceforth, we consistently use the letter σ for a predicate lifting of the branching behavior T and λ for that of the transition behavior F. We call σ *"path quantifier"* and λ *"next-time operator."*[5]

[3] Here a *boolean operator* means a map on a complete lattice constructed from operators $\bot, \top, \vee$ and $\wedge$.

[4] Recall the Yoneda lemma: there is a bijective correspondence between natural transformations from $\Omega^{(-)}$ to Ω^{G-} and elements of $\Omega^{G\Omega}$.

[5] Although we treat only unary next-time operators for simplicity, we can easily extend our framework to contain 0-ary next-time operators. Such 0-ary modalities are used to include atomic predicates to our syntax (see Definition 3.7) as in [19,33]. We will freely exploit this extension when we talk about concrete examples.

Example 2.5 (predicate liftings). First, note that $\mathbf{2} \in \mathbf{Set}$ and $(\mathbf{2}, \mathcal{P}\mathbf{2}) \in \mathbf{SB}$ are logical value objects.

1. There is a trivial lifting $\mathrm{id}_{\Omega^X} : \Omega^X \to \Omega^X$ of the identify functor Id. More generally, there is a canonical predicate lifting $\mathrm{Pred}(F)$ for each polynomial functor F [24, Lemma 6.1.3]. For an arity functor $F = \coprod_{\alpha \in A} X^{|\alpha|}$, the lifting $\mathrm{Pred}(F)$ is induced from the map $[\wedge^{|\alpha|}]_{\alpha \in A} \colon \coprod_{\alpha \in A} \Omega^{|\alpha|} \to \Omega$, where $[\_]$ denotes a cotuple of the coproduct and $\wedge^{|\alpha|} \colon \Omega^{|\alpha|} \to \Omega$ denotes $|\alpha|$-ary conjunction. Thus, $\mathrm{Pred}(F)(Q)$ for a predicate $Q \in \Omega^X$ is given by $\mathrm{Pred}(F)(Q) = [\wedge^{|\alpha|} \circ Q^{|\alpha|}]_{\alpha \in A}$.
2. (non-determinism) Liftings $\mathcal{P}^+_\Diamond, \mathcal{P}^+_\Box$ of $\mathcal{P}^+$ are respectively induced by the maps $\Diamond, \Box \colon \mathcal{P}\mathbf{2} \to \mathbf{2}$ (i.e., $\mathcal{P}^+_\Diamond(P) = \Diamond \circ \mathcal{P}^+(P)$ and $\mathcal{P}^+_\Box(P) = \Box \circ \mathcal{P}^+(P)$, recall Definition 2.4). These maps $\Diamond, \Box$ are defined as follows: for $S \in \mathcal{P}^+\mathbf{2}$, $\Diamond(S) = 1$ if and only if $S = \{0,1\}, \{1\}$ and $\Box(S) = 1$ if and only if $S = \{1\}$.
3. (reliability) The Giry monad $\mathcal{G}_1$ has liftings $\mathcal{G}_{1,\geq q}, \mathcal{G}_{1,>q}$ which are induced by the "larger-than-q-or-equal" and "larger-than-q" maps $\geq_q, >_q \colon \mathcal{G}_1(\mathbf{2}, \mathcal{P}\mathbf{2}) \cong ([0,1], \Sigma_{[0,1]}) \to (\mathbf{2}, \mathcal{P}\mathbf{2})$. The map $\geq_q$ is defined by $\geq_q (r) = 1$ if and only if $r \geq q$, and the map $>_q$ is also defined likewise.

3 Coalgebraic Path-Based Temporal Logics: $\mu\mathcal{L}$, CCTL

3.1 Coalgebraic Abstraction of Systems

We first set up a semantic domain of coalgebraic path-based logics $\mu\mathcal{L}$ and CCTL, dubbed *BT situation*. It is categorical data that includes branching and transition types T and F, a coalgebra of these types, path quantifiers, and next-time operators.

Definition 3.1 (BT situation). A *branching-transition situation* (*BT situation*, in short) is given by a tuple $(\mathbb{C}, T, F, c, \Omega, \Sigma, \Lambda)$ where:

1. $\mathbb{C}$ is a concrete, finitely complete, and countably cocomplete category,
2. $T \colon \mathbb{C} \to \mathbb{C}$ is a commutative monad,
3. $F \colon \mathbb{C} \to \mathbb{C}$ is a polynomial functor,
4. $c \colon X \to TFX$ is a TF-coalgebra,
5. $\Omega \in \mathbb{C}$ is a logical value object (see Definition 2.3),
6. Σ is a set of predicate liftings of T, called *path quantifiers*,
7. Λ is a set of predicate liftings of F, called *next-time operators*.

Example 3.2 (BT situation). Table 2 defines our examples of BT situation. Note that our instantiations $\mathcal{S}_{\mathrm{ND}}$ and $\mathcal{S}_{\mathrm{R}}$ still have a genericity of F.

1. (non-determinism) In $\mathcal{S}_{\mathrm{ND}}$, $\mathrm{Pred}(F)$ and $\mathcal{P}^+_\Diamond, \mathcal{P}^+_\Box$ are the liftings as in Example 2.5. A $\mathcal{P}^+F$-coalgebra is a (generalized) left-total Kripke frame. Besides the classical Kripke frames when $F = \mathrm{Id}$, F-genericity allows other variants: labeled Kripke frames when $F = \mathcal{P}(\mathrm{AP}) \times \mathrm{Id}$ (here AP is the set of atomic propositions) and Kripke structures with termination when $F = \mathbf{1} + \mathrm{Id}$.

Table 2. Examples of BT situation

	parameters	$\mathcal{S}_{\text{ND}}$	$\mathcal{S}_{\text{R}}$
category	$\mathbb{C}$	**Set**	**SB**
branching type	a monad T	$\mathcal{P}^{+}$	$\mathcal{G}_1$
transition type	a polynomial F	F	F
system	$c\colon X \to TFX$	a Kripke frame	a Markov chain
truth values	$\Omega \in \mathbb{C}$	$\mathbf{2}$	$(\mathbf{2}, \mathcal{P}\mathbf{2})$
path quantifiers	$\{\sigma\}_{\sigma \in \Sigma}$	$\{\mathcal{P}^{+}_{\Diamond}, \mathcal{P}^{+}_{\Box}\}$	$\{\mathcal{G}_{1,\geq q}, \mathcal{G}_{1,>q}\}_{q \in [0,1]}$
next-time operators	$\{\lambda\}_{\lambda \in \Lambda}$	$\{\text{Pred}(F)\}$	$\{\text{Pred}(F)\}$

2. (reliability) In $\mathcal{S}_{\text{R}}$, $\mathcal{G}_1$ is the Giry monad, $\text{Pred}(F)$ and $\mathcal{G}_{1,\geq q}, \mathcal{G}_{1,>q}$ are as in Example 2.5. A $\mathcal{G}_1 F$-coalgebra is a (generalized) Markov chain, which coincides with a classical one when the state space is given by the discrete space $(X, \mathcal{P}X)$ for a countable set X and F is given by Id or $\mathcal{P}(\text{AP}) \times \text{Id}$.
3. (qualitative reliability) We also define a BT situation $\mathcal{S}_{\text{qR}}$ for *qualitative* reliability by restricting the set of path quantifiers of $\mathcal{S}_{\text{R}}$ to $\{\mathcal{G}_{1,\geq 1}, \mathcal{G}_{1,>0}\}$.

3.2 Maximal Traces as Computation Paths

We recall concepts of maximal trace map and maximal execution map of TF-coalgebras. The latter is an abstraction of the classical notion of computation trees and will be used in the formulation of our path-based semantics.

We first recall Jacobs' formulation of maximal trace [22] on the Kleisli category of the monad T [24]. Let $J\colon \mathbb{C} \to \mathcal{K}\ell(T)$ be the canonical left adjoint of the monad T. This J sends an object of $\mathbb{C}$ to itself and a map $f\colon A \to B$ to $\eta_B \circ f$. Given a distributive law $\xi\colon FT \Rightarrow TF$, we have the induced functor $\overline{F}\colon \mathcal{K}\ell(T) \to \mathcal{K}\ell(T)$ that sends a Kleisli arrow $f\colon A \rightarrow\!\!\!\!\!\rightarrow B$ to $\xi_B \circ Ff\colon FA \rightarrow\!\!\!\!\!\rightarrow FB$.

Definition 3.3 (maximal trace, [22,37]). Suppose that each homset of the Kleisli category $\mathcal{K}\ell(T)$ carries an order $\sqsubseteq$. A functor F and a monad T constitute *a maximal trace situation* if

1. F has a final coalgebra $\zeta\colon Z \to FZ$,
2. a distributive law ξ of F over T exists,
3. for every $\overline{F}$-coalgebra $c\colon X \rightarrow\!\!\!\!\!\rightarrow \overline{F}X$, there exists the *greatest* map $u\colon X \rightarrow\!\!\!\!\!\rightarrow Z$ satisfying $J\zeta \odot u = \overline{F}u \odot c$ w.r.t. the order $\sqsubseteq$, where $\odot$ is the Kleisli composition.

The greatest map u in condition 3 is called the *maximal trace map* of c and is denoted by $\text{tr}(c)$.

Condition 1 and 2 are automatically satisfied in our semantic domain, BT situation $\mathcal{S}$, since every polynomial functor has its final coalgebra, and there is a canonical distributive law of polynimial F over commutative T (cf. [24, Prop. 5.2.12]).

We will use the maximal trace map for the polynomial functor $F_X := X \times F$. Note that the auxiliary coefficient X is added to capture the current state.

Let $\zeta = \langle \zeta_1, \zeta_2 \rangle : Z_X \to X \times F(Z_X)$ be the final coalgebra of F_X. We call the object Z_X the *path space* of X with type F, and the maps ζ_1 and ζ_2 the *head operator* and the *tail operator* on the path space Z_X.

We can render any TF-coalgebra $c\colon X \to TFX$ into a TF_X-coalgebra $c' = \mathsf{dst}_{X,FX} \circ \langle \eta_X, c \rangle : X \to T(X \times FX)$. We call the maximal trace w.r.t. this TF_X-coalgebra c' the *maximal execution map* for the TF-coalgebra $c\colon X \to TFX$.

Definition 3.4 (BT situation with maximal execution). A *BT situation with maximal execution* is a BT situation $\mathcal{S}$ with the maximal execution map $\mathrm{tr}(c')$ for the TF-coalgebra $c\colon X \to TFX$.

In the remainder of this paper, we fix a BT situation $\mathcal{S} = (\mathbb{C}, T, F, c, \Omega, \Sigma, \Lambda)$ with maximal execution $\mathrm{tr}(c')$.

Example 3.5 (examples of maximal executions). In the examples below, for the sake of simplicity, we fix F to be Id on **Set** or **SB**.

1. (non-determinism) The final coalgebra of $(\mathrm{Id}_{\mathbf{Set}})_X = X \times \mathrm{Id}_{\mathbf{Set}}$ for **Set** is the set X^ω of streams, where ω is the set of finite ordinals. The existence of maximal executions for $\mathcal{P}^+$ is assured by an adaptation of [37, Prop. 4.1]. Concretely, the maximal execution map $\mathrm{tr}(c')\colon X \to \mathcal{P}^+X^\omega$ maps x to $\{\pi \in X^\omega \mid \pi_0 = x \text{ and } \forall n \in \omega. \pi_{n+1} \in c(\pi_n)\}$.
2. (reliability) The final coalgebra of $(\mathrm{Id}_{\mathbf{SB}})_{(X,\Sigma_X)} = (X, \Sigma_X) \times \mathrm{Id}_{\mathbf{SB}}$ for **SB** is the measurable set $(X^\omega, \Sigma_{X^\omega})$ of streams. Its measurable structure Σ_{X^ω} is generated by the cylinder sets $\mathrm{Cyl}(t) = \{\pi \in X^\omega \mid \pi \text{ has the prefix } t\}$ for every finite path t. The existence of maximal execution for $\mathcal{P}^+$ is assured by [37, Prop. 5.2]. Concretely, the maximal execution map $\mathrm{tr}(c')\colon (X, \Sigma_X) \to \mathcal{G}_1(X^\omega, \Sigma_{X^\omega})$ is given by
$$\mathrm{tr}(c')(x)\big(\mathrm{Cyl}(t)\big) = c(x)(x_1) \cdot c(x_1)(x_2) \cdot \ldots \cdot c(x_{n-1})(x_n)$$
where $t = x x_1 x_2 \ldots x_{n-1} x_n$. For a detailed description, see [36, Def. E.9].

3.3 The Logics $\mu\mathcal{L}$ and CCTL

We first recall the coalgebraic logic $\mu\mathcal{L}$ [5]. Its syntax is given by coalgebra-generic *path formulas* and *state formulas*. The following definition is a slight adaptation of the original $\mu\mathcal{L}$.[6]

[6] The notations $\mu\mathcal{L}_F$, $\mu\mathcal{L}$, $[\lambda_F]$ and $[\lambda]$ of the original $\mu\mathcal{L}$ [5] correspond to PFml, SFml, ♡ and ♠, respectively, in our presentation. We also omit variables in SFml.

Definition 3.6 (state formulas and path formulas). Let Σ, Λ be sets, and Γ be a ranked alphabet. Two sets $\mathsf{PFml}_{\Gamma,\Lambda,\Sigma}$ and $\mathsf{SFml}_{\Gamma,\Lambda,\Sigma}$ (or simply $\mathsf{PFml}, \mathsf{SFml}$) are defined by the following mutual induction:

$$\varphi \in \mathsf{PFml} ::= u \mid \boxdot_\gamma(\varphi_1, \dots, \varphi_{|\gamma|}) \mid \heartsuit_\lambda \varphi \mid \mu u.\varphi \mid \nu u.\varphi \mid \psi$$
$$\psi \in \mathsf{SFml} ::= \boxdot_\gamma\,(\psi_1, \dots, \psi_{|\gamma|}) \mid \spadesuit_\sigma \varphi,$$

where u is a proposition variable, $\gamma \in \Gamma$, $\lambda \in \Lambda$ and $\sigma \in \Sigma$. Furthermore, we assume φ in $\spadesuit_\sigma \varphi \in \mathsf{SFml}$ is closed, i.e., φ has no proposition variables.

The symbols $\boxdot_\gamma$, $\heartsuit_\lambda$ and $\spadesuit_\sigma$ correspond to boolean operators, next-time operators and path quantifiers, respectively.

We can then define the new logic CCTL as a fragment of SFml by restricting the forms of fixpoint formulas.

Definition 3.7 (CCTL). Let Σ, Λ be sets and Γ be a ranked alphabet with subsets $\Gamma_\mu, \Gamma_\nu \subset \Gamma$. The set $\mathsf{CCTL}_{\Gamma_\mu,\Gamma_\nu}$ (whose subscripts we will sometimes omit) is the subset of SFml defined by the following grammar:

$$\begin{aligned}\psi \in \mathsf{CCTL}_{\Gamma_\mu,\Gamma_\nu} ::= {} & \boxdot_\gamma\,(\psi_1, \dots, \psi_{|\gamma|}) \mid \spadesuit_\sigma \heartsuit_\lambda \psi \\ & \mid \spadesuit_\sigma(\mu u.\, \boxdot_{\gamma_\mu}\,(\psi_1, \dots, \psi_{|\gamma_\mu|-1}, \heartsuit_\lambda u)) \\ & \mid \spadesuit_\sigma(\nu u.\, \boxdot_{\gamma_\nu}\,(\psi_1, \dots, \psi_{|\gamma_\nu|-1}, \heartsuit_\lambda u))\end{aligned}$$

where u is a proposition variable, $\gamma \in \Gamma$, $\lambda \in \Lambda$, $\sigma \in \Sigma$, $\gamma_\mu \in \Gamma_\mu$ and $\gamma_\nu \in \Gamma_\nu$.

The operators γ_μ and γ_ν in the fixpoint formula are called *μ-schemes* and *ν-schemes*, res pectively. These are used to recover temporal operators (EF, AF, etc. in classical CTL) and are crucial in characterizing $\mathsf{CCTL}_{\Gamma_\mu,\Gamma_\nu}$ within the mu-calculus.

Example 3.8. In the literature, the modality symbols $\spadesuit_{\mathcal{P}^+_\diamond}$ and $\spadesuit_{\mathcal{P}^+_\Box}$ in CTL are respectively denoted by E and A. The modality symbols $\spadesuit_{\mathcal{G}_{1,\geq q}}$ and $\spadesuit_{\mathcal{G}_{1,>q}}$ in PCTL are respectively denoted by $\mathbb{P}_{\geq q}$ and $\mathbb{P}_{>q}$ [1]. The modality symbol $\heartsuit_{\mathrm{Pred}(F)}$ in both CTL and PCTL is often denoted by X. In both CTL and PCTL, their sets of μ-schemes and ν-schemes are respectively given by $\{(\_ \vee (\_ \wedge \_))\}$ and $\{(\_ \wedge (\_ \vee \_))\}$. The least/greatest fixpoint formulas made of $(\_ \vee (\_ \wedge \_))/(\_ \wedge (\_ \vee \_))$ is often denoted by U/W. [7]

The relationships between $\mathsf{SFml}, \mathsf{PFml}, \mathsf{CCTL}$ can be summarized as follows:

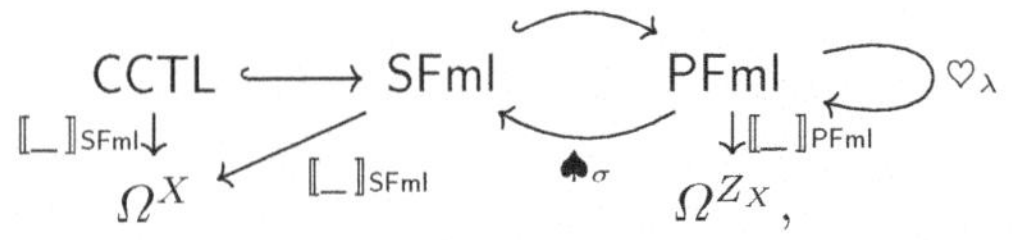

where the semantics $[\![\_]\!]_{\mathsf{SFml}}$ and $[\![\_]\!]_{\mathsf{PFml}}$ is defined below [5].

[7] Another (equivalent) choice of Γ_μ and Γ_ν is possible: we can put $\Gamma_\mu = \{\vee, (\_ \vee (\_ \wedge \_))\}$ and $\Gamma_\nu = \{\wedge, (\_ \wedge (\_ \vee \_))\}$, and the least/greatest fixpoint formula made of $\vee/\wedge$ is denoted by F/G.

Definition 3.9 (semantics of PFml and SFml formulas). For each PFml formula φ with free variables $u_1, \ldots, u_m$, and each SFml formula ψ, their interpretation $[\![\varphi]\!]_{\mathsf{PFml}}\colon (\Omega^{Z_X})^m \to \Omega^{Z_X}$ and $[\![\psi]\!]_{\mathsf{SFml}}\colon \Omega^X$ are defined in the following mutually inductive manner: for $\vec{V} = V_1, \ldots, V_m$ with $V_i\colon X \to \Omega$,

$$\begin{aligned}
[\![u_i]\!]_{\mathsf{PFml}}(\vec{V}) &:= V_i, \\
[\![\boxdot_\gamma(\varphi_1, \ldots, \varphi_{|\gamma|})]\!]_{\mathsf{PFml}}(\vec{V}) &:= \gamma\big(\, [\![\varphi_1]\!]_{\mathsf{PFml}}(\vec{V}),\ \ldots,\ [\![\varphi_{|\gamma|}]\!]_{\mathsf{PFml}}(\vec{V})\,\big), \\
[\![\heartsuit_\lambda \varphi]\!]_{\mathsf{PFml}}(\vec{V}) &:= [\![\heartsuit_\lambda]\!]\big(\, [\![\varphi_1]\!]_{\mathsf{PFml}}(\vec{V}),\ \ldots,\ [\![\varphi_n]\!]_{\mathsf{PFml}}(\vec{V})\,\big), \\
[\![\mu u.\, \varphi]\!]_{\mathsf{PFml}}(\vec{V}) &:= \big(\, \mu\big(\, [\![\varphi]\!]_{\mathsf{PFml}}(\vec{V}, _)\colon \Omega^{Z_X} \to \Omega^{Z_X}\,\big)\big), \\
[\![\nu u.\, \varphi]\!]_{\mathsf{PFml}}(\vec{V}) &:= \big(\, \nu\big(\, [\![\varphi]\!]_{\mathsf{PFml}}(\vec{V}, _)\colon \Omega^{Z_X} \to \Omega^{Z_X}\,\big)\big), \\
[\![\psi]\!]_{\mathsf{PFml}}(\vec{V}) &:= \zeta_1^*([\![\psi]\!]_{\mathsf{SFml}}), \\
[\![\boxdot_\gamma(\psi_1, \ldots, \psi_{|\gamma|})]\!]_{\mathsf{SFml}} &:= \gamma\big(\, [\![\psi_1]\!]_{\mathsf{SFml}},\ \ldots,\ [\![\psi_{|\gamma|}]\!]_{\mathsf{SFml}}\,\big), \\
[\![\spadesuit_\sigma \varphi]\!]_{\mathsf{SFml}} &:= [\![\spadesuit_\sigma]\!]([\![\varphi]\!]_{\mathsf{PFml}}),
\end{aligned}$$

where

$$\begin{aligned}
[\![\heartsuit_\lambda]\!] &:= \zeta_2^* \circ \lambda_{Z_X}\colon \Omega^{Z_X} \to \Omega^{Z_X}, \\
[\![\spadesuit_\sigma]\!] &:= \big(\mathrm{tr}(c')\big)^* \circ \sigma_{Z_X}\colon \Omega^{Z_X} \to \Omega^X.
\end{aligned}$$

In this interpretation, f^* denotes the pullback of a map f, and μ, ν denote the least/greatest fixpoint of the monotone function $[\![\varphi]\!]_{\mathsf{PFml}}(\vec{V}, \_)\colon \Omega^{Z_X} \to \Omega^{Z_X}$.

Definition 3.10 (path-based semantics of CCTL). The *path-based semantics* of a CCTL formula ψ is given by $[\![\psi]\!]_{\mathsf{SFml}}$. Especially, the interpretations of the restricted fixpoints $\spadesuit_\sigma(\mu u.\, \boxdot_{\gamma_\mu} (\psi_1, \ldots, \psi_{|\gamma_\mu|-1}, \heartsuit_\lambda u))$ and $\spadesuit_\sigma(\nu u.\, \boxdot_{\gamma_\nu} (\psi_1, \ldots, \psi_{|\gamma_\nu|-1}, \heartsuit_\lambda u))$ are given by

$$\begin{aligned}
[\![\spadesuit_\sigma(\mu u.\, \boxdot_{\gamma_\mu} (\psi_1, \ldots, \psi_{|\gamma_\mu|-1}, \heartsuit_\lambda u))]\!]_{\mathsf{PFml}} &= [\![\spadesuit_\sigma]\!](\mu \Phi_{\lambda, \gamma_\mu, (\varphi_1, \ldots, \varphi_{|\gamma_\mu|})}) \\
[\![\spadesuit_\sigma(\nu u.\, \boxdot_{\gamma_\nu} (\psi_1, \ldots, \psi_{|\gamma_\nu|-1}, \heartsuit_\lambda u))]\!]_{\mathsf{PFml}} &= [\![\spadesuit_\sigma]\!](\nu \Phi_{\lambda, \gamma_\nu, (\varphi_1, \ldots, \varphi_{|\gamma_\nu|})})
\end{aligned}$$

where

$$\Phi_{\lambda, \gamma, (\varphi_1, \ldots, \varphi_{|\gamma|})} := \gamma([\![\varphi_1]\!]_{\mathsf{PFml}}, \ldots, [\![\varphi_{|\gamma|}]\!]_{\mathsf{PFml}}, [\![\heartsuit_\lambda]\!](\_))\colon \Omega^{Z_X} \to \Omega^{Z_X}$$

whose subscripts we will sometimes omit.

Example 3.11 (instantiations of CCTL, cf. Example 3.2). Using the BT situation $\mathcal{S}_{\mathrm{ND}}$, we can obtain classical CTL semantics [12]. The instantiated operators $[\![\mathsf{E}]\!]$ and $[\![\mathsf{A}]\!]$ respectively map a predicate Q (on computation paths) to the predicates

$$\begin{aligned}
&\{x \in X \mid \text{there is a computation path } \pi \text{ of } x \text{ with } \pi \in Q\}, \\
&\{x \in X \mid \text{every computation path } \pi \text{ of } x \text{ belongs to } Q\}.
\end{aligned}$$

The operator $[\![\mathsf{X}]\!]$ maps a path predicate Q to the path predicate

$$\{\pi \in X^\omega \mid \text{the tail of } \pi \text{ belongs to } Q\}.$$

Using the BT situation $\mathcal{S}_{\mathrm{R}}$, we can also obtain the PCTL semantics [18]. The instantiated operator $[\![\mathbb{P}_{\geq q}]\!]$ maps a path predicate Q to

$$\left\{x \in X \,\middle\|\, \begin{array}{l}\text{the probability of computation paths of } x \text{ belonging to } Q\\ \text{is greater than or equal } q\end{array}\right\}.$$

4 Fixpoint Characterization of CCTL

The aim of this section is to give an alternative *step-wise* semantics of CCTL, and prove its equivalence to the path-based semantics (Definition 3.10). The equivalence, *fixpoint characterization*, is crucial in obtaining our polynomial time model-checking algorithm of CCTL formulas in Sect. 5.

4.1 A Coalgebraic μ-calculus μ^{CCTL}

We first introduce a fragment μ^{CCTL} of the coalgebraic μ-calculus [19,38]. The fragment instantiates the coalgebraic μ-calculus using composite modalities $\spadesuit_\sigma\heartsuit_\lambda$, and restricts formulas inside fixpoints to be in a particular form.

Definition 4.1 (the μ-calculus μ^{CCTL}). Let Σ, Λ be sets, and Γ be a ranked alphabet. We define the μ-calculus $\mu^{\mathsf{CCTL}}_{\Gamma_\mu,\Gamma_\nu}$ by the following grammar:

$$\begin{aligned}\theta \in \mu^{\mathsf{CCTL}}_{\Gamma_\mu,\Gamma_\nu} ::= &\ \boxdot_\gamma(\theta_1,\ldots,\theta_{|\gamma|}) \mid \spadesuit_\sigma\heartsuit_\lambda\theta\\ &\mid \mu u.\, \boxdot_{\gamma_\mu}(\theta_1,\ldots,\theta_{|\gamma_\mu|-1},\spadesuit_\sigma\heartsuit_\lambda u)\\ &\mid \nu u.\, \boxdot_{\gamma_\nu}(\theta_1,\ldots,\theta_{|\gamma_\nu|-1},\spadesuit_\sigma\heartsuit_\lambda u)\end{aligned}$$

where u is a proposition variable, $\gamma \in \Gamma$, $\lambda \in \Lambda$, $\sigma \in \Sigma$, and $\gamma_\mu \in \Gamma_\mu, \gamma_\nu \in \Gamma_\nu$.

Note here our $\mu^{\mathsf{CCTL}}_{\Gamma_\mu,\Gamma_\nu}$ has no open formula since any occurrence of variables is bound immediately.

Definition 4.2 (semantics of μ^{CCTL} formulas). For each $\mu^{\mathsf{CCTL}}_{\Gamma_\mu,\Gamma_\nu}$ formula θ, its interpretation $[\![\theta]\!]_{\mu^{\mathsf{CCTL}}} \in \Omega^X$ is defined by:

$$\begin{aligned}[\![\boxdot_\gamma(\theta_1,\ldots,\theta_n)]\!]_{\mu^{\mathsf{CCTL}}} &:= \gamma\big([\![\theta_1]\!]_{\mu^{\mathsf{CCTL}}},\ \ldots,\ [\![\theta_n]\!]_{\mu^{\mathsf{CCTL}}}\big),\\ [\![\spadesuit_\sigma\heartsuit_\lambda\theta]\!]_{\mu^{\mathsf{CCTL}}} &:= [\![\spadesuit_\sigma\heartsuit_\lambda]\!]\big([\![\theta]\!]_{\mu^{\mathsf{CCTL}}}\big),\\ [\![\mu u.\, \boxdot_{\gamma_\mu}(\theta_1,\ldots,\theta_{|\gamma_\mu|-1},\spadesuit_\sigma\heartsuit_\lambda u)]\!]_{\mu^{\mathsf{CCTL}}} &:= \mu\,\Psi_{(\sigma,\lambda),\gamma_\mu,(\theta_1,\ldots,\theta_{|\gamma_\mu|-1})},\\ [\![\nu u.\, \boxdot_{\gamma_\nu}(\theta_1,\ldots,\theta_{|\gamma_\nu|-1},\spadesuit_\sigma\heartsuit_\lambda u)]\!]_{\mu^{\mathsf{CCTL}}} &:= \nu\,\Psi_{(\sigma,\lambda),\gamma_\nu,(\theta_1,\ldots,\theta_{|\gamma_\nu|-1})},\end{aligned}$$

where we denote monotone functions

$$\begin{aligned}[\![\spadesuit_\sigma\heartsuit_\lambda]\!] &:= c^* \circ \sigma_{FX} \circ \lambda_X \colon \Omega^X \to \Omega^X,\\ \Psi_{(\sigma,\lambda),\gamma,(\theta_1,\ldots,\theta_{|\gamma|-1})} &:= \gamma([\![\theta_1]\!]_{\mu^{\mathsf{CCTL}}},\ldots,[\![\theta_{|\gamma|-1}]\!]_{\mu^{\mathsf{CCTL}}},[\![\spadesuit_\sigma\heartsuit_\lambda]\!](\_))\colon \Omega^X \to \Omega^X,\end{aligned}$$

whose subscripts we will sometimes omit.

Example 4.3. In $\mathcal{S}_{\mathrm{ND}}$, the operator $[\![\mathsf{EX}]\!]$ maps a predicate P on states to the predicate

$$\{x \in X \mid \text{there is a successor } x' \text{ of } x \text{ with } x' \in Q\}.$$

The operator $[\![\mathsf{AX}]\!]$ is also defined similarly. In $\mathcal{S}_{\mathrm{R}}$, the operator $[\![\mathbb{P}_{\geq q}\mathsf{X}]\!]$ maps a predicate P to the predicate

$$\left\{x \in X \| \begin{array}{l}\text{the probability of successors of } x \text{ belonging to } Q\\ \text{is greater than or equal to } q\end{array}\right\}.$$

4.2 Step-Wise Semantics of CCTL and Fixpoint Characterization

To define the step-wise semantics of CCTL, we first define a bijective translation between μ^{CCTL} formulas and CCTL formulas.

Definition 4.4 (translation of μ^{CCTL} into CCTL). We define a translation ι of $\mu^{\mathsf{CCTL}}_{\Gamma_\mu,\Gamma_\nu}$ formulas θ into $\mathsf{CCTL}_{\Gamma_\mu,\Gamma_\nu}$ formulas by

$$\begin{aligned}
\iota\big(\boxdot_\gamma(\theta_1,\dots,\theta_{|\gamma|})\big) &:= \boxdot_\gamma(\iota\theta_1,\dots,\iota\theta_{|\gamma|}),\\
\iota\big(\spadesuit_\sigma\heartsuit_\lambda\theta\big) &:= \spadesuit_\sigma\heartsuit_\lambda(\iota\theta),\\
\iota\big(\mu u.\,\boxdot_{\gamma_\mu}(\theta_1,\dots,\theta_{|\gamma_\mu|-1},\spadesuit_\sigma\heartsuit_\lambda u)\big) &:= \spadesuit_\sigma\big(\mu u.\,\boxdot_{\gamma_\mu}(\iota\theta_1,\dots,\iota\theta_{|\gamma_\mu|-1},\heartsuit_\lambda u)\big),\\
\iota\big(\nu u.\,\boxdot_{\gamma_\nu}(\theta_1,\dots,\theta_{|\gamma_\nu|-1},\spadesuit_\sigma\heartsuit_\lambda u)\big) &:= \spadesuit_\sigma\big(\nu u.\,\boxdot_{\gamma_\nu}(\iota\theta_1,\dots,\iota\theta_{|\gamma_\nu|-1},\heartsuit_\lambda u)\big).
\end{aligned}$$

The translation ι is a bijection between μ^{CCTL} formulas and CCTL formulas. We call the inverse map ι^{-1} the *(fixpoint) encoding* of CCTL into μ^{CCTL}. Via this encoding, the semantics of μ^{CCTL} induces another semantics of CCTL, the step-wise semantics of CCTL.

Definition 4.5 (step-wise semantics). The *step-wise semantics* of each CCTL-formula ψ is given by $[\![\iota^{-1}\psi]\!]_{\mu^{\mathsf{CCTL}}}$.

We will prove the so-called *fixpoint characterization*, which is the equivalence of the path-based semantics (Definition 3.10) and the step-wise semantics (Definition 4.5) of CCTL. The classical fixpoint characterization theorem [13] for CTL asserts, for example, the following equivalence (1). The LHS below is the (path-based) interpretation of the CTL formula $\mathsf{E}(\mu u.\theta \vee \mathsf{X}u)$, and the RHS below is the (step-wise) interpretation of the $\mathbf{L}\boldsymbol{\mu}$ formula that encodes the formula $\mathsf{E}(\mu u.\theta \vee \mathsf{X}u)$.

$$[\![\mathsf{E}(\mu u.\theta \vee \mathsf{X}u)]\!] = [\![\mu u.\theta \vee \mathsf{EX}u]\!]. \tag{1}$$

Figure 1 illustrates the critical difference between these two interpretations. To verify the CTL formula $\mathsf{E}(\mu u.\theta\vee\mathsf{X}u)$, the path-based semantics (Fig. 1a) searches for a computation path along which the property θ eventually occurs. In contrast, the step-wise semantics (Fig. 1b) searches in a breadth-first manner for a state validating the property θ in the computation tree.

We generalize this classical result to our coalgebraic setting:

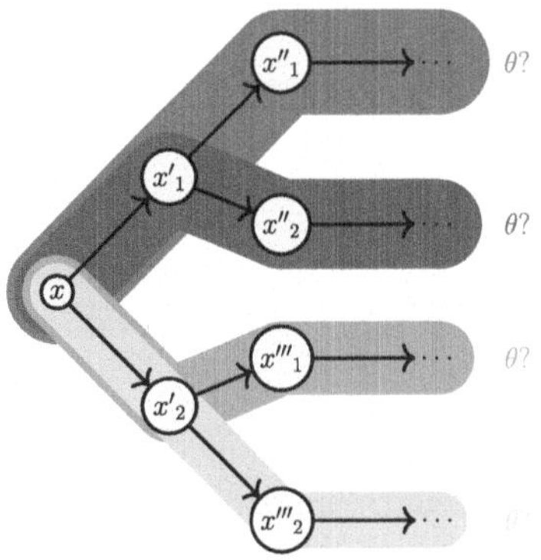

(a) Via the path-based semantics

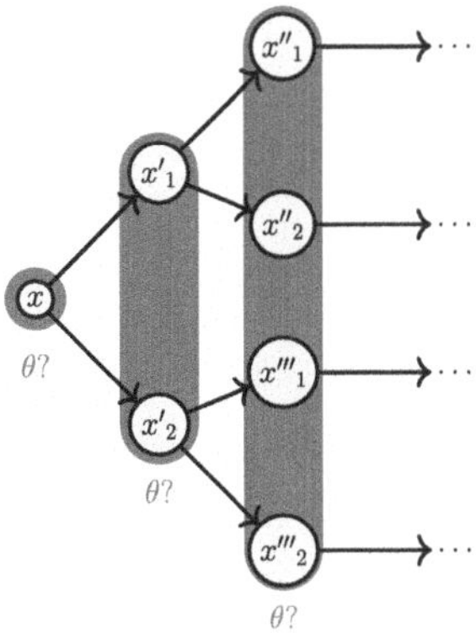

(b) Via the step-wise semantics

Fig. 1. Two equivalent interpretations of the CTL formula $\mathsf{E}(\mu u.\theta \vee \mathsf{X}u)$.

Theorem 4.6 (fixpoint characterization). *If the BT situation $\mathcal{S}$ with maximal execution satisfies Assumption 4.7, we have $[\![\theta]\!]_{\mu\mathsf{CCTL}} = [\![\iota\theta]\!]_{\mathsf{SFml}}$ for every μ^{CCTL} formula θ, and $[\![\iota^{-1}\psi]\!]_{\mu\mathsf{CCTL}} = [\![\psi]\!]_{\mathsf{SFml}}$ for every* CCTL *formula ψ.*

In this theorem, we identify sufficient conditions on the BT situation in categorical terms so that the fixpoint characterization holds.

Assumption 4.7 (the main assumption).

1. T is an affine monad,
2. the maximal trace $\mathrm{tr}(c')$ satisfies

$$\begin{array}{ccc} X \times TZ_X & \xrightarrow{\mathsf{st}_{X,Z_X}} & T(X \times Z_X) \\ {\scriptstyle \langle \mathrm{id}_X, \mathrm{tr}(c')\rangle}\uparrow & & \uparrow{\scriptstyle T\langle \zeta_1, \mathrm{id}_{Z_X}\rangle} \\ X & \xrightarrow{\mathrm{tr}(c')} & TZ_X, \end{array} \tag{2}$$

3. for every $\sigma \in \Sigma$, $\mathrm{ev}_\sigma = \sigma_\Omega(\mathrm{id}_\Omega) \colon T\Omega \to \Omega$ is an Eilenberg-Moore T-algebra,
4. for every $\sigma \in \Sigma$, $\lambda \in \Lambda$, and for every μ-scheme $\gamma_\mu \in \Gamma_\mu$ and ν-scheme $\gamma_\nu \in \Gamma_\nu$, we have

$$[\![\spadesuit_\sigma]\!](\mu\Phi_{\lambda,\gamma_\mu,\iota\vec{\theta}_{|\gamma_\mu|}}) \sqsubseteq \mu\Psi_{(\sigma,\lambda),\gamma_\mu,\vec{\theta}_{|\gamma_\mu|}}, \tag{3}$$

$$[\![\spadesuit_\sigma]\!](\nu\Phi_{\lambda\gamma_\nu,\iota\vec{\theta}_{|\gamma_\nu|}}) \sqsupseteq \nu\Psi_{(\sigma,\lambda),\gamma_\nu,\vec{\theta}_{|\gamma_\nu|}}, \tag{4}$$

   for every tuple of $\mu^{\mathsf{CCTL}}_{\Gamma_\mu,\Gamma_\nu}$ formulas $\vec{\theta}_{|\gamma|} = (\theta_1, \ldots, \theta_{|\gamma|})$, where Ψ, Φ are the operators defined in Definition 3.10 and Definition 4.2,
5. for every $\gamma \in \Gamma_\mu \cup \Gamma_\nu$ and $\sigma \in \Sigma$, $\gamma \colon \Omega^{|\gamma|} \to \Omega$ is bilinear [25, Section 1] with respect to the T-algebra $\mathrm{ev}_\sigma \colon T\Omega \to \Omega$, i.e.,

$$\begin{array}{ccccc} \Omega^n \times T\Omega & \xrightarrow{\mathsf{st}_{\Omega^n,\Omega}} & T(\Omega^n \times \Omega) & \xrightarrow{T\gamma} & T\Omega \\ {\scriptstyle \mathrm{id}_{\Omega^n} \times \mathrm{ev}_\sigma}\downarrow & & & & \downarrow{\scriptstyle \mathrm{ev}_\sigma} \\ \Omega^n \times \Omega & & \xrightarrow{\gamma} & & \Omega \end{array} \tag{5}$$

where $n = |\gamma| - 1$. In the case $|\gamma| = 0$, the above diagram becomes

$$\begin{array}{ccccc} \mathbf{1} \times T\mathbf{1} & \xrightarrow{\mathsf{st}_{\mathbf{1},\mathbf{1}}} & T(\mathbf{1} \times \mathbf{1}) & \xrightarrow{T\gamma} & T\Omega \\ {\scriptstyle \mathrm{id}_{\mathbf{1}} \times !_{T\mathbf{1}}}\downarrow & & & & {\scriptstyle \mathrm{ev}_\sigma}\downarrow \\ \mathbf{1} \times \mathbf{1} & & \xrightarrow{\quad\quad\gamma\quad\quad} & & \Omega, \end{array} \tag{6}$$

6. for every $\sigma \in \Sigma$ and $\lambda \in \Lambda$, the map $\mathrm{ev}_\lambda \circ \mathrm{inj}_\alpha \colon \Omega^{|\alpha|} \to \Omega$ is bilinear w.r.t. ev_σ, where $\mathrm{inj}_\alpha \colon \Omega^{|\alpha|} \to \coprod_{\alpha \in A} \Omega^{|\alpha|}$ is the injection of the index α.

Let us explain each condition in Assumption 4.7.

1. This condition asserts absence of deadlock states. Technically, it is needed here to ensure the compatibility of the strength map of T with the first projection (that is, $T\pi_1 \circ \mathsf{st}_{X,Y} = \eta_X \circ \pi_2$), which, in turn, ensures that the original $T \circ F$-coalgebra structure can be recovered from its execution map $\mathrm{tr}(c')$ (that is, Lemma 4.8).
2. This condition is quite technical but harmless and used to prove one of our key results, Proposition 4.9. A similar condition can be found in [23], as *strong affine-ness*. Indeed, we can show every strongly affine monad satisfies condition 2. Since both $\mathcal{P}^+$ and $\mathcal{G}_1$ are strongly affine, condition 2 is satisfied by both $\mathcal{S}_{\mathrm{ND}}$ and $\mathcal{S}_{\mathrm{R}}$ (see Table 2).
3. This condition, especially the associativity of the Eilenberg-Moore T-algebra ev_σ, enables us to reduce many-fold branching (i.e., several applications of the path quantifier σ) to single branching (i.e., just one application).
4. This condition states that the path quantifier σ preserves the least/greatest fixpoints of the operators Ψ, Φ. In the logical perspective, the inequality (3) means "any path-based witness can be reached in step-wise manner," and the inequality (4) means "step-wise validity guarantees path-based validity."
5. This condition expresses the bilinearity of μ-schemes $\gamma_\mu \in \Gamma_\mu$ and ν-schemes $\gamma_\nu \in \Gamma_\nu$: each application of a path quantifier $\spadesuit_\sigma$ on a formula of the form $\boxdot_\gamma(\vec{\psi}, \heartsuit_\lambda \varphi)$ is calculated by passing $\spadesuit_\sigma$ *inside*, as $\boxdot_\gamma(\vec{\psi}, \spadesuit_\sigma \heartsuit_\lambda \varphi)$.
6. This condition captures the coherence between path quantifiers $\sigma \in \Sigma$ and next-time operators $\lambda \in \Lambda$. If we choose the canonical predicate lifting $\mathrm{Pred}(F)$ as λ, this condition is a consequence of condition 5, because $\mathrm{Pred}(F)$ is constructed from conjunction, i.e., $\mathrm{ev}_\lambda \circ \mathrm{inj}_\alpha = \wedge$; see Example 2.5 (1).

Before starting the proof of Theorem 4.6, we introduce two important results.

The first one (Lemma 4.8) is a consequence of condition 1 of Assumption 4.7, and states that taking the head (ζ_1) of the tail (ζ_2) of paths starting from a state x yields successors of x.

Lemma 4.8. $T(F\zeta_1 \circ \zeta_2) \circ \mathrm{tr}(c') = c$.

The second one (Proposition 4.9) is a coalgebraic generalization of the *expansion law* [1] of CTL. When instantiated to the CTL formula $\mathsf{E}(p_1 \mathsf{U} p_2)$, it means

$$[\![\mathsf{E}(p_1 \mathsf{U} p_2)]\!] = p_2 \vee (p_1 \wedge [\![\mathsf{EX}]\!][\![\mathsf{E}(p_1 \mathsf{U} p_2)]\!]).$$

Analogous to the classical one, our coalgebraic expansion law is critically used in the induction in the proof of the fixpoint characterization. It depends on all conditions of Assumption 4.7 but condition 4.

Proposition 4.9 (coalgebraic expansion law). *Let $\sigma \in \Sigma$, $\lambda \in \Lambda$, and μ-schemes $\gamma_\mu \in \Gamma_\mu$ and ν-schemes $\gamma_\nu \in \Gamma_\nu$. We have*

$$[\![\spadesuit_\sigma]\!](\mu \Phi_{\lambda,\gamma_\mu,\iota\vec{\theta}_{|\gamma_\mu|-1}}) \sqsupseteq \Psi_{(\sigma,\lambda),\gamma_\mu,\vec{\theta}_{|\gamma_\mu|-1}}([\![\spadesuit_\sigma]\!](\mu \Phi_{\lambda,\gamma_\mu,\iota\vec{\theta}_{|\gamma_\mu|-1}})) \tag{7}$$

for $\theta_1, \ldots, \theta_{|\gamma_\mu|-1}$ with $[\![\iota\theta_i]\!]_{\mathsf{SFml}} \sqsupseteq [\![\theta_i]\!]_{\mu^{\mathsf{CCTL}}}$ for $i = 1, \ldots, |\gamma_\mu| - 1$, and

$$[\![\spadesuit_\sigma]\!](\nu \Phi_{\lambda,\gamma_\nu,\iota\vec{\theta}_{|\gamma_\nu|-1}}) \sqsubseteq \Psi_{(\sigma,\lambda),\gamma_\nu,\vec{\theta}_{|\gamma_\nu|-1}}([\![\spadesuit_\sigma]\!](\nu \Phi_{\lambda,\gamma_\nu,\iota\vec{\theta}_{|\gamma_\nu|-1}})) \tag{8}$$

for $\theta_1, \ldots, \theta_{|\gamma_\nu|-1}$ with $[\![\iota\theta_i]\!]_{\mathsf{SFml}} \sqsubseteq [\![\theta_i]\!]_{\mu^{\mathsf{CCTL}}}$ for $i = 1, \ldots, |\gamma_\nu| - 1$. Furthermore, if $[\![\iota\theta_i]\!]_{\mathsf{SFml}} = [\![\theta_i]\!]_{\mu^{\mathsf{CCTL}}}$ for every subformula θ_i, the inequalities 7 and 8 are both equalities.

Proof (Sketch of Theorem 4.6). Since ι is a bijection between μ^{CCTL} and CCTL, it suffices to show

$$[\![\theta]\!]_{\mu^{\mathsf{CCTL}}} = [\![\iota\theta]\!]_{\mathsf{SFml}} \tag{9}$$

for every $\theta \in \mu^{\mathsf{CCTL}}$. We prove Eq. (9) by induction on the construction of θ.

For $\theta = \boxdot_\gamma(\theta_1, \ldots, \theta_{|\gamma|})$, Eq. (9) is straightforward.

For $\theta = \spadesuit_\sigma \heartsuit_\lambda \theta'$, by I.H. and naturality of λ and σ, we obtain

$$[\![\iota(\spadesuit_\sigma \heartsuit_\lambda \theta')]\!]_{\mathsf{SFml}} = \big(T(F\zeta_1 \circ \zeta_2) \circ \mathrm{tr}(c')\big)^* \circ \sigma_{FX} \circ \lambda_X([\![\theta']\!]_{\mu^{\mathsf{CCTL}}}).$$

Thus, by Lemma 4.8 and Definition 4.2, we have

$$[\![\iota(\spadesuit_\sigma \heartsuit_\lambda \theta')]\!]_{\mathsf{SFml}} = c^* \circ \sigma_{FX} \circ \lambda_X([\![\theta']\!]_{\mu^{\mathsf{CCTL}}}) = [\![\spadesuit_\sigma \heartsuit_\lambda \theta']\!]_{\mu^{\mathsf{CCTL}}}.$$

Next, we move on to the case $\theta = \mu u.\ \boxdot_{\gamma_\mu} (\theta_1, \ldots, \theta_{|\gamma_\mu|-1}, \spadesuit_\sigma \heartsuit_\lambda u)$. Firstly, we hypothesize $\theta_1, \ldots, \theta_{|\gamma_\mu|-1}$ with $[\![\iota\theta_i]\!]_{\mathsf{SFml}} = [\![\theta_i]\!]_{\mu^{\mathsf{CCTL}}}$ for $i = 1, \ldots, |\gamma_\mu| - 1$. Under the notation introduced in Definition 3.10 and Definition 4.2, we have

$$[\![\mu\, u.\ \boxdot_{\gamma_\mu} (\theta_1, \ldots, \theta_{|\gamma_\mu|}, \spadesuit_\sigma \heartsuit_\lambda u)]\!]_{\mu^{\mathsf{CCTL}}} = \mu \Psi_{\gamma_\mu,\vec{\theta}_{|\gamma_\mu|}},$$
$$[\![\iota(\mu u.\ \boxdot_{\gamma_\mu} (\theta_1, \ldots, \theta_{|\gamma_\mu|}, \spadesuit_\sigma \heartsuit_\lambda u))]\!]_{\mathsf{SFml}} = [\![\spadesuit_\sigma]\!] \mu \Phi_{\gamma_\mu,\iota\vec{\theta}_{|\gamma_\mu|}}.$$

Thus, the last task is to prove $[\![\spadesuit_\sigma]\!](\mu\Phi_{\gamma,\vec{\psi}}) = \mu\Psi_{\gamma,\vec{\psi}}$. The direction LHS $\sqsubseteq$ RHS is already assumed in condition 4 of Assumption 4.7.

We show the other direction, LHS $\sqsupseteq$ RHS. To prove this, we recall the Knaster-Tarski fixpoint theorem [35]: the least fixpoint of a monotone function on a complete lattice is exactly the minimal of all pre-fixpoints of the function. Since LHS is a pre-fixpoint of the operator Ψ by Proposition 4.9, we conclude LHS $\sqsupseteq$ RHS by the Knaster-Tarski fixpoint theorem.

The proof for the case $\theta = \nu u.\ \boxdot_{\gamma_\nu} (\theta_1, \ldots, \theta_{|\gamma_\nu|-1}, \spadesuit_\sigma \heartsuit_\lambda u)$ is similar to the least fixpoint case since condition 4 of Assumption 4.7 is symmetric to μ and ν. $\square$

Examining the above proof, we can also obtain a *partial* fixpoint characterization.

Proposition 4.10 (partial fixpoint characterization). *Under the same assumption of Theorem 4.6 (Assumption 4.7) but without condition 4, we have*

1. $[\![\theta]\!]_{\mu\mathsf{CCTL}} = [\![\iota\theta]\!]_{\mathsf{SFml}}$ *for a formula* θ *without any* μ *or* ν*,*
2. $[\![\theta]\!]_{\mu\mathsf{CCTL}} \sqsubseteq [\![\iota\theta]\!]_{\mathsf{SFml}}$ *for a formula* θ *with only* μ*'s,*
3. $[\![\theta]\!]_{\mu\mathsf{CCTL}} \sqsupseteq [\![\iota\theta]\!]_{\mathsf{SFml}}$ *for a formula* θ *with only* ν*'s.*

4.3 Examples and Non-examples of Assumption 4.7

The non-deterministic BT situation $\mathcal{S}_{\mathrm{ND}}$ satisfies Assumption 4.7, as expected.

Proposition 4.11. *$\mathcal{S}_{ND}$ satisfies Assumption 4.7 with $\Gamma_\mu = \{(\_\vee(\_\wedge\_))\}$ and $\Gamma_\nu = \{(\_\wedge(\_\vee\_))\}$. Thus, $\mathcal{S}_{ND}$ enjoys the fixpoint characterization (Theorem 4.6).*

Proof (Sketch). The conditions of Assumption 4.7 other than 4 can be shown by literal calculation. Condition 4 is instantiated in $\mathcal{S}_{\mathrm{ND}}$ as

$$[\![\mathsf{E}(\theta_1 \mathsf{U} \theta_2)]\!] \subseteq \mu u.\, \theta_2 \vee (\theta_1 \wedge [\![\mathsf{EX}]\!] u)$$
$$[\![\mathsf{E}(\theta_1 \mathsf{W} \theta_2)]\!] \supseteq \nu u.\, \theta_1 \wedge (\theta_2 \vee [\![\mathsf{EX}]\!] u)$$

for E (the A case is given likewise). Proof of these inequalities is presented in [1, Thm. 6.23]. Note that although there the state set is assumed to be finite, this assumption can be lifted: the proof uses the expansion law, but the law can be obtained by checking the conditions of Assumption 4.7 other than 4 (recall our proof of the coalgebraic expansion law does not depend on condition 4). The rest of the proof in [1] can be done without the finiteness assumption.

□

On the other hand, the probabilistic BT situations ($\mathcal{S}_{\mathrm{R}}$ and $\mathcal{S}_{\mathrm{qR}}$) fail to satisfy some conditions of Assumption 4.7, and hence to have the fixpoint characterization.

Fact 4.12. *$\mathcal{S}_R$ and $\mathcal{S}_{qR}$ do not satisfy Assumption 4.7.*

Firstly, $\mathcal{S}_{\mathrm{R}}$ does not satisfy condition 3 of Assumption 4.7, i.e., the requirement for the $\mathcal{G}_1$-modality $\sigma = \geq_q : \mathcal{G}_1(\mathbf{2}, \mathcal{P}\mathbf{2}) \to (\mathbf{2}, \mathcal{P}\mathbf{2})$ to be an Eilenberg-Moore $\mathcal{G}_1$-algebra in **SB**. Indeed, the modality breaks the associativity condition of Eilenberg-Moore $\mathcal{G}_1$-algebras. The associativity means the following diagram commutes for every $\rho \in \mathcal{G}_1(\mathcal{G}_1(\mathbf{2}, \mathcal{P}\mathbf{2})) \cong \mathcal{G}_1([0,1], \Sigma_{[0,1]})$, where $\Sigma_{[0,1]}$ is the Borel set generated from the usual topology of $[0,1]$:

$$\begin{array}{ccc}
\mathcal{G}_1(\mathcal{G}_1(\mathbf{2},\mathcal{P}\mathbf{2})) \cong \mathcal{G}_1([0,1],\Sigma_{[0,1]}) & \xrightarrow{\mathcal{G}_1(\geq_q)} & \mathcal{G}_1(\mathbf{2},\mathcal{P}\mathbf{2}) \cong ([0,1],\Sigma_{[0,1]}) \\
\downarrow{\scriptstyle \mu_{(\mathbf{2},\mathcal{P}\mathbf{2})}} & & \downarrow{\scriptstyle \geq_q} \\
\mathcal{G}_1(\mathbf{2},\mathcal{P}\mathbf{2}) \cong ([0,1],\Sigma_{[0,1]}) & \xrightarrow[\geq_q]{} & (\mathbf{2},\mathcal{P}\mathbf{2}).
\end{array}$$

The commutativity of this diagram can be further rephrased as follows: the condition $\rho([q,1]) \geq q$ is equivalent to $\int_{r\in[0,1]} \rho(r)\, dr \geq q$ for every measure ρ. However, by taking a real number q other than 0 or 1, this equivalence fails. Thus, the associativity condition of Eilenberg-Moore $\mathcal{G}_1$-algebras also fails for q other than 0 or 1.

This suggests that by restricting the modality parameter q to 0 or 1, we can make condition 3 hold. This restriction is realized by the BT situation $\mathcal{S}_{\mathrm{qR}}$ (see Example 3.2).

Nevertheless, for $\mathcal{S}_{\mathrm{qR}}$, condition 4 of Assumption 4.7 is violated. The violation can be seen in a simple counterexample shown in Fig. 2 (found in [3]). While the PCTL formula $\mathbb{P}_{\geq 1}(\mu u.p \vee \mathsf{X}u)$ is interpreted as $\{x, y\}$ in this example, the encoded probabilistic mu-formula $\mu u.p \vee \mathbb{P}_{\geq 1}\mathsf{X}u$ is interpreted as $\{y\}$. Thus, we have $\mathbb{P}_{\geq 1}(\mu u.p \vee \mathsf{X}u) \sqsupset \mu u.p \vee \mathbb{P}_{\geq 1}\mathsf{X}u$, which breaks condition 4 of Assumption 4.7.

Nonetheless, we also have the following positive result.

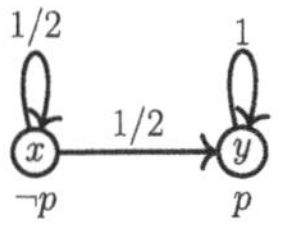

Fig. 2. A counterexample Markov chain.

Proposition 4.13. *$\mathcal{S}_{qR}$ with its state space $(X, \mathcal{P}X)$ for a countable set X satisfies the other conditions of Assumption 4.7 than condition 4 with $\Gamma_\mu = \{(\_ \vee (\_ \wedge \_))\}$ and $\Gamma_\nu = \{(\_ \wedge (\_ \vee \_))\}$ Thus, $\mathcal{S}_{qR}$ with countable $(X, \mathcal{P}X)$ enjoys the partial fixpoint characterization (Proposition 4.10).*

Remark 4.14. We saw we can not construct a step-wise semantics of PCTL equivalent to its path-based one via our encoding (Definition 4.4). In fact, we can make a stronger statement: *any* fixpoint encoding of PCTL into the probabilistic mu-calculus does not preserve semantics. Indeed, PCTL does not have the finite model property [3], whereas the probabilistic mu-calculus does [7]. One example of PCTL formula with no finite model is $\mathbb{P}_{>0}\mathsf{G}(\neg p \wedge \mathbb{P}_{>0}\mathsf{F}p)$ for any atomic predicate p.

5 A Polynomial-Time Model-Checking Algorithm for CCTL

Thanks to the fixpoint characterization, we can obtain a polynomial-time model-checking algorithm $\mathsf{MC}_{\mathcal{S}}^{\mathsf{CCTL}}$ for CCTL. It is based on the standard model-checking algorithm for CTL [8]. Nevertheless, the algorithm $\mathsf{MC}_{\mathcal{S}}^{\mathsf{CCTL}}$ is described in categorical terms, with the following additional conditions on the BT situation $\mathcal{S}$.

Assumption 5.1.

1. The ambient category $\mathbb{C}$ is concrete [27].
2. The underlying set of X is finite, with its size denoted by $|X|$.
3. The underlying set of Ω is $\mathbf{2}$.

Algorithm 1 A CCTL model-checking algorithm $\mathsf{MC}_{\mathcal{S}}^{\mathsf{CCTL}}$.

Input: A CCTL formula ψ.
Output: An Ω-predicate $U \in \Omega^X$. ▷ where $\mathcal{S} = (\mathbb{C}, T, F, c, \Omega, \Sigma, \Lambda)$.

procedure CHECK(θ)
  switch θ **do**

    case $\boxdot_\gamma(\theta_1, \dots, \theta_{|\gamma|})$
      return $\gamma(\mathrm{CHECK}(\theta_1), \dots, \mathrm{CHECK}(\theta_{|\gamma|}))$
    end case

    case $\spadesuit_\sigma \heartsuit_\lambda \theta'$
      return $[\![\spadesuit_\sigma \heartsuit_\lambda]\!](\mathrm{CHECK}(\theta'))$
    end case

    case $\mu u.\ \boxdot_\gamma\ (\theta_1, \dots, \theta_{|\gamma_\mu|-1}, \spadesuit_\sigma \heartsuit_\lambda u)$
      $U := \bot$; $V := \gamma_\mu(\mathrm{CHECK}(\theta_1), \dots, \mathrm{CHECK}(\theta_{|\gamma_\mu|-1}), [\![\spadesuit_\sigma \heartsuit_\lambda]\!](\bot))$
      while $U \neq V$ **do**
        $U := V$
        $V := \gamma_\mu(\mathrm{CHECK}(\theta_1), \dots, \mathrm{CHECK}(\theta_{|\gamma_\mu|-1}), [\![\spadesuit_\sigma \heartsuit_\lambda]\!](U))$
      end while
      return U
    end case

    case $\nu u.\ \boxdot_{\gamma_\nu}\ (\theta_1, \dots, \theta_{|\gamma_\nu|-1}, \spadesuit_\sigma \heartsuit_\lambda u)$
      $U := \top$; $V := \gamma_\nu(\mathrm{CHECK}(\theta_1), \dots, \mathrm{CHECK}(\theta_{|\gamma_\nu|-1}), [\![\spadesuit_\sigma \heartsuit_\lambda]\!](\top))$
      while $U \neq V$ **do**
        $U := V$
        $V := \gamma_\nu(\mathrm{CHECK}(\theta_1), \dots, \mathrm{CHECK}(\theta_{|\gamma_\nu|-1}), [\![\spadesuit_\sigma \heartsuit_\lambda]\!](U))$
      end while
      return U
    end case

end procedure
return CHECK($\iota^{-1}\psi$)

By Assumption 5.1, we can identify Ω-predicates with subsets of the underlying set of X and the maps γ and $[\![\spadesuit_\sigma \heartsuit_\lambda]\!]$ with corresponding predicate transformers.

Given the BT situation $\mathcal{S}$ and a specification $\psi \in \mathsf{CCTL}$, the algorithm $\mathsf{MC}_{\mathcal{S}}^{\mathsf{CCTL}}$ calculates $[\![\psi]\!]_{\mathsf{SFml}}$, which is the interpretation of ψ. The calculation steps are shown in Algorithm 1. Firstly, the CCTL formula ψ is encoded into a μ^{CCTL} formula $\iota^{-1}\psi$ (cf. Definition 4.4). Next, the μ^{CCTL} formula $\iota^{-1}\psi$ is passed to the procedure CHECK($\iota^{-1}\psi$), which is the core of $\mathsf{MC}_{\mathcal{S}}^{\mathsf{CCTL}}$. The procedure call calculates $[\![\iota^{-1}\psi]\!]_{\mu^{\mathsf{CCTL}}}$ in a step-wise manner. The calculation result coincides with $[\![\psi]\!]_{\mathsf{SFml}}$ by the fixpoint characterization (Theorem 4.6).

The procedure CHECK(θ) is a simplification of an existing model-checking algorithm for the coalgebraic μ-calculus $\mathbf{C}\boldsymbol{\mu}$ [19]. In the body of CHECK(θ), one out of four cases is chosen according to the structure of θ. The first two cases, one for boolean operators and one for modalities, are straightforward. In the least fixpoint case, we exploit the Cousot-Cousot fixpoint theorem [9], which approximates the least fixpoint by an ascending chain in Ω^X starting from the least element $\bot$. The greatest fixpoint case is similar to the least fixpoint case.

Termination of CHECK(θ), and hence $\mathsf{MC}_{\mathcal{S}}^{\mathsf{CCTL}}$ as a whole, is a direct consequence of our finiteness assumption in Assumption 5.1. The encoding ι^{-1} is also terminating. Correctness, particularly that of the two while loops (at Line 11 and Line 19), follows from the Cousot-Cousot fixpoint theorem.

Proposition 5.2 (termination and correctness of $\mathsf{MC}_{\mathcal{S}}^{\mathsf{CCTL}}$). *For a given* CCTL *formula* ψ*, the algorithm* $\mathsf{MC}_{\mathcal{S}}^{\mathsf{CCTL}}$ *terminates and returns* $[\![\psi]\!]_{\mathsf{SFml}}$.

To estimate the complexity bound of our algorithm $\mathsf{MC}_{\mathcal{S}}^{\mathsf{CCTL}}$, we abstract the time to compute each modality $\spadesuit_\sigma \heartsuit_\lambda$. Our formulation here follows [20, Def. 2].

Definition 5.3 (one-step satisfaction problem [20, Def. 2]). The *one-step satisfaction problem* w.r.t. σ and λ for a state $x \in X$ and an Ω-predicate U is to decide whether $x \in [\![\spadesuit_\sigma \heartsuit_\lambda]\!](U)$ or not. We denote the time to solve this problem by $t\big((\sigma, \lambda), x, U\big)$ and define $t(\sigma, \lambda) = \max_{x \in X, U \in \Omega^X} t\big((\sigma, \lambda), x, U\big)$.

We show $\mathsf{MC}_{\mathcal{S}}^{\mathsf{CCTL}}$ is at most polynomial time under moderate assumptions.

Proposition 5.4 (complexity bound of $\mathsf{MC}_{\mathcal{S}}^{\mathsf{CCTL}}$). *Let* $|\psi|$ *be the number of subformulas in* ψ*, and* N *be a constant that bounds the time to execute the boolean operations used in* ψ*. The complexity of* $\mathsf{MC}_{\mathcal{S}}^{\mathsf{CCTL}}$ *is given by*

$$O\Big(|\psi| \cdot |X| \cdot \big(N + t(\sigma, \lambda) + 2 \cdot t(\sigma, \lambda) \cdot N\big) + |\psi|\Big).$$

When $t(\sigma, \lambda)$ *is polynomial to the size* $|X|$*, so is the complexity of* $\mathsf{MC}_{\mathcal{S}}^{\mathsf{CCTL}}$.

Example 5.5 (fixpoint model checking for CTL). The instance $\mathsf{MC}_{\mathcal{S}_{\mathrm{ND}}}^{\mathsf{CCTL}}$ corresponds to the well-known model-checking algorithm for CTL via fixpoints [8]. Since the time $t(\sigma, \lambda)$ is bounded by $|X|$ as in [20, Example 3], Proposition 5.4 recovers the known quadratic complexity bound of the classical algorithm.

6 Conclusion and Future Work

We formulated a new path-based coalgebraic logic CCTL (Definition 3.7), as an abstraction of classical CTL. We introduced an encoding of CCTL formulas into step-wise μ^{CCTL} formulas, which captures the categorical essence of the standard encoding of CTL into $\mathbf{L}\boldsymbol{\mu}$. This encoding is proven to preserve the semantics (Theorem 4.6) under some semantic conditions (Assumption 4.7) formulated in purely categorical terms. We saw these conditions distinguish classical CTL, which enjoys the fixpoint characterization (Proposition 4.11), and PCTL, which violates some conditions and enjoys only limited results (Proposition 4.13). Our coalgebraic fixpoint characterization yielded a naive model-checking algorithm $\mathsf{MC}_{\mathcal{S}}^{\mathsf{CCTL}}$ of CCTL, whose complexity is analyzed to be polynomial (Proposition 5.4).

The genericity of our framework of CCTL will allow several interesting extensions: n-ary next-time operators and non-boolean logical connectives could be smoothly incorporated. By changing the branching type T, our framework is expected to not only encompass other known examples like quantitative variants of CTL [4,30] but also yield novel efficient path-based logics. We will investigate monotone neighborhood frames [16] and aim to establish "Monotone Neighborhood CTL" which would provide an efficient path-based language for Parikh's game logic [17,31]. We will also explore $[0, 1]$-valued probabilistic path-based logics and corresponding probabilistic mu-calculus validating the fixpoint characterization. Such path-based logics would resemble the quantitative LTL [6].

We could also extend our encoding ι^{-1} to the coalgebraic path-based logic $\mu\mathcal{L}$, as an abstraction of classical exponential encodings of CTL* into the mu-calculus [2,10].

Acknowledgement. PR.K. and I.H. are supported by JST ERATO HASUO Metamathematics for Systems Design Project (No. JPMJER1603). R.K. is also supported by JST, the establishment of university fellowships towards the creation of science technology innovation (No. JPMJFS2123). I.H. is also supported by JST ASPIRE Grant No. JPMJAP2301. C.C. is supported by the Leverhulme Trust Research Project Grant RPG-2020-232, UK. K.M. is supported by JSPS, KAKENHI Project No. 22K17850, Japan.

References

1. Baier, C., Katoen, J.: Principles of Model Checking. MIT Press, Cambridge (2008)
2. Bhat, G., Cleaveland, R.: Efficient model checking via the equational μ-calculus. In: Proceedings, 11th Annual IEEE Symposium on Logic in Computer Science, New Brunswick, New Jersey, USA, 27–30 July 1996, pp. 304–312. IEEE Computer Society (1996). https://doi.org/10.1109/LICS.1996.561358
3. Brázdil, T., Forejt, V., Kretínský, J., Kucera, A.: The satisfiability problem for probabilistic CTL. In: Proceedings of the Twenty-Third Annual IEEE Symposium on Logic in Computer Science, LICS 2008, 24–27 June 2008, Pittsburgh, PA, USA, pp. 391–402. IEEE Computer Society (2008). https://doi.org/10.1109/LICS.2008.21
4. Chechik, M., Devereux, B., Easterbrook, S.M., Gurfinkel, A.: Multi-valued symbolic model-checking. ACM Trans. Softw. Eng. Methodol. **12**(4), 371–408 (2003). https://doi.org/10.1145/990010.990011
5. Cîrstea, C.: Maximal traces and path-based coalgebraic temporal logics. Theor. Comput. Sci. **412**(38), 5025–5042 (2011). https://doi.org/10.1016/j.tcs.2011.04.025
6. Cirstea, C.: Linear-time logics – a coalgebraic perspective (2023)
7. Cîrstea, C., Kupke, C., Pattinson, D.: EXPTIME tableaux for the coalgebraic μ-calculus. In: Grädel, E., Kahle, R. (eds.) CSL 2009. LNCS, vol. 5771, pp. 179–193. Springer, Heidelberg (2009). https://doi.org/10.1007/978-3-642-04027-6_15
8. Clarke, E.M., Grumberg, O., Kroening, D., Peled, D.A., Veith, H.: Model Checking, 2nd edn. MIT Press, Cambridge (2018). https://mitpress.mit.edu/books/model-checking-second-edition
9. Cousot, P., Cousot, R.: Constructive versions of Tarski's fixed point theorems. Pac. J. Math. **82**(1), 43–57 (1979)
10. Dam, M.: CTL* and ECTL* as fragments of the modal mu-calculus. Theor. Comput. Sci. **126**(1), 77–96 (1994). https://doi.org/10.1016/0304-3975(94)90269-0

11. Doob, J.L.: Measure Theory. Springer, New York (1994). https://doi.org/10.1007/978-1-4612-0877-8
12. Emerson, E.A., Clarke, E.M.: Using branching time temporal logic to synthesize synchronization skeletons. Sci. Comput. Program. **2**(3), 241–266 (1982). https://doi.org/10.1016/0167-6423(83)90017-5
13. Emerson, E.A., Halpern, J.Y.: Decision procedures and expressiveness in the temporal logic of branching time. J. Comput. Syst. Sci. **30**(1), 1–24 (1985). https://doi.org/10.1016/0022-0000(85)90001-7
14. Emerson, E.A., Halpern, J.Y.: "sometimes" and "not never" revisited: on branching versus linear time temporal logic. J. ACM **33**(1), 151–178 (1986). https://doi.org/10.1145/4904.4999
15. Giry, M.: A categorical approach to probability theory. In: Banaschewski, B. (ed.) Categorical Aspects of Topology and Analysis. LNM, vol. 915, pp. 68–85. Springer, Heidelberg (1982). https://doi.org/10.1007/BFb0092872
16. Hansen, H.H., Kupke, C.: A coalgebraic perspective on monotone modal logic. In: Adámek, J., Milius, S. (eds.) Proceedings of the Workshop on Coalgebraic Methods in Computer Science, CMCS 2004, Barcelona, Spain, 27–29 March 2004. Electronic Notes in Theoretical Computer Science, vol. 106, pp. 121–143. Elsevier (2004). https://doi.org/10.1016/j.entcs.2004.02.028
17. Hansen, H.H., Kupke, C., Marti, J., Venema, Y.: Parity games and automata for game logic (extended version). CoRR abs/1709.00777 (2017). http://arxiv.org/abs/1709.00777
18. Hansson, H., Jonsson, B.: A logic for reasoning about time and reliability. Formal Aspects Comput. **6**(5), 512–535 (1994). https://doi.org/10.1007/BF01211866
19. Hasuo, I., Shimizu, S., Cîrstea, C.: Lattice-theoretic progress measures and coalgebraic model checking. In: Bodík, R., Majumdar, R. (eds.) Proceedings of the 43rd Annual ACM SIGPLAN-SIGACT Symposium on Principles of Programming Languages, POPL 2016, St. Petersburg, FL, USA, 20–22 January 2016, pp. 718–732. ACM (2016). https://doi.org/10.1145/2837614.2837673
20. Hausmann, D., Schröder, L.: Game-based local model checking for the coalgebraic mu-calculus. In: Fokkink, W.J., van Glabbeek, R. (eds.) 30th International Conference on Concurrency Theory, CONCUR 2019, 27–30 August 2019, Amsterdam, the Netherlands. LIPIcs, vol. 140, pp. 35:1–35:16. Schloss Dagstuhl - Leibniz-Zentrum für Informatik (2019). https://doi.org/10.4230/LIPIcs.CONCUR.2019.35
21. Jacobs, B.: Semantics of weakening and contraction. Ann. Pure Appl. Log. **69**(1), 73–106 (1994). https://doi.org/10.1016/0168-0072(94)90020-5
22. Jacobs, B.: Trace semantics for coalgebras. In: Adámek, J., Milius, S. (eds.) Proceedings of the Workshop on Coalgebraic Methods in Computer Science, CMCS 2004, Barcelona, Spain, 27–29 March 2004. Electronic Notes in Theoretical Computer Science, vol. 106, pp. 167–184. Elsevier (2004). https://doi.org/10.1016/j.entcs.2004.02.031
23. Jacobs, B.: Affine monads and side-effect-freeness. In: Hasuo, I. (ed.) CMCS 2016. LNCS, vol. 9608, pp. 53–72. Springer, Cham (2016). https://doi.org/10.1007/978-3-319-40370-0_5
24. Jacobs, B.: Introduction to Coalgebra: Towards Mathematics of States and Observation. Cambridge Tracts in Theoretical Computer Science. Cambridge University Press, Cambridge (2016). https://doi.org/10.1017/CBO9781316823187
25. Kock, A.: Bilinearity and cartesian closed monads. Math. Scand. **29**, 161–174 (1971)
26. Kozen, D.: Results on the propositional mu-calculus. Theor. Comput. Sci. **27**, 333–354 (1983). https://doi.org/10.1016/0304-3975(82)90125-6

27. MacLane, S.: Categories for the Working Mathematician. Graduate Texts in Mathematics, vol. 5. Springer, New York (1971)
28. Milner, R.: Communication and Concurrency. PHI Series in Computer Science. Prentice Hall, Hoboken (1989)
29. Moss, L.S.: Coalgebraic logic. Ann. Pure Appl. Log. **96**(1–3), 277–317 (1999). https://doi.org/10.1016/S0168-0072(98)00042-6
30. Pan, H., Li, Y., Cao, Y., Ma, Z.: Model checking computation tree logic over finite lattices. Theor. Comput. Sci. **612**, 45–62 (2016). https://doi.org/10.1016/J.TCS.2015.10.014
31. Parikh, R.: The logic of games and its applications. In: Selected Papers of the International Conference on "Foundations of Computation Theory" on Topics in the Theory of Computation, pp. 111–139. Elsevier North-Holland, Inc., USA (1985)
32. Pattinson, D.: Coalgebraic modal logic: soundness, completeness and decidability of local consequence. Theor. Comput. Sci. **309**(1–3), 177–193 (2003). https://doi.org/10.1016/S0304-3975(03)00201-9
33. Schröder, L., Pattinson, D.: PSPACE bounds for rank-1 modal logics. ACM Trans. Comput. Log. **10**(2), 13:1–13:33 (2009). https://doi.org/10.1145/1462179.1462185
34. Schubert, C.: Terminal coalgebras for measure-polynomial functors. In: Chen, J., Cooper, S.B. (eds.) TAMC 2009. LNCS, vol. 5532, pp. 325–334. Springer, Heidelberg (2009). https://doi.org/10.1007/978-3-642-02017-9_35
35. Tarski, A.: A lattice-theoretical fixpoint theorem and its applications. Pac. J. Math. **5**, 285–309 (1955). https://api.semanticscholar.org/CorpusID:13651629
36. Urabe, N., Hasuo, I.: Categorical Büchi and parity conditions via alternating fixed points of functors. In: Cîrstea, C. (ed.) CMCS 2018. LNCS, vol. 11202, pp. 214–234. Springer, Cham (2018). https://doi.org/10.1007/978-3-030-00389-0_12
37. Urabe, N., Hasuo, I.: Coalgebraic infinite traces and kleisli simulations. Log. Methods Comput. Sci. **14**(3) (2018). https://doi.org/10.23638/LMCS-14(3:15)2018
38. Venema, Y.: Automata and fixed point logic: a coalgebraic perspective. Inf. Comput. **204**(4), 637–678 (2006). https://doi.org/10.1016/J.IC.2005.06.003

A Categorical Approach to Coalgebraic Fixpoint Logic

Ezra Schoen[1(✉)], Clemens Kupke[1], Jurriaan Rot[2], and Ruben Turkenburg[2]

[1] University of Strathclyde, Glasgow, Scotland
{ezra.schoen,clemens.kupke}@strath.ac.uk
[2] Radboud Universiteit, Nijmegen, The Netherlands
jrot@cs.ru.nl, ruben.turkenburg@ru.nl

Abstract. We define a framework for incorporating alternation-free fixpoint logics into the dual-adjunction setup for coalgebraic modal logics. We achieve this by using order-enriched categories. We give a least-solution semantics as well as an initial algebra semantics, and prove they are equivalent. We also show how to place the alternation-free coalgebraic μ-calculus in this framework, as well as PDL and a logic with a probabilistic dynamic modality.

1 Introduction

Coalgebra provides a versatile framework for representing different types of state-based dynamic systems in a uniform way [26]. At the heart of the coalgebraic theory lie the semantic notions of behaviour and behavioural equivalence. It is well-known that modal logics provide the appropriate syntactic tools to specify and reason about labelled transition systems in a fully abstract way, i.e., in a way that ensures that the language precisely characterises behavioural equivalence [14]. This is the basis for research into coalgebraic modal logics, i.e., modal logics that are developed parametric in the type of the coalgebra that the logic is supposed to be interpreted on. Many different modal logics, such as monotone modal logic, graded modal logic and various probabilistic modal logics, have been shown to be instances of coalgebraic modal logics [5,20,27]. By placing those logics in a common framework, one is able to provide generic proofs of expressivity, soundness and completeness of the logics that can then be instantiated to each of the logics, thus avoiding the need for proving those results for each logic individually. Mathematically, the close connection between coalgebras and their corresponding coalgebraic modal logics has been represented in the elegant framework of a dual adjunction that links B-coalgebras over a category $\mathcal{C}$ to L-algebras over a category $\mathcal{D}$ [16,19,23]. This category $\mathcal{D}$ should be thought of as the category of algebras for the propositional logic of predicates, whereas L encodes the modal operators of the logic. The "one-step" semantics of the

This research is partially supported by the Leverhulme Trust Research Project Grant RPG-2020-232 and NWO grant No. OCENW.M20.053.

Published by Springer Nature Switzerland AG 2024
B. König and H. Urbat (Eds.): CMCS 2024, LNCS 14617, pp. 23–43, 2024.
https://doi.org/10.1007/978-3-031-66438-0_2

modal operators is then provided by a certain type of natural transformation connecting L and B across the adjunction.

The biggest stumbling block for studying coalgebraic modal logics abstractly is probably the inherent "one-stepness" of the theory: the transition structure typically only allows to look one step ahead in the model. Consequently, modal operators are usually one-step and axioms of the logics are assumed to be non-iterative [9] i.e., not allowing nesting of modal operators (exceptions such as [7] confirm the rule). Fixpoint operators pose a problem in this context as they specify properties that can look *arbitrarily deep* into the model. There are several existing approaches to adding fixpoint operators to coalgebraic modal logics [4, 28, 29] and even a coalgebraic model-checking tool for those logics [12]. However, none of the existing approaches to fixpoint logics provide a *categorical* treatment of the fixpoints within the above outlined dual adjunction framework. This means that existing coalgebraic fixpoint logics are developed on the category of sets and that results such as invariance under behavioural equivalence cannot be proven in an abstract, diagrammatic way.

The main contribution of this paper is to extend the dual adjunction framework to incorporate alternation-free fixpoint logics. We will first introduce the key concept of an *unfolding system* that contains as ingredients a one-step logic, a functor corresponding to the fixpoint operators and a natural transformation that we call unfolding and that is used to represent the unfolding of fixpoints. In order to ensure existence of fixpoints we will assume that both the category $\mathcal{D}$ and the functors on $\mathcal{D}$ corresponding to the logic are **Poset**-enriched. We will define the semantics of a given unfolding system as the least solution of an unfolding operation. After we present the abstract framework for fixpoint logics in Sect. 3 we will demonstrate how to place several examples in the framework: a positive modal logic with a transitive closure modality, a probabilistic version of a similar logic whose fixpoint operator resembles iteration in PPDL [10, 17] and finally the positive fragment of PDL [25] without tests. We will then prove key technical results: a diagrammatic proof of "adequacy" of the given fixpoint logic and, in Sect. 4, the fact that the semantics can equivalently be obtained as more standard initial algebra semantics.

In the final part of the paper, consisting of Sect. 5 and Sect. 6, we show how to place the alternation-free fragment of the coalgebraic μ-calculus (in the sense of [4]) into our framework. This is done in two stages: first, in Sect. 5, we translate the positive fragment of the coalgebraic μ-calculus into a syntax that allows us to place the logic into our framework by representing the logic and its semantics as an unfolding system. In Sect. 6 we then provide a general recipe for adding negations to a fixpoint logic represented in our framework. The latter will in particular show that the full alternation-free fragment of the coalgebraic μ-calculus can be represented using a suitable functor L and the associated initial L-algebra semantics. Finally, in Sect. 7, we conclude with some ideas for ongoing and future work.

2 Preliminaries

We will assume familiarity with basic category theory, coalgebra and modal logic.

Notation. Throughout the paper we will use $\mathscr{P}$ to denote the covariant powerset functor. The functor P denotes the left adjoint of the dual adjunction we will be working with. In many concrete instances of this adjunction, P will thus denote the contravariant power set functor.

Fixpoints. We will heavily rely on Kleene's fixpoint theorem and its generalisation by Cousot & Cousot [6] that states that the least fixpoint of a monotone function $f : L \to L$ on a complete lattice L exists and can be obtained as the limit of the sequence $f^0 = \bot$, $f^{i+1} = f(f^i)$ for an arbitrary ordinal i and $f^j = \bigvee_{i<j} f^i$ for limit ordinals j. We will not make explicit use of the dual statement concerning greatest fixpoints.

2.1 One-Step Logics

We model coalgebraic modal logics via the dual adjunction approach, cf. e.g. [16]. Throughout the paper we will assume to work in a setting where we are given:

- A category $\mathcal{C}$ of *spaces*, which are the carriers for coalgebras;
- A category $\mathcal{D}$ of *algebras* for some underlying 'propositional' logic;
- A dual adjunction $P : \mathcal{C} \to \mathcal{D}^{\mathrm{op}}$ and $Q : \mathcal{D}^{\mathrm{op}} \to \mathcal{C}$ with $P \dashv Q$, i.e., for all $X \in \mathcal{C}$, $A \in \mathcal{D}$ we have $\mathrm{Hom}_{\mathcal{C}}(X, QA) \cong \mathrm{Hom}_{\mathcal{D}}(A, PX)$ and this isomorphism is natural in both X and A;
- and an endofunctor $B : \mathcal{C} \to \mathcal{C}$ specifying the coalgebra type.

The adjunctions $P \dashv Q$ we consider are *logical connections*, as in [21]. That is to say, P and Q are both of the form $\mathrm{Hom}(-, \Omega)$, where Ω is a so-called 'ambimorphic' object, living in both $\mathcal{C}$ and $\mathcal{D}$.

Definition 1. *Given a functor $B : \mathcal{C} \to \mathcal{C}$, a pair (L_0, δ) consisting of a functor $L_0 : \mathcal{D} \to \mathcal{D}$ and a natural transformation $\delta : L_0 P \to PB$ is called a* one-step logic for B.

The natural transformation captures the "one-step semantics" of the logic. Crucially, while we will be able to place fixpoint logics into the dual adjunction framework, we will see that those logics cannot be described as one-step logics. This reflects the fact that fixpoint logics are inherently multi-step, as fixpoint operators can be used to inspect the model arbitrarily deep.

Examples of One-Step Logics

Example 1. We take $\mathcal{C}$ to be the category **Sets**, and $\mathcal{D}$ to be the category **DL** of distributive lattices. One half of the adjunction is formed by the powerset functor $P : \mathbf{Sets} \to \mathbf{DL}^{\mathrm{op}}$, which can be seen as exponentiation 2^-. Similarly, for $Q : \mathbf{DL}^{\mathrm{op}} \to \mathbf{Sets}$ we use $\mathrm{Hom}_{\mathbf{DL}}(-, 2)$. It is well known that $P \dashv Q$, forming the logical connection using 2 as the ambimorphic object.

For our behavior functor, we fix a set Prop of propositional letters, and define $B : \mathbf{Sets} \to \mathbf{Sets}$ as $BX = \mathscr{P}\,\mathrm{Prop} \times \mathscr{P}X$. For our one-step logic, we let $L_0 : \mathbf{DL} \to \mathbf{DL}$ be given by $L_0\mathbb{A} = \text{Free}\left(\{\Diamond a \mid a \in A\} \cup \{p \mid p \in \text{Prop}\}\right)/{\approx}$ where Free generates the free distributive lattice on a set of generators, and $\approx$ is the least congruence satisfying $\Diamond a \approx (\Diamond a \wedge \Diamond b)$ whenever $a \leq b$.

We note for future reference that quotienting out $\approx$ exactly ensures that $\Diamond a \leq \Diamond b$ whenever $a \leq b$. We obtain a one-step semantics $\delta : L_0P \to PB$ via

$$\delta : \begin{cases} p & \mapsto \{\langle m, U\rangle \mid p \in m\} \\ \Diamond v & \mapsto \{\langle m, U\rangle \mid v \cap u \neq \varnothing\} \end{cases}$$

and extending freely. This yields the expected semantics of (positive) modal logic as follows: consider a B-coalgebra (X, γ), let (Ψ, α) be the initial L_0-algebra (to be thought of as algebra of formulas) and consider the L_0-algebra $P\gamma \circ \delta : L_0PX \to PX$. The initial algebra map $[\![-]\!]$ will satisfy the following

$$\begin{aligned} [\![\alpha(\Diamond a)]\!] &= P\gamma(\delta_X(L[\![-]\!](\Diamond a))) = \{x \in X \mid \gamma(x) \in \delta_X(\Diamond[\![a]\!])\} \\ &= \{x \in X \mid \gamma(x) \in \{\langle m, U\rangle \in BX \mid U \cap [\![a]\!] \neq \emptyset\}\} \end{aligned}$$

which expresses precisely that $\Diamond a$ will be true at those states that have at least one successor "satisfying" a.

Example 2. For $\mathcal{C}$ we again take the category of sets. For $\mathcal{D}$, we take distributive lattices that come equipped with subconvex combinations. By this, we mean that if $\lambda_1, \ldots, \lambda_n \in [0, 1]$ with $\sum_i \lambda_i < 1$, then for each $a_1, \ldots, a_n \in \mathbb{A}$, we obtain an element

$$\sum_i \lambda_i a_i = a \in \mathbb{A}$$

Moreover, we demand that

$$1 \cdot a = a, \quad \sum_i \lambda_i (\sum_j \mu_{i,j} a_{i,j}) = \sum_{i,j} (\lambda_i \cdot \mu_{i,j}) a_{i,j}$$

and

$$\lambda a + \mu(b \vee c) = (\lambda a + \mu b) \vee (\lambda a + \mu b)$$

and similar for $\wedge$. Let **SCL** be the category with objects subconvex lattices, and morphisms the maps preserving both subconvex and lattice structure.

A key example of a subconvex lattice is given by $[0, 1]$, with max and min as lattice operations, and subconvex structure given in the obvious way. Note also that if $\mathbb{A}$ is a subconvex lattice, then so is $\mathbb{A}^X$ with pointwise structure; hence, we obtain our main examples as $[0, 1]^X$, with X any set.

Note also that if $f : X \to Y$ is any function, then $f^* : [0, 1]^Y \to [0, 1]^X$ is a subconvex lattice homomorphism. Hence, we obtain a functor $P : \mathbf{Sets} \to \mathbf{SCL}^{\mathrm{op}}$ given by $X \mapsto [0, 1]^X$. Vice versa, we clearly have a morphism $Q : \mathbf{SCL}^{\mathrm{op}} \to \mathbf{Sets}$ given by $\mathbb{A} \mapsto \mathrm{Hom}_{\mathbf{SCL}}(\mathbb{A}, [0, 1])$. It is easy to see that there is an adjunction $P \dashv Q$, since they are both of the form $\mathrm{Hom}(-, [0, 1])$. Finally, for our fixpoint extensions later on, note that **SCL** is enriched in posets in the obvious way, and that PX is a complete lattice for all X.

For a set X, let $\Delta(X)$ be the set of finitely supported subdistributions on X:

$$\Delta(X) := \{\mu : X \to [0,1] \mid \mu(x) = 0 \text{ all but finitely many } x, \sum_{x \in X} \mu(x) \leq 1\}.$$

Δ is an endofunctor on **Sets**, where Δf maps a distribution μ on X to μ_f with $\mu_f(y) = \sum_{f(x)=y} \mu(x)$. Now fix a set A of labels, and let B be the functor $BX = [0,1]^A \times \Delta(X)$. We can think of a B-coalgebra as a probabilistic 1-player game, where in a given state, the player may select a label A to obtain a 'payout', or probabilistically transition to a next state; but if the player takes the probabilistic transition, there is a possibility of failure, since the probabilities need not add up to 1.

In this context, we may posit the following one-step logic L_0 : **SCL** $\to$ **SCL**: for a given subconvex lattice $\mathbb{A}$, we let $L_0\mathbb{A}$ be the free subconvex lattice generated by $\{\lozenge a \mid a \in \mathbb{A}\} \cup \{p \mid p \in A\}$, quotiented by the equations

$$\sum_i \lambda_i \lozenge a_i \approx \lozenge(\sum_i \lambda_i a_i)$$

$$\lozenge a \wedge \lozenge b \approx \lozenge a \qquad \text{whenever } a \wedge b = a$$

Intuitively, $\lozenge a$ should be read as 'the expected value of a'; this is why we demand that $\lozenge$ acts linearly and preserves the order, but does *not* necessarily preserve lattice structure (as $\mathbb{E}[\max(X,Y)]$ is usually strictly greater than both $\mathbb{E}X$ and $\mathbb{E}Y$). Using the intuition of 'expectation', we obtain a one-step semantics $\delta : L_0 P \to PB$ by

$$\delta(\lozenge u) : \langle \pi, \mu \rangle \mapsto \sum_{x \in X} \mu(x) u(x) =: \mathbb{E}_\mu(u)$$

$$\delta(p) : \langle \pi, \mu \rangle \mapsto \pi(p)$$

and extending freely; it is easy to verify that δ respects the equations for L_0, so this indeed is a well-defined subconvex lattice morphism $L_0 P \to PB$. Naturality is also easy to verify.

2.2 Enriched Categories

We review the concepts from enriched categories that we use in this paper. Since we only consider categories enriched in **Poset**, some things simplify. In particular, for our purposes we do not need the general case of weighted (co)limits, which allows us to stick close to unenriched category theory. For a more in-depth treatment of enriched categories see [15].

Definition 2. *A* **Poset***-enriched category $\mathcal{C}$ is a category $\mathcal{C}$, together with a partial order $\leq = \leq_{A,B}$ on each homset* $\mathrm{Hom}_\mathcal{C}(A,B)$*, such that*

$$- \circ - : \mathrm{Hom}_\mathcal{C}(B,C) \times \mathrm{Hom}_\mathcal{C}(A,B) \to \mathrm{Hom}_\mathcal{C}(A,C)$$

is an order-preserving map for all A, B, C.

A key example of a **Poset**-enriched category is **Poset** itself: one can order morphisms 'pointwise', that is, $f \leq_{A,B} g$ iff $f(x) \leq_B g(x)$ for all $x \in A$. In fact, all examples of **Poset**-enriched categories in this paper are ordered pointwise in a similar way. That is, we consider categories $\mathcal{D}$ where objects are ordered structures, and the morphisms respect the orders; we then enrich $\mathcal{D}$ in **Poset** by ordering the morphisms pointwise. There is also a notion of enriched functor.

Definition 3. *Let $\mathcal{C}, \mathcal{D}$ be* **Poset**-*enriched categories. A* **Poset**-enriched functor *is a functor $F : \mathcal{C} \to \mathcal{D}$ such that for all parallel arrows $f, g : A \to B$ in $\mathcal{C}$, we have $f \leq g \implies Ff \leq Fg$.*

Finally, we have a notion of enriched coproduct:

Definition 4. *Let $\mathcal{C}$ be a* **Poset**-*enriched category, and let A, B be objects in $\mathcal{C}$. We get a functor $F : \mathcal{C} \to$* **Poset** *given by $FX = \mathrm{Hom}_{\mathcal{C}}(A, X) \times \mathrm{Hom}_{\mathcal{C}}(B, X)$. We call an object C of $\mathcal{C}$ an* enriched coproduct of A and B *if there is a natural isomorphism of functors $F \cong \mathrm{Hom}_{\mathcal{C}}(C, -)$.*

That is to say, an enriched coproduct of A and B is a coproduct $A + B$ such that the (natural) isomorphism $\mathrm{Hom}_{\mathcal{C}}(A, X) \times \mathrm{Hom}_{\mathcal{C}}(B, X) \cong \mathrm{Hom}_{\mathcal{C}}(A+B, X)$ is an isomorphism of *posets*, not merely sets.

3 Semantics of Fixpoint Logics

3.1 Unfolding Systems

To define a fixpoint logic, we take a 'two-tiered' approach: one starts with a 'base' modal (one-step) logic, to which one adds fixpoint modalities. Hence we will work in a setting where we are given a one-step logic as in Definition ,1, together with an enrichment that supports fixpoint operators.

Enriching the Dual Adjunction. To define the semantics of a fixpoint formula as a 'least solution', it is necessary to be able to compare predicates (i.e., elements of objects in $\mathcal{D}$). Since we intend a fully diagrammatic approach, we generalize comparing elements to comparing *maps* $X \to Y$ in $\mathcal{D}$. This is a generalization indeed: if $\mathcal{D}$ is a concrete category where each object is an ordered set, then maps can be ordered pointwise. To be able to compare maps, we fix an enrichment of $\mathcal{D}$ in **Poset**. We need some additional assumptions:

Assumption 1 *(i) We assume that $\mathrm{Hom}_{\mathcal{D}}(X, PY)$ is a complete lattice for each $Y \in \mathcal{C}$, and that for each morphism $f : Y' \to Y$, the map $Pf \circ - : \mathrm{Hom}_{\mathcal{D}}(X, PY) \to \mathrm{Hom}_{\mathcal{D}}(X, PY')$ preserves the lattice structure and arbitrary directed joins.*
(ii) Since P is a left adjoint, it preserves colimits; however, we require the stronger condition that P maps coproducts to enriched *coproducts.*

Remark 1. A natural candidate for $\mathcal{D}$ is the category **BA** of Boolean algebras, as we have the powerset-ultrafilter adjunction $P \dashv \mathbf{uf}$ between **Sets** and $\mathbf{BA}^{\mathrm{op}}$. Note, however, that if we enrich **BA** by pointwise order on the morphisms, the morphisms end up being ordered discretely. For, if $f \leq g$ in $\mathrm{Hom}_{\mathbf{BA}}(\mathbb{A}, \mathbb{B})$, then for each $x \in \mathbb{A}$ we have $f(\neg x) \leq g(\neg x)$. Hence, $g(x) = \neg g(\neg x) \leq \neg f(\neg x) = f(x)$, showing $g \leq f$ as well.

It may be possible to enrich **BA** via a different, non-pointwise strategy; however, it can be shown that there is no enrichment of **BA** that makes $\mathrm{Hom}(X, PY)$ a complete lattice for all X, Y; hence, we will work with the category **DL** of distributive lattices, rather than **BA**. This could be seen as an analogue of the constraint on fixpoint equations, that fixpoint variables only occur *positively*. In Sect. 6, we outline how negations can be added into the picture outside of the fixpoint equations.

Logical Functors. In order to be able to enrich a given logic with fixpoints we will assume to be given a one-step logic (L_0, δ) for B and posit a functor $L : \mathcal{D} \to \mathcal{D}$ representing the fixpoint modalities. We assume moreover that both L_0 and L are **Poset**-enriched functors, and that L has an initial algebra. Throughout the paper we will write Φ for the initial L-algebra, with structure map $\alpha : L\Phi \to \Phi$ (note that α is an isomorphism); it should be thought of as the 'algebra of formulas'.

Remark 2. Intuitively the assumption that the functor L_0 is enriched means that all modal operators in the base logic are monotone.

We will now define a (alternation-free) fixpoint logic as an extension of a basic one-step logic. A key element is the so-called unfolding operation for the fixpoints.

Definition 5. *Let $\mathcal{D}$ be a* **Poset***-enriched category. Let $B : \mathcal{C} \to \mathcal{C}$ be a functor, and let $L_0 : \mathcal{D} \to \mathcal{D}$ be a* **Poset***-enriched functor. An* unfolding system *for L_0 is a pair $(\mathsf{u} : L \to \mathrm{id} + L_0 L, \delta : L_0 P \to PB)$ where $L : \mathcal{D} \to \mathcal{D}$ is a* **Poset***-enriched functor that has an initial algebra Φ, u is a natural transformation and (L_0, δ) is a one-step logic for B in the sense of Definition 1.*

The natural transformation $\mathsf{u} : L \to \mathrm{id} + L_0 L$ provides the key to define the semantics of fixpoint modalities.

Intuitively, if we think of an element ℓ of $L\mathbb{A}$ as being a loop of some type, we see that there are two 'branches' to unfolding ℓ; we may exit the loop, which yields an element of $\mathbb{A}$, or we may take a step in the model (represented by a modality taken from L_0), after which we continue with a loop.

In more detail, a generic element of $L\mathbb{A}$ may look like $\ell(x_1, \ldots, x_n)$, with ℓ some fixpoint modality, and $\bar{x} = (x_1, \ldots, x_n) \in \mathbb{A}^n$. We want ℓ to satisfy a fixpoint equation

$$\ell(\bar{x}) = \sigma(\bar{x}, g_1 \cdot \ell_1(\bar{x}), \ldots, g_k \cdot \ell_k(\bar{x}))$$

Here σ is some 'propositional' expression, and each ℓ_j occurs *guarded* by a one-step modality g_i. The transformation u replaces the LHS of each such expression with the RHS; and the RHS lives in $\mathbb{A} + L_0 L\mathbb{A}$.

3.2 Definition of the Semantics

We are now ready to define the semantic map $[\![-]\!]$.

Definition 6. *Let* (u, δ) *be an unfolding system for* B *and let* (X, γ) *be a* B*-coalgebra. The semantic map* $[\![-]\!]_\gamma : \Phi \to PX$ *on* (X, γ) *is defined as the least map* $t : \Phi \to PX$ *making the following diagram commute:*

$$
\begin{array}{ccc}
\Phi & \xrightarrow{\quad t \quad} & PX \\
\downarrow{\scriptstyle \alpha^{-1}} & & \uparrow{\scriptstyle [PX, P\gamma]} \\
L\Phi & & PX + PBX \\
\downarrow{\scriptstyle \mathsf{u}} & & \uparrow \\
\Phi + L_0 L\Phi & & {\scriptstyle \delta} \\
\downarrow{\scriptstyle \mathrm{id} + L_0\alpha} & & \\
\Phi + L_0\Phi & \xrightarrow{t + L_0 t} & PX + L_0 PX
\end{array}
$$

When the X *is clear from the context we often will drop the subscript and simply denote the semantic map by* $[\![-]\!]$.

Intuitively, the left-hand side of the above diagram takes a formula, and unfolds the top-level fixpoint modalities, to obtain a one-step L_0-formula over Φ. Using the semantic map, this can be interpreted as a one-step L_0-formula over PX, which we can reduce to a predicate in PX using the one-step semantics δ, and the coalgebra structure γ.

In order to see that $[\![-]\!]$ is well-defined, we note that the above formulation is equivalent to saying that $[\![-]\!]$ is the least fixpoint of the operator

$$
\begin{aligned}
\Xi_\gamma : \mathrm{Hom}_{\mathcal{D}}(X, PX) &\to \mathrm{Hom}_{\mathcal{D}}(X, PX) \\
t &\mapsto [PX, P\gamma] \circ \delta \circ (t + L_0 t) \circ (\mathrm{id} + L_0\alpha) \circ \mathsf{u} \circ \alpha^{-1}
\end{aligned} \tag{1}
$$

That Ξ_γ is a monotone operator can be seen as follows:

1. as $\mathcal{D}$ is order-enriched, both pre- and post-composition of morphisms are monotone, and
2. the operation that maps t to $t + L_0 t$ is monotone.

The first fact is immediate from the definition of enriched categories. The second fact follows from the operation $t \mapsto L_0 t$ being monotone (L_0 is an enriched functor) and because the cotupling operation is monotone as we have

$$\mathrm{Hom}(X + Y, Z) \cong \mathrm{Hom}(X, Z) \times \mathrm{Hom}(Y, Z)$$

by definition of the coproduct and because the above isomorphism is an isomorphism of posets. As Ξ is monotone and as $\mathrm{Hom}_{\mathcal{D}}(X, PX)$ is a complete lattice by assumption, the least fixpoint $[\![-]\!]$ of Ξ exists.

Remark 3. While the definition of $[\![-]\!]$ is in terms of a least fixpoint, the same setup can be used to define greatest fixpoints as well, by inverting the order on the homsets. However, we are as yet restricted to only a single *type* of fixpoint.

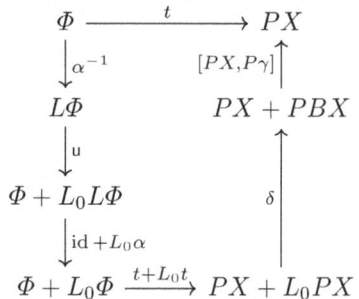

3.3 Examples

Example 3. As a first example, we give a simple logic of transitive closure in Kripke frames.

We start with the one-step logic as given in Example 1. To obtain an unfolding system, we take $L : \mathbf{DL} \to \mathbf{DL}$ similarly to L_0, but using $\Diamond^*$ rather than $\Diamond$:

$$L\mathbb{A} = \text{Free}\left(\{\Diamond^* a \mid a \in A\} \cup \{p \mid p \in \text{Prop}\}\right) / {\approx}$$

where $\approx$ is as in Example 1. We get an unfolding transformation $\mathsf{u} : L \to \text{id} + L_0 L$ by

$$\mathsf{u} : \begin{cases} p & \mapsto \text{inr}(p) \\ \Diamond^* a & \mapsto \text{inl}(a) \vee \text{inr}(\Diamond\Diamond^* a) \end{cases}$$

To spell out the concrete description of the above logic, we have formulas defined via

$$\phi, \psi ::= p \in \text{Prop} \mid \top \mid \bot \mid \phi \wedge \psi \mid \phi \vee \psi \mid \Diamond^* \phi$$

For a given coalgebra seen as a Kripke model $\mathfrak{M} = (W, R, V)$, consider the usual definition of satisfaction of formulas, given by

$$x \Vdash p \text{ iff } x \in V(p)$$
$$x \Vdash \Diamond^* \phi \text{ iff } \exists v_1, \dots, v_n \text{ such that } x R v_1 R \dots R v_n \text{ and } v_n \Vdash \phi$$

We claim that our approach yields the same semantics, in the sense that

$$[\![\phi]\!] = \{x \in W \mid x \Vdash \phi\}$$

To see this, first note that the map $t : \Phi \to PW$ given by $\phi \mapsto \{x \in W \mid x \Vdash \phi\}$ is a solution to the diagram in Definition 6; hence, $[\![\phi]\!] \subseteq t(\phi)$, as $[\![-]\!]$ is the least solution. For the other direction, let $t^{(k)}$ be the map that interprets $\Diamond^*$ as 'reachability *in at most k steps*'; by induction on k, one can show that each $t^{(k)}$ is below all solutions to the diagram, and hence each $t^{(k)} \leq [\![-]\!]$. Since t is clearly the supremum of the $t^{(k)}$, we also have $t(\phi) \subseteq [\![\phi]\!]$ for all formulas ϕ.

Example 4. The above example may be extended to cover Propositional Dynamic Logic (PDL). We take the same base adjunction, but adjust B and L_0 to include a set of modalities. That is, for a set Π of 'atomic programs', we let $BX = \mathscr{P}\,\text{Prop} \times (\mathscr{P}X)^{\Pi}$, and take $L_0 : \mathbf{DL} \to \mathbf{DL}$ to be

$$L_0\mathbb{A} = \text{Free}(\text{Prop} \cup \{\langle\pi\rangle a \mid \pi \in \Pi, a \in \mathbb{A}\}) / {\approx}$$

with obvious action on morphisms. Then just as before, we obtain a one-step semantics as

$$\delta : \begin{cases} p & \mapsto \{\langle m, u\rangle \mid p \in m\} \\ \langle\pi\rangle v & \mapsto \{\langle m, u\rangle \mid v \cap u(\pi) \neq \varnothing\} \end{cases}$$

For PDL, we define (composite) programs via the following grammar:

$$\alpha ::= \pi \in \Pi \mid \epsilon \mid \alpha \cup \alpha \mid \alpha ; \alpha \mid \alpha^*$$

We write $\bar{\Pi}$ for the set of programs. There is a function $g : \bar{\Pi} \to \bar{\Pi}$ such that $g(\alpha)$ is equivalent to α, and $g(\alpha)$ is of the form $\sum_i \pi_i; \alpha_i$ or $(\sum_i \pi_i; \alpha_i) \cup \epsilon$. Existence of g can be proven as the normal form in [3, Thm. 4.4] using the Brzozowski derivative.

We let $L : \mathbf{DL} \to \mathbf{DL}$ be given by

$$L\mathbb{A} := \text{Free}(\text{Prop} \cup \{\langle \alpha \rangle a \mid \alpha \in \bar{\Pi}, a \in \mathbb{A}\})/\approx$$

with obvious action on morphisms. We define $\mathsf{u} : L \to \text{id} + L_0 L$ as

$$\mathsf{u}(p) := \text{inr}(p)$$
$$\mathsf{u}(\langle \alpha \rangle a) := \begin{cases} \text{inr}\left(\bigvee_i \langle \pi_i \rangle \langle \alpha_i \rangle a\right) & g(\alpha) = \sum_i \pi_i; \alpha_i \\ \text{inr}\left(\bigvee_i \langle \pi_i \rangle \langle \alpha_i \rangle a\right) \vee \text{inl}(a) & g(\alpha) = (\sum_i \pi_i; \alpha_i) \cup \epsilon \end{cases}$$

Concretely, we see that $g(\pi^*) = \pi; \pi^* \cup \epsilon$, and hence (omitting coproduct inclusions) we have

$$\mathsf{u}(\langle \pi^* \rangle a) = \langle \pi \rangle \langle \pi^* \rangle a \vee a$$

Just like $\Diamond^*$ above, it can be shown that $[\![\langle \alpha \rangle \phi]\!]$ has the usual denotation.

Example 5. As a third example, we give a quantitative logic for one-player games.

In Example 2, we gave a simple quantitative modal logic. We may define a functor L by setting

$$L\mathbb{A} := \text{Free}(\{\sigma_q a, \Diamond^* a \mid a \in \mathbb{A}, q \in [0,1]\})$$

(where Free denotes the 'free subconvex lattice'-functor) and unfolding the generators as

$$\mathsf{u}(\sigma_q a) = qa + (1-q)\Diamond \sigma_q a$$
$$\mathsf{u}(\Diamond^* a) = a \vee \Diamond \Diamond^* a$$

Under this unfolding, $[\![\sigma_q \phi]\!]$ will assign to a state x the expected outcome of a q-probabilistic strategy, where with probability q, the player chooses the payout corresponding to ϕ, and with probability $(1-q)$ she decides to play on.

On the other hand $[\![\Diamond^* \phi]\!]$ assigns to a state x the expected outcome of the *optimal* strategy, where the player chooses to play on only if the outcome she can expect from playing on is greater than the payout she can get now.

3.4 Invariance Under Behavioural Equivalence

A fundamental property of coalgebraic modal logic is that the semantics of the logic is invariant under behavioural equivalence. This follows from invariance under coalgebra morphisms which can be concisely expressed as in (2) below. The proposition shows that invariance under behavioural equivalence also holds in the presence of fixpoint operators.

Proposition 1. *For all coalgebra morphisms $f : (X_1, \gamma_1) \to (X_2, \gamma_2)$ we have*

$$Pf \circ [\![-]\!]_{\gamma_2} = [\![-]\!]_{\gamma_1} \tag{2}$$

Proof. Recall the definition of Ξ from (1) on page 8. Using naturality of δ it is a matter of routine checking that

$$\Xi_{\gamma_1} = \mathrm{Hom}(-, Pf) \circ \Xi_{\gamma_2}.$$

As a consequence we have $Pf \circ [\![-]\!]_{\gamma_2} \geq [\![-]\!]_{\gamma_1}$ because

$$\Xi_{\gamma_1}(Pf \circ [\![-]\!]_{\gamma_2}) = Pf \circ \Xi_{\gamma_2}([\![-]\!]_{\gamma_2}) = Pf \circ [\![-]\!]_{\gamma_2},$$

i.e., $Pf \circ [\![-]\!]_{\gamma_2}$ is a fixpoint of Ξ_{γ_1} and $[\![-]\!]_{\gamma_1}$ is the least such fixpoint.

We will now show by ordinal induction that for all $i \in \mathrm{ORD}$ we have

$$\Xi^i_{\gamma_1} \geq \mathrm{Hom}(-, Pf) \circ \Xi^i_{\gamma_2}$$

where $\Xi^i_{\gamma_1}$ and $\Xi^i_{\gamma_2}$ are the approximants defined in Sect. 2.

Case $i = j + 1$. Then

$$\begin{aligned} \Xi^i_{\gamma_1} &= \Xi_{\gamma_1}(\Xi^j_{\gamma_1}) \geq \Xi_{\gamma_1}(\mathrm{Hom}(-, Pf) \circ \Xi^j_{\gamma_2}) \\ &= \mathrm{Hom}(-, Pf) \circ \Xi_{\gamma_2}(\Xi^j_{\gamma_2}) = \mathrm{Hom}(-, Pf) \circ \Xi^i_{\gamma_2} \end{aligned}$$

Case i is a limit ordinal. Then

$$\begin{aligned} \Xi^i_{\gamma_1} &= \bigvee_{j<i} \Xi^j_{\gamma_1} \geq \bigvee_{j<i} \mathrm{Hom}(-, Pf) \circ \Xi^j_{\gamma_2} \\ &\overset{\text{Ass. 1}}{=} \mathrm{Hom}(-, Pf) \circ \bigvee_{j<i} \Xi^j_{\gamma_2} = \mathrm{Hom}(-, Pf) \circ \Xi^i_{\gamma_2} \end{aligned}$$

The claim follows now as $\mathcal{D}$ is locally small which implies the hom sets $\mathrm{Hom}(\Phi, PX_i)$ for $i = 1, 2$ are sets, and thus $[\![-]\!]_{\gamma_i} = \Xi^i_{\gamma_i}$ for a suitably large ordinal i.

3.5 Guarded Equations of Arbitrary Depth

The framework presented so far seems to place some restrictions on the shape of the fixpoint equations; consider the following toy logic:

$$\phi ::= \top \mid \bot \mid \phi \vee \phi \mid \phi \wedge \phi \mid \Diamond^{2*}\phi$$

with the intended interpretation of $\Diamond^{2*}$ in a transition system $(X, \to)$ being the least solution to $[\![\Diamond^{2*}\phi]\!] = [\![\phi]\!] \vee \{x \in X \mid \exists y, z : x \to y \to z \text{ and } z \in [\![\Diamond^{2*}\phi]\!]\}$ The natural guarding 1-step logic is positive modal logic; but this would yield an unfolding transformation $\mathsf{u} : \Diamond^{2*}a \mapsto a \vee \Diamond\Diamond\Diamond^{2*}a$ and the RHS is not of the form $\mathbb{A} + L_0 L\mathbb{A}$. A possible remedy would be to add an extra operator $\Diamond^{2*+1}$ to the logic, which would allow the depth-one transformation

$$\Diamond^{2*}a \mapsto a \vee \Diamond\Diamond^{2*+1}a \qquad\qquad \Diamond^{2*+1}a \mapsto \Diamond\Diamond^{2*}a$$

but this requires an extension of the logic, and is somewhat ad-hoc.

A second issue is that in the current presentation, *all* modalities are treated as fixpoint modalities. But most fixpoint logics contain a mixture of fixpoint and one-step modalities. In fact both fixpoint modalities of higher depth and one-step modalities (which may be seen as 'depth 0 fixpoint modalities') can be accommodated by our setup, in a uniform way.

Proposition 2. *Assume that* $L + L_0$ *generates an algebraically free monad* T. *Let* $\mathsf{u} : L \to \mathrm{id} + L_0T$ *be a natural transformation. Then we obtain a natural transformation* $\mathsf{u}^* : T \to \mathrm{id} + L_0T$ *such that* u *factors through* u^*.

Note that the assumption is satisfied whenever L and L_0 are given via a finitary presentation, i.e. as a quotient of a free algebra on some finitary operations. The above proposition allows us to apply our results even in the less restrictive case where the codomain of u is expanded to allow formulas of arbitrary depth, and including both one-step and fixpoint modalities. For instance, in the above toy logic, we get a transformation $\mathsf{u} : L \to \mathrm{id} + L_0L_0L$ sending $\Diamond^{2*}a$ to $a \vee \Diamond\Diamond\Diamond^{2*}a$, and clearly L_0L is a subfunctor of the algebraically-free monad T on $L_0 + L$.

Proof. We know that if T is the algebraically free monad on $L_0 + L$, then $T\mathbb{A}$ is a free $L_0 + L$-algebra on $\mathbb{A}$. Hence, to obtain u^*, we need to exhibit $\mathbb{A} + L_0T\mathbb{A}$ as an algebra

$$k : \mathbb{A} + L_0(\mathbb{A} + L_0T\mathbb{A}) + L(\mathbb{A} + L_0T\mathbb{A}) \to \mathbb{A} + L_0T\mathbb{A}$$

The first component can simply be included; on the second component, we have a natural map $\mathbb{A} + L_0T\mathbb{A} \xrightarrow{\mathrm{inl}} \mathbb{A} + L_0T\mathbb{A} + LT\mathbb{A} \to T\mathbb{A}$ using the free $(L_0 + L)$-algebra structure on $T\mathbb{A}$; applying L_0 to this map yields a map $L_0(\mathbb{A} + L_0T\mathbb{A}) \to L_0T\mathbb{A}$; finally, for the third component, we get

$$L(\mathbb{A} + L_0T\mathbb{A}) \xrightarrow{\mathsf{u}} \mathbb{A} + L_0T\mathbb{A} + L_0T(\mathbb{A} + L_0T\mathbb{A}) \xrightarrow{\text{algebra}} \mathbb{A} + L_0T\mathbb{A} + L_0TT\mathbb{A}$$

$$\xrightarrow{\text{monad}} \mathbb{A} + L_0T\mathbb{A} + L_0T\mathbb{A} \xrightarrow{\text{codiagonal}} \mathbb{A} + L_0T\mathbb{A}$$

The resulting algebra map $\mathsf{u}^* : T\mathbb{A} \to \mathbb{A} + L_0T\mathbb{A}$ is natural in $\mathbb{A}$ as it is induced by the composition of natural transformations.

4 Initial Algebra Semantics

The definition of the semantic map above is somewhat different from the usual definition in terms of initial algebra semantics. The reason for this is that a coalgebra map $\gamma : X \to BX$ does not directly give rise to an L-algebra structure on PX. With some additional work, we can exhibit the semantic map in this way as well. This section is devoted to presenting an initial algebra semantics, and proving it equivalent to the least-solution semantics.

Before we start, however, we will give our reasons for choosing the above construction as fundamental. Firstly, the definition of $[\![-]\!]$ as a 'least solution' corresponds more closely to intuitions about the semantics of (least) fixpoint formulas. Secondly, it yields a direct proof technique via the ordinal approximation sequence. The proof of invariance under behavioral equivalence from Sect. 3.4 would be more involved if one were to work with the initial algebra semantics.

For the main result of this section, the proof of equivalence between initial and least-solution semantics, we need the following very mild categorical assumption.

Assumption 2. *We assume that in $\mathcal{D}$ pre-composition with morphisms distributes over arbitrary joins, i.e., for $g : \mathbb{A} \to \mathbb{A}'$, an index set I and $f_i : \mathbb{A}' \to \mathbb{A}''$ for all $i \in I$ we have*

$$\bigvee_{i\in I}(f_i \circ g) = (\bigvee_{i\in I} f_i) \circ g$$

where the equality means that if the join on the right side of the equation exists, the join on the left side also exists and in this case both morphisms are equal.

It can be easily verified that Assumption 2 is always satisfied in case the poset structure on $\mathrm{Hom}_{\mathcal{D}}(\mathbb{A}, \mathbb{A}')$ is induced pointwise, as is usually the case.

For the definition of the initial algebra semantics we first recall that any L-algebra structure on an object $\mathbb{B}$ induces a morphism from the initial L-algebra.

Definition 7. *For an L-algebra $\beta : L\mathbb{B} \to \mathbb{B}$ we let $\ulcorner m \urcorner : \varPhi \to \mathbb{B}$ denote the unique L-algebra morphism from the initial L-algebra $(\varPhi, \alpha)$ to $(\mathbb{B}, \beta)$.*

We now define a suitable L-algebra structure on the algebra of predicates.

Definition 8. *Let $\gamma : X \to BX$ be a coalgebra. We define $m_\gamma : LPX \to PX$ as the least map m making*

$$\begin{array}{ccc}
LPX & \xrightarrow{\quad m \quad} & PX \\
 & & \uparrow{\scriptstyle [PX,P\gamma]} \\
\downarrow{\scriptstyle u} & & PX + PBX \\
 & & \uparrow{\scriptstyle PX+\delta} \\
PX + L_0LPX & \xrightarrow{PX+L_0m} & PX + L_0PX
\end{array}$$

commute. We define $\| - \| : \varPhi \to PX$ as $\ulcorner m_\gamma \urcorner$, i.e., as the unique L-algebra morphism; that is, the unique map making the following diagram commute

$$\begin{array}{ccc}
\varPhi & \xrightarrow{\|-\|} & PX \\
{\scriptstyle \alpha}\uparrow & & \uparrow{\scriptstyle m_\gamma} \\
L\varPhi & \xrightarrow{L\|-\|} & LPX
\end{array}$$

Similarly as for the semantic map $[\![-]\!]$ it is easy to see that m_γ - and thus $\|-\|$ - is well-defined as m_γ is the least fixpoint of the *monotone* operator m given by $\mathrm{m}(m) := [PX, P\gamma \circ \delta] \circ (PX + L_0 m) \circ \mathsf{u}$. We wish to prove that $\|-\| = [\![-]\!]$. We first consider the easier direction of the statement.

Lemma 1. *We have* $\|-\| \geq [\![-]\!]$.

Proof (Sketch). The lemma can be proven by showing that $\|-\|$ makes the definition square of $[\![-]\!]$ commute. As $[\![-]\!]$ is the least such map, the claim follows.

In order to prove the other inequality, we first prove the following lemma.

Lemma 2. *Let* $\{\mathrm{m}^i\}_{i \in \mathrm{Ord}}$ *denote the approximants of* m_γ *in* $\mathrm{Hom}_{\mathcal{D}}(LPX, PX)$. *We have that* $\ulcorner \mathrm{m}^i \urcorner \leq [\![-]\!]$ *for all ordinals* $i \in \mathrm{Ord}$.

Proof. We first prove that for all ordinals i we have

$$\mathrm{m}^i \circ L[\![-]\!] \circ \alpha^- \leq [\![-]\!] \tag{3}$$

In other words, we show that $[\![-]\!]$ is a pre-fixpoint of the following monotone operator on $\mathrm{Hom}_{\mathcal{D}}(\Phi, PX)$ given by$g \mapsto \mathrm{m}^i \circ Lg \circ \alpha^{-1}$. The claim of the lemma follows that by the fact that $\ulcorner \mathrm{m}^i \urcorner$ is the unique fixpoint of the above operator and thus also the smallest pre-fixpoint. The proof of (3) is by ordinal induction on i. For $i = 0$ the claim is trivial. For the successor ordinal case we calculate:

$$\begin{aligned}
[\![-]\!] &\overset{\text{Def.}}{=} [PX, P\gamma \circ \delta] \circ ([\![-]\!] + L_0 L[\![-]\!]) \circ (\Phi + L_0 \alpha) \circ \mathsf{u} \circ \alpha^- \\
&\overset{(*)}{\geq} [PX, P\gamma \circ \delta] \circ (PX + L_0 \mathrm{m}^i) \circ ([\![-]\!] + L_0 L[\![-]\!]) \circ \mathsf{u} \circ \alpha^- \\
&\overset{\text{nat of } \mathsf{u}}{=} [PX, P\gamma \circ \delta] \circ (PX + L_0 \mathrm{m}^i) \circ \mathsf{u} \circ L[\![-]\!] \circ \alpha^- \\
&\overset{\text{Def of } \mathrm{m}^{i+1}}{=} \mathrm{m}^{i+1} \circ L[\![-]\!] \circ \alpha^-
\end{aligned}$$

Here (*) is a consequence of the I.H. and the fact that composition in $\mathcal{D}$ is monotonic. For the limit case consider $\mathrm{m}^i = \bigvee_{j<i} \mathrm{m}^j$. We have:

$$[\![-]\!] \overset{\text{I.H.}}{\geq} \bigvee_{j<i} \left(\mathrm{m}^j \circ L[\![-]\!] \circ \alpha^-\right) \overset{\text{Ass.}}{\geq} \left(\bigvee_{j<i} \mathrm{m}^j\right) \circ L[\![-]\!] \circ \alpha^-$$

where for the second inequality we made use of Assumption 2 that pre-composition of morphisms in $\mathcal{D}$ distributes over joins.

We are now ready to prove the main result of this section.

Theorem 1. *Let* (X, γ) *be a* B*-coalgebra. and let* (u, δ) *be an unfolding system for* B. *The semantic map* $[\![-]\!]_\gamma : \Phi \to PX$ *coincides with the initial algebra map* $\|-\|_X$ *from* (Φ, α) *to* (PX, m_γ)

Proof. By Lemma 1 we have $[\![-]\!]_\gamma \leq \|-\|_X$. For the converse direction note that we have $m_\gamma = \mathrm{m}^i$ for some $i \in \mathrm{Ord}$. Therefore we can apply Lemma 2 as follows: $\|-\|_{PX} = \ulcorner m_\gamma \urcorner = \ulcorner \mathrm{m}^i \urcorner \overset{\text{Lem. 2}}{\leq} [\![-]\!]_\gamma$.

5 Correctness for Positive Alternation-Free Coalgebraic Fixpoint Logics

In this section, we show how our approach can express the alternation-free fragment of the coalgebraic μ-calculus as defined in [4]. More specifically, we give a pair of logical functors (L, L_0) and an unfolding system (u, δ) such that the initial L-algebra Φ contains 'enough' formulas of the μ-calculus, and the induced semantics $[\![-]\!]$ coincides with the asusual semantics for the μ-calculus.

Before we can give the unfolding system, we need to define a special syntax for the μ-calculus. The reason for this is that our setup is most suited for logics with fixpoint *modalities*; however, in the μ-calculus, there is no syntactical distinction between formulas, and fixpoint modalities applied to a formula.

In order to get the μ-calculus into a more amenable form, we use a presentation similar to the flat coalgebraic fixpoint logics from [28]. This approach can also be compared to PDL, which similarly has a syntactic distinction between *formulas* and *programs*, corresponding to our fixpoint schemes.

Fix a countable set V of 'parametric variables'. Let Λ be modal similarity type, i.e. a countable set of modal operators of finite arity. Let $x \notin V$ be a designated 'fixpoint variable'. Then we define by mutual induction the formulas ϕ and fixpoint schemes γ as follows:

$$\gamma ::= v \in V \mid x \mid \phi \mid \gamma \vee \gamma \mid \gamma \wedge \gamma \mid \heartsuit(\bar{\gamma}) \mid \sharp_\gamma(\bar{\gamma}/\bar{v}) \mid \flat_\gamma(\bar{\gamma}/\bar{v})$$
$$\phi ::= \top \mid \bot \mid \phi \vee \phi \mid \phi \wedge \phi \mid \neg\phi \mid \heartsuit(\bar{\phi}) \mid \sharp_\gamma(\bar{\phi}/\bar{v}) \mid \flat_\gamma(\bar{\phi}/\bar{v})$$

where the notation $\bar{\gamma}, \bar{\phi}$ indicates that operators might have lists of fixpoint schemes/formulas as arguments. The intended reading of $\sharp_\gamma$ and $\flat_\gamma$ is as taking the least and greatest fixpoints respectively of the operator defined by γ. We also define the set of *guarded* fixpoint schemes via

$$\gamma_g ::= v \in V \mid \gamma_g \vee \gamma_g \mid \gamma_g \wedge \gamma_g \mid \heartsuit(\bar{\gamma})$$

That is, a guarded fixpoint scheme is a propositional combination of parametric variables and one-step modalities applied to (non-guarded) fixpoint schemes.

Note that negations may only occur in formulas, not fixpoint schemes; hence, the fixpoint variable x only ever occurs positively. Note also that on formulas, the 'greatest fixpoint operator' $\flat_\gamma$ may be defined in terms of $\sharp_\gamma$ via $\flat_\gamma(\bar{\phi}/\bar{v}) := \neg\sharp_\gamma(\overline{\neg\phi}/\bar{v})$. We write CFL for the above coalgebraic fixpoint logic and CFS for the set of fixpoint schemes. We will also write $\mathrm{CFS}_\sharp$ for the set of fixpoint schemes not containing $\flat$, $\mathrm{CFL}_\sharp$ for the set of formulas not containing $\flat$, and $\mathrm{CFL}_\sharp^+$ for the set of formulas not containing $\flat$ or $\neg$.

Definition 9. *Let B be a* **Sets**-*functor, and fix for every modality $\heartsuit$ of arity n a monotone predicate lifting $(\!|\heartsuit|\!) : P^n \to PB$. Then for a B-coalgebra $\sigma : X \mapsto BX$, we define by mutual recursion the semantics $[\![\phi]\!]$ of a formula, and $[\![\gamma]\!]_p^\xi$ of a fixpoint scheme, where $p : V \to PX$ is a valuation of the parametric variables, and $\xi \subseteq PX$ is the value of the fixpoint variable*

$$
\begin{array}{ll}
[\![\phi]\!]^\xi_p := [\![\phi]\!] & \\
[\![v]\!]^\xi_p := p(v) & [\![\top]\!] := S \\
[\![x]\!]^\xi_p := \xi & [\![\bot]\!] := \varnothing \\
[\![\gamma \vee \delta]\!]^\xi_p := [\![\gamma]\!]^\xi_p \cup [\![\delta]\!]^\xi_p & [\![\phi \vee \psi]\!] := [\![\phi]\!] \cup [\![\psi]\!] \\
[\![\gamma \wedge \delta]\!]^\xi_p := [\![\gamma]\!]^\xi_p \cap [\![\delta]\!]^\xi_p & [\![\phi \wedge \psi]\!] := [\![\phi]\!] \cap [\![\psi]\!] \\
[\![\heartsuit(\bar{\gamma})]\!]^\xi_p := \sigma^{-1}\left((\!|\heartsuit|\!)([\![\bar{\gamma}]\!]^\xi_p)\right) & [\![\heartsuit(\bar{\phi})]\!] := \sigma^{-1}\left((\!|\heartsuit|\!)([\![\bar{\phi}]\!])\right) \\
[\![\sharp_\gamma(\bar{\delta}/\bar{v})]\!]^\xi_p := \mu\left(U \mapsto [\![\gamma]\!]^U_{[\![\bar{\delta}]\!]^\xi_p/\bar{v}}\right) & [\![\sharp_\gamma(\bar{\phi}/\bar{v})]\!] := \mu\left(U \mapsto [\![\gamma]\!]^U_{[\![\bar{\phi}]\!]/\bar{v}}\right) \\
[\![\flat_\gamma(\bar{\delta}/\bar{v})]\!]^\xi_p := \nu\left(U \mapsto [\![\gamma]\!]^U_{[\![\bar{\delta}]\!]^\xi_p/\bar{v}}\right) & [\![\flat_\gamma(\bar{\phi}/\bar{v})]\!] := \nu\left(U \mapsto [\![\gamma]\!]^U_{[\![\bar{\phi}]\!]/\bar{v}}\right)
\end{array}
$$

We claim that CFL is equivalent to the coalgebraic μ-calculus (CMC), and $\mathrm{CFL}_\sharp$ is equivalent to the alternation-free fragment of the coalgebraic μ-calculus (AFCMC). We will illustrate by showing how to express two μ-calculus formulas in CFL.

Example 6. As a simple example, consider the PDL formula $\Diamond^* p$. In the μ-calculus, this formula may be expressed as $\phi := \mu X.p \vee \Diamond X$. We can express ϕ in CFL as follows: $\sharp_\gamma(p/v)$, where $\gamma(v;x) := v \vee \Diamond x$. Note that this is not exactly the formula generated by the translation; in fact we have

$$
trl(\phi) = \sharp_\delta(), \qquad \text{where } \delta(x) = p \vee \Diamond x
$$

We have chosen $\sharp_\gamma(p/v)$ as our intuitive translation, since it highlights how the fixpoint *modality* $\Diamond^*$ may be expressed as the modality $\sharp_\gamma$.

Example 7. As a second example, we look at the formula

$$
\phi = \mu X.p \wedge \mu Y.((q \wedge \Diamond Y) \vee (r \wedge \Box X)).
$$

This example exhibits intrinsic nesting of fixpoints. It is guarded as a formula of the μ-calculus; however, in CFL, guardedness also requires that all fixpoint *quantifiers* (other than the first) appear directly under a modality. So, let ψ be the formula bound by Y - i.e., $\psi = (q \wedge \Diamond Y) \vee (r \wedge \Box X)$. Clearly, ϕ is equivalent to

$$
\mu X.(p \wedge ((q \wedge \Diamond(\mu Y.\psi)) \vee (r \wedge \Box X)))
$$

and in this formula, all but the outer quantifier appear guarded. Now translation is a straightforward affair: consider

$$
\begin{aligned}
&\sharp_\gamma() \text{ where} \\
&\quad \gamma(x) := p \wedge ((q \wedge \Diamond \sharp_\delta(x)) \vee (r \wedge \Box x)) \\
&\quad \delta(v;x) := (q \wedge \Diamond x) \vee (r \wedge \Box v)
\end{aligned}
$$

Then $\binom{\sharp_\gamma()}{\sharp_\delta(\sharp_\gamma()/v)}$ is the (mutual) least fixed point of

$$
\begin{pmatrix} X \\ Y \end{pmatrix} \mapsto \begin{pmatrix} p \wedge ((q \wedge \Diamond Y) \vee (r \wedge \Box X)) \\ (q \wedge \Diamond Y) \vee (r \wedge \Box X) \end{pmatrix}
$$

which shows that $\sharp_\gamma()$ has the same interpretation as ϕ.

Next, we show how $\mathrm{CFL}_\sharp^+$ fits into the dual-adjunction picture. We take $\mathcal{C}$ to be the category **Sets**, and $\mathcal{D}$ to be the category **DL** of distributive lattices. **DL** is enriched over **Poset** by the pointwise order. For the adjunction $P \dashv Q$ we take the adjunction from Example 1 in Sect. 2.1.

Our behavior functor $B : \mathbf{Sets} \to \mathbf{Sets}$ was given, together with a set of predicate liftings $\{(\!|\heartsuit|\!) \mid \heartsuit \in \Lambda\}$. Similarly to the one-step logic from Example 1, we take L_0 to be given by

$$L_0\mathbb{A} := \mathrm{Free}(\{\heartsuit(a_1, \dots, a_n) \mid a_i \in A, \heartsuit \in \Lambda \text{ of arity } n\})/{\approx}$$

where $\approx$ is the least congruence such that if $a \leq b$, then $\heartsuit a \leq \heartsuit b$. Now $(\!|-|\!)$ can be used to define a one-step semantics $\delta : L_0 P \to PB$ (cf. [20, Ex. 3.11]). Next, we set

$$L\mathbb{A} := \mathrm{Free}(\{\sharp_\gamma(\bar{a}/\bar{v}) \mid \gamma \in \mathrm{CFS}_\sharp^\mu, \gamma \text{ guarded}\})/{\approx}$$

Note that $L_0 + L$ generates an algebraically-free monad T, the monad of $L_0 + L$-*terms*. Since each γ is an $L_0 + L$-term in free variables $\bar{v} \cup x$, we obtain a natural transformation $\mathsf{u} : L \to T$ via $\mathsf{u} : \sharp_\gamma(\bar{a}/\bar{v}) \mapsto \gamma(\bar{a}/\bar{v}, \sharp_\gamma(\bar{a}/\bar{v})/x)$. Since we stipulate that γ is guarded, we know that u will in fact factor through $\mathrm{id} + L_0 T$. Hence, using Proposition 2, we are able to define the desired semantic maps $[\![-]\!]$.

Proposition 3. *Let $\sigma : X \to BX$ be a B-coalgebra, and let $[\![-]\!] : \Phi \to PX$ be the semantic map induced by the unfolding system (δ, u). Then $[\![\phi]\!]$ is the subset of X defined as in definition 9.*

Proof (Sketch). The argument for this is similar to the one given in Example 3: let $t : \Phi \to PX$ be the explicitly defined semantics. Then t is clearly a solution to the diagram from definition 6, and hence $t \geq [\![-]\!]$. For the other direction, we can set up an approximation process for t, and prove by induction that each approximant is contained in all solutions.

Now since we have shown how to faithfully translate the μ-calculus into CFL, this allows us to interpret the negation-free, ν-free fragment of the coalgebraic μ-calculus. In Sect. 6, we will sketch how to extend this to the full alternation-free fragment of the coalgebraic μ-calculus.

6 Negations

So far, we have only been working in the positive fragments of the modal logics under consideration. However, many modal (fixpoint) logics come equipped with operations that are not order preserving - think of negation $\neg$ in Boolean algebras, or subtraction in quantitative algebras. We will use the term 'negation(s)' as a stand-in for such operations generally. In this section, we sketch how the semantics on a positive fragment may be extended to the full logic with negations.

Let $\mathcal{D}^\neg$ be a category of 'algebras-with-negations', equipped with a free-forgetful adjunction $F : \mathcal{D} \to \mathcal{D}^\neg$, $U : \mathcal{D}^\neg \to \mathcal{D}$ with $F \dashv U$. This yields a endofunctors $L_0^\neg, L^\neg : \mathcal{D}^\neg \to \mathcal{D}^\neg$ given by $L^\neg = FLU, L_0^\neg = FL_0U$. We also assume that we have an adjunction $P^\neg \dashv Q^\neg$ with $P^\neg : \mathcal{C} \to (\mathcal{D}^\neg)^{\mathrm{op}}$, and moreover assume that $UP^\neg = P$. Using this, the map m from Sect. 4 has type $m : LUP^\neg X \to UP^\neg X$. Hence, we can transpose it along $F \dashv U$ to get

$$m^\neg : FLUP^\neg = L^\neg P^\neg X \to P^\neg X$$

So, if we write $\Phi^\neg$ for the initial $L^\neg$-algebra, we obtain an initial algebra semantics for formulas with negations, which respects the already defined semantics for negation-free formulas.

As an example, consider the logic CFL from Sect. 5. As written there, we can only interpret $\mathrm{CFL}_\sharp^+$, corresponding to the fragment of CMC without greatest fixpoints and negations. However, if we take $\mathcal{D}^\neg$ to be **BA**, the category of Boolean algebras, we see that the adjunction $P_{\mathbf{DL}} \dashv Q$ factors as $UP_{\mathbf{BA}} \dashv \mathbf{uf}F$, where $U : \mathbf{BA} \to \mathbf{DL}$ is forgetful, $F : \mathbf{DL} \to \mathbf{BA}$ is free, $P_{\mathbf{DL}}$ and $P_{\mathbf{BA}}$ are the powerset functors into **DL** and **BA**, respectively, and $P_{\mathbf{BA}} \dashv \mathbf{uf}$ is the well-known powerset-ultrafilter adjunction.

By the above discussion, we also get an interpretation of $L^\neg$-formulas in $P_{\mathbf{BA}}X$ for any coalgebra $\xi : X \to BX$. The concrete analogue of this shift consists of adding in negations to the syntax of formulas (but not fixpoint schemes)

$$\begin{aligned}\gamma &::= v \in V \mid x \mid \phi \mid \gamma \vee \gamma \mid \gamma \wedge \gamma \mid \heartsuit(\bar{\gamma}) \mid \sharp_\gamma(\bar{\gamma}/\bar{v})\\ \phi &::= \top \mid \bot \mid \phi \vee \phi \mid \phi \wedge \phi \mid \heartsuit(\bar{\phi}) \mid \sharp_\gamma(\bar{\phi}/\bar{v}) \mid \neg\phi\end{aligned}$$

yielding $\mathrm{CFL}_\sharp$, which corresponds to the full alternation-free fragment of the coalgebraic μ-calculus.

The picture is similar to that in [1]. There, the authors define 'positive fragments' of one-step logics (however restricted to the logical connection between **Sets** and **BA**). For our purposes, we would prefer to start with a (positive) logic, and consider 'negative extensions'. There is also the interesting question of expanding the notion of a 'positive fragment'/'negative extension' beyond the pair **BA**/**DL**, to more quantitative logics.

It is important to note that, while this section shows how to extend the initial algebra semantics of the positive logic to a logic with negation, it is not clear whether we can define the semantics for the logic with negation directly, i.e., based on the unfolding system as in Definition 6.

7 Conclusion

We have provided a new categorical framework for studying coalgebraic fixpoint logics based on the dual adjunction framework. The framework provides both a least-solution and an initial algebra semantics. We exemplified the framework using a number of different examples such as the positive modal logic of transitive closure, its probabilistic variant, positive PDL and the positive alternation-free fragment of the coalgebraic μ-calculus. We also showed how to add negations to these logics, but hasten to remark that in this case we only have an initial algebra semantics. As a first simple example of how the framework can be used we provided a generic proof of adequacy of our fixpoint logics, i.e., the semantics of any logic that fits in the framework is invariant under behavioural equivalence.

Our framework is based on adding recursion to coalgebraic modal logic where the one-step semantics is given by a type of distributive law. In [24], instead, it is shown how to add a form of recursion to *abstract GSOS specifications*—a different type of distributive laws that can be used to study structural operational semantics at a high level of generality—using "unfolding" natural transformations, similar to the current approach. An interesting theoretical direction may be to try and unify these extensions; a good starting point may be the steps-and-adjunctions framework studied in [25].

The main contribution of this paper is the new categorical framework for fixpoint logics. In the future we hope to exploit this in several ways: Firstly, we will study proof systems [22] and reasoning procedures [13] for alternation-free fixpoint logics and will lift soundness and completeness proofs to our framework. Secondly, the alternation-free fragment of the modal μ-calculus has interesting model-theoretic properties [8] and we will explore whether those can be recovered within our framework. Another important question – suggested by two of the anonymous referees – is whether we can use our framework to provide a completeness proof for the Segerberg axioms for PDL [18] or their coalgebraic generalisation from [11]. These logics certainly do fit into our framework, the key question to solve will be to formally connect the induction axiom or the least fixpoint rule to the requirement that the semantic map is defined as a least fixpoint. Finally, the standard dual adjunction framework can be used to define certain filtrations [2]. We have reasons to believe that our framework of unfolding systems can be used to show that a (fixpoint) logic has filtrations. This opens an avenue for studying filtrations and abstraction techniques for fixpoint logics in a categorical way.

Acknowledgments. The authors would like to thank the anonymous referees for instructive comments and interesting questions for future work, as well as for suggesting the formula in Example 7.

References

1. Balan, A., Kurz, A., Velebil, J.: Positive fragments of coalgebraic logics. In: Heckel, R., Milius, S. (eds.) CALCO 2013. LNCS, vol. 8089, pp. 51–65. Springer, Heidelberg (2013). https://doi.org/10.1007/978-3-642-40206-7_6
2. Barlocco, S., Kupke, C., Rot, J.: Coalgebra learning via duality. In: Bojańczyk, M., Simpson, A. (eds.) FoSSaCS 2019. LNCS, vol. 11425, pp. 62–79. Springer, Cham (2019). https://doi.org/10.1007/978-3-030-17127-8_4
3. Brzozowski, J.A.: Derivatives of regular expressions. J. ACM **11**(4), 481–494 (1964)
4. Cîrstea, C., Kupke, C., Pattinson, D.: EXPTIME tableaux for the coalgebraic mu-calculus. Log. Methods Comput. Sci. **7**(3) (2011)

5. Cîrstea, C., Kurz, A., Pattinson, D., Schröder, L., Venema, Y.: Modal logics are coalgebraic. In: BCS International Academic Conference, pp. 128–140. British Computer Society (2008)
6. Cousot, P., Cousot, R.: Constructive versions of Tarski's fixed point theorems. Pac. J. Math. **82**(1), 43–57 (1979)
7. Dahlqvist, F.: Coalgebraic completeness-via-canonicity. In: Hasuo, I. (ed.) CMCS 2016. LNCS, vol. 9608, pp. 174–194. Springer, Cham (2016). https://doi.org/10.1007/978-3-319-40370-0_11
8. Facchini, A., Venema, Y., Zanasi, F.: A characterization theorem for the alternation-free fragment of the modal μ-calculus. In: LICS, pp. 478–487. IEEE Computer Society (2013)
9. Forster, J., Schröder, L.: Non-iterative modal logics are coalgebraic. In: AiML, pp. 229–248. College Publications (2020)
10. Gu, T., Silva, A., Zanasi, F.: Hennessy-Milner results for probabilistic PDL. In: MFPS. Electronic Notes in Theoretical Computer Science, vol. 352, pp. 283–304. Elsevier (2020)
11. Hansen, H.H., Kupke, C.: Weak completeness of coalgebraic dynamic logics. In: FICS. EPTCS, vol. 191, pp. 90–104 (2015)
12. Hausmann, D., Humml, M., Prucker, S., Schröder, L., Strahlberger, A.: Generic model checking for modal fixpoint logics in COOL-MC. In: Dimitrova, R., Lahav, O., Wolff, S. (eds.) VMCAI 2024. LNCS, vol. 14499, pp. 171–185. Springer, Cham (2024). https://doi.org/10.1007/978-3-031-50524-9_8
13. Hausmann, D., Schröder, L., Egger, C.: Global caching for the alternation-free μ-calculus. In: CONCUR. LIPIcs, vol. 59, pp. 34:1–34:15. Schloss Dagstuhl - Leibniz-Zentrum für Informatik (2016)
14. Hennessy, M., Milner, R.: On observing nondeterminism and concurrency. In: de Bakker, J., van Leeuwen, J. (eds.) ICALP 1980. LNCS, vol. 85, pp. 299–309. Springer, Heidelberg (1980). https://doi.org/10.1007/3-540-10003-2_79
15. Kelly, M.: The basic concepts of enriched category theory. Reprints in Theory and Applications of Categories [electronic only] (2005)
16. Klin, B.: Coalgebraic modal logic beyond sets. In: MFPS. Electronic Notes in Theoretical Computer Science, vol. 173, pp. 177–201. Elsevier (2007)
17. Kozen, D.: A probabilistic PDL. J. Comput. Syst. Sci. **30**(2), 162–178 (1985)
18. Kozen, D., Parikh, R.: An elementary proof of the completness of PDL. Theor. Comput. Sci. **14**, 113–118 (1981)
19. Kupke, C., Kurz, A., Pattinson, D.: Algebraic semantics for coalgebraic logics. In: CMCS. Electronic Notes in Theoretical Computer Science, vol. 106, pp. 219–241. Elsevier (2004)
20. Kupke, C., Pattinson, D.: Coalgebraic semantics of modal logics: An overview. Theor. Comput. Sci. **412**(38), 5070–5094 (2011)
21. Kurz, A., Velebil, J.: Enriched logical connections. Appl. Categorical Struct. **21**(4), 349–377 (2013)
22. Marti, J., Venema, Y.: A focus system for the alternation-free μ-calculus. In: Das, A., Negri, S. (eds.) TABLEAUX 2021. LNCS (LNAI), vol. 12842, pp. 371–388. Springer, Cham (2021). https://doi.org/10.1007/978-3-030-86059-2_22
23. Pavlovic, D., Mislove, M., Worrell, J.B.: Testing semantics: connecting processes and process logics. In: Johnson, M., Vene, V. (eds.) AMAST 2006. LNCS, vol. 4019, pp. 308–322. Springer, Heidelberg (2006). https://doi.org/10.1007/11784180_24
24. Rot, J., Bonsangue, M.M.: Structural congruence for bialgebraic semantics. J. Log. Algebr. Methods Program. **85**(6), 1268–1291 (2016)

25. Rot, J., Jacobs, B., Levy, P.B.: Steps and traces. J. Log. Comput. **31**(6), 1482–1525 (2021)
26. Rutten, J.J.M.M.: Universal coalgebra: a theory of systems. Theor. Comput. Sci. **249**(1), 3–80 (2000)
27. Schröder, L., Pattinson, D.: PSPACE bounds for rank-1 modal logics. ACM Trans. Comput. Log. **10**(2), 13:1–13:33 (2009)
28. Schröder, L., Venema, Y.: Flat coalgebraic fixed point logics. In: Gastin, P., Laroussinie, F. (eds.) CONCUR 2010. LNCS, vol. 6269, pp. 524–538. Springer, Heidelberg (2010). https://doi.org/10.1007/978-3-642-15375-4_36
29. Venema, Y.: Automata and fixed point logic: a coalgebraic perspective. Inf. Comput. **204**(4), 637–678 (2006)

Preorder-Constrained Simulations for Program Refinement with Effects

Koko Muroya[1(✉)], Takahiro Sanada[1], and Natsuki Urabe[2]

[1] RIMS, Kyoto University, Kyoto, Japan
{kmuroya,tsanada}@kurims.kyoto-u.ac.jp
[2] National Institute of Informatics, Tokyo, Japan
urabenatsuki@nii.ac.jp

Abstract. We propose a notion of *preorder-constrained simulation*. It is parameterised by a preorder ("observation preorder") on traces, so that it can uniformly characterise quantitative notions of program refinement for different effects, such as exception, nondeterminism and I/O. Preorder-constrained simulation is additionally parameterised by a positive number ("look-ahead bound"), and forms a generative spectrum governed by the look-ahead bound. We analyse the complexity of determining preorder-constrained similarity, and show that preorder-constrained simulation can be enhanced by the so-called up-to technique.

1 Introduction

It is often important to have two programs behave the same in programming, for example, in code refactoring and compiler optimisation. *Observational equivalence* [21] is the standard notion that asserts the same observable behaviour between two programs.

Among known proof techniques for observational equivalence are *coinductive* techniques. These techniques reduce the problem of observational equivalence to *stepwise* comparison based on operational semantics. An important example is *applicative bisimilarity* [1]. It is tailored to the ordinary reduction semantics, syntactically constructing a binary relation that characterises observational equivalence. Together with its extension known as *environmental bisimilarity* [17], applicative bisimilarity is applicable to a wide range of side effects, such as general state [17], I/O [30] and continuation [32].

Other coinductive techniques that are for abstract machines have emerged recently. Examples are what we shall call *counting simulation* [22], and *improvement* [4] which can be seen as an instance of counting simulation. Notably, counting simulation can not only assert the same behaviour between programs, but also compare efficiency of two programs in terms of the number of execution steps. Despite this strength, it is only known to work on "deterministic" effects that yield at most one result per program.

It is desirable to obtain a coinductive proof technique for observational equivalence that is applicable to various effects and capable of quantitative comparison of efficiency. We have two kinds of techniques with different strengths: (1)

Published by Springer Nature Switzerland AG 2024
B. König and H. Urbat (Eds.): CMCS 2024, LNCS 14617, pp. 44–64, 2024.
https://doi.org/10.1007/978-3-031-66438-0_3

$(1, \mathbf{Q})$-similarity $\lesssim_{1,\mathbf{Q}}$ $\longrightarrow$ $(2, \mathbf{Q})$-similarity $\lesssim_{2,\mathbf{Q}}$ $\longrightarrow \cdots \longrightarrow$ $\mathbf{Q}$-trace inclusion $\sqsubseteq_{\mathbf{Q}}$

Fig. 1. A generative spectrum, parameterised by the observation preorder $\mathbf{Q}$

Table 1. Instances of the two ends of the generative spectrum (see Sect. 4 for details)

observation preorder $\mathbf{Q}$	$(1, \mathbf{Q})$-simulation	$\mathbf{Q}$-trace inclusion $\sqsubseteq_{\mathbf{Q}}$
$=$	standard simulation	finite trace inclusion
$=_{\mathsf{rem}_{\{\tau\}}}$	weak simulation	weak trace inclusion
$\dot{Q}$	(new instances)	refinement $\preceq^{Q}_{\mathsf{err}}$ for exception
$\dot{Q} \cap =_{\mathsf{rem}_{\{\tau\} \cup \overline{\Omega_{\mathsf{nd}}}}}$		refinement $\preceq^{Q}_{\mathsf{nd}}$ for nondeterminism
$\dot{Q} \cap =_{\mathsf{rem}_{\{\tau\}}}$		refinement $\preceq^{Q}_{\mathsf{io}}$ for I/O

applicative bisimilarity and environmental bisimilarity accommodating various effects, and (2) counting simulation and improvement which are capable of quantitative comparison. Our goal here is to enhance the techniques (2) so that they can accommodate a wider range of effects including nondeterminism.

One of the challenges of accommodating various effects is that a notion of observational equivalence, in particular that of observation, varies between effects. For example, nondeterministic choice is regarded as internal and unobservable. A program $\mathtt{or}(\underline{1}, \underline{1})$ with a binary nondeterministic choice operator $\mathtt{or}$ would be identified with a program $\underline{1}$. The choice (with the same result) is *ignored*. On the other hand, the choice that is made according to a 1-bit input is regarded as external and observable. A program $\mathtt{in}(\underline{1}, \underline{1})$, which results in $\underline{1}$ regardless of the value of the 1-bit input, would *not* be identified with the program $\underline{1}$. The received input value (0 or 1) and the induced choice (between $\underline{1}$ and $\underline{1}$) is *observed*. We need to be able to both ignore and observe effectful choices.

We propose a notion of *preorder-constrained simulation* that is applicable to effects such as nondeterminism and I/O, in addition to the "deterministic" effects such as exception. It is notably parameterised by an *observation preorder*, a preorder on traces (or words). By altering the observation preorder, we can characterise quantitative notions of observational refinement, which is the asymmetric version of observational equivalence, for both internal and ignored effects, and external and observed effects.

Preorder-constrained simulations are additionally parameterised by a positive number dubbed *look-ahead bound*. It determines the degree of awareness of branching. By altering the look-ahead bound, one can obtain a spectrum illustrated in Fig. 1. The "limit" of the spectrum is a novel generalisation of trace inclusion (i.e. $\mathbf{Q}$-trace inclusion $\sqsubseteq_{\mathbf{Q}}$) that is also parameterised by the observation preorder. The spectrum is generative in the sense that it yields various concrete spectra by instantiating the observation preorder $\mathbf{Q}$. Some instances are shown in Table 1, whose details will be evident in Sect. 4.

We analyse the complexity of determining preorder-constrained similarity $\lesssim_{M,\mathbf{Q}}$ in the finite setting, and show that the complexity is polynomial time. The complexity analysis is via a game-theoretic characterisation of preorder-constrained simulation that is similar to *buffered simulation game* [12,13]. We also show that preorder-constrained simulation can be enhanced by the so-called *up-to* technique [26]. We discuss and identify sufficient conditions that make the up-to technique work, in terms of observation preorders.

Our contributions can be summarised as follows.

- (Sect. 4) The notion of preorder-constrained simulation, which is a new variant of simulation for characterising observational refinement between programs (Corollary 1). Preorder-constrained simulations enjoy soundness (Theorem 1) with respect to a novel generalisation of trace inclusion, and also monotonicity (Corollary 2) that yields a generative spectrum (Fig. 1).
- (Sect. 5) A game-theoretic characterisation of preorder-constrained simulation, with complexity analysis. The games also form a generative spectrum (Fig. 12).
- (Sect. 6) Integration of the so-called *up-to* technique [26], in terms of observation preorders.

Proofs of statements marked by (†) are formalised in Agda, and available at https://github.com/urabenatsuki/preorder-constr-sim-agda.

2 Preliminaries

Let $\mathbb{N}$ be the set of natural numbers, and $\mathbb{N}_+$ be the set of positive numbers. For a set X, X^* denotes the set $\{x_1 \dots x_n \mid n \in \mathbb{N}; x_1, \dots, x_n \in X\}$ of finite words over X. We let ε denote the empty word, X^+ denote $X^* \setminus \{\varepsilon\}$, and $|w|$ denote the length of a finite word $w \in X^*$.

Given $X' \subseteq X$, the *filtered equality* $=_{\mathsf{rem}_{X'}}$ on the set X^* is defined by $w =_{\mathsf{rem}_{X'}} w'$ if w and w' are the same except for symbols in X'. For example, $a\tau b\sigma\sigma c =_{\mathsf{rem}_{\{\tau,\sigma\}}} \sigma ab\tau c\tau =_{\mathsf{rem}_{\{\tau,\sigma\}}} abc$.

A preorder $Q \subseteq \mathbb{N} \times \mathbb{N}$ is said to be *s-closed* if it is closed under summation, i.e. $kQl \wedge k'Ql' \implies (k+k')Q(l+l')$. For a set Σ, a preorder $\mathbf{Q} \subseteq \Sigma^* \times \Sigma^*$ is said to be *c-closed* if it is closed under concatenation, i.e. $w_1\mathbf{Q}w_2 \wedge w_1'\mathbf{Q}w_2' \implies (w_1w_1')\mathbf{Q}(w_2'w_2')$.

2.1 Nondeterministic Automata

A *nondeterministic automaton* (NA) is a quadruple $\mathcal{A} = (X, \Sigma, \rightsquigarrow, F)$ consisting of a set X called a *state space*, a set Σ called an *alphabet*, a *transition relation* $\rightsquigarrow \subseteq X \times \Sigma \times X$ and a set $F \subseteq X$ of *accepting states*. We write $x \overset{a}{\rightsquigarrow} x'$ when $(x, a, x') \in \rightsquigarrow$, and $x \overset{w}{\twoheadrightarrow} x'$ for $w = a_1 \cdots a_n$ when there exist $x_0 \dots x_n \in X^+$ such that $x_0 = x$, $x_n = x'$ and $x_{i-1} \overset{a_i}{\rightsquigarrow} x_i$ for each $i \in \{1, \dots, n\}$. In particular, $x \overset{\varepsilon}{\twoheadrightarrow} x$. We write $x \not\rightsquigarrow$ and say x is *stuck*, if there exist no $x' \in X$ and $a \in \Sigma$

$$
\begin{aligned}
t &::= x \mid t\ t \mid \lambda x.t \mid t[x \leftarrow t] \mid \underline{n} \mid t + t \mid t \times t \mid f(t, \ldots, t) && \text{(Terms } \mathbf{T}_\Omega\text{)} \\
v &::= \lambda x.t \mid \underline{n} && \text{(Values)} \\
L &::= \langle\rangle \mid L[x \leftarrow t] && \text{(Answer contexts)} \\
E &::= \langle\rangle \mid E\ t \mid L\langle v\rangle\ E \mid E[x \leftarrow t] \mid E + t \mid L\langle v\rangle + E \mid E \times t \mid L\langle v\rangle \times E && \text{(Evaluation contexts)}
\end{aligned}
$$

$$
\frac{}{L\langle \lambda x.t\rangle\ L'\langle v\rangle \xrightarrow{\tau} L\langle t[x \leftarrow L\langle v\rangle]\rangle} \qquad \frac{}{E\langle x\rangle[x \leftarrow L\langle v\rangle] \xrightarrow{\tau} L\langle E\langle v\rangle[x \leftarrow v]\rangle}
$$

$$
\frac{\bullet \in \{+, \times\}}{L\langle \underline{n}\rangle \bullet L'\langle \underline{m}\rangle \xrightarrow{\tau} \underline{n \bullet m}} \qquad \frac{}{f(t_0, \ldots, t_{\mathrm{ar}(f)-1}) \xrightarrow{f_i} t_i} \qquad \frac{t \xrightarrow{\ell} u \quad \ell \in \{\tau\} \cup \overline{\Omega}}{E\langle t\rangle \xrightarrow{\ell} E\langle u\rangle} \qquad \frac{}{L\langle \underline{n}\rangle \xrightarrow{n} \checkmark}
$$

Fig. 2. Reduction semantics as an NA $\mathcal{A}_\Omega = (\mathbf{T}_\Omega \cup \{\checkmark\},\ \{\tau\} \cup \mathbb{N} \cup \overline{\Omega},\ \rightarrow,\ \{\checkmark\})$

such that $x \overset{a}{\rightsquigarrow} x'$. An NA is said to be *branching-free* if its transition relation $\rightsquigarrow \subseteq X \times \Sigma \times X$ satisfies the following: for any $x \in X$, if there exist two pairs $(a_1, x_1), (a_2, x_2) \in \Sigma \times X$ such that $x \overset{a_1}{\rightsquigarrow} x_1$ and $x \overset{a_2}{\rightsquigarrow} x_2$, then $(a_1, x_1) = (a_2, x_2)$.

The *(finite) language* of an NA $\mathcal{A}$ is a function $L^*_{\mathcal{A}} : X \to \Sigma^*$ defined by $L^*_{\mathcal{A}}(x) := \{w \in \Sigma^* \mid \exists x' \in F.\ x \overset{w}{\rightsquigarrow\!\!\!\!\rightsquigarrow} x'\}$. In particular, $\varepsilon \in L^*_{\mathcal{A}}(x)$ when $x \in F$. We omit the subscript and write $L^*(x)$ for $L^*_{\mathcal{A}}(x)$ when no confusion is likely.

2.2 A Linear Substitution Calculus with Algebraic Effects

We recall *algebraic effects* [23] and present their reduction semantics as an NA. Algebraic effects are specified by a *signature* Ω. It is a set of *algebraic operations* f, each of which comes with an *arity* $\mathrm{ar}(f) \in \mathbb{N}$. We write $f\colon n$ when $\mathrm{ar}(f) = n$.

Example 1 (algebraic operations).

1. $\Omega_{\mathsf{err}} = \{\mathtt{err}\colon 0\}$ for raising an error.
2. $\Omega_{\mathsf{nd}} = \{\mathtt{or}\colon 2\}$ for nondeterministic choice between two operands.
3. $\Omega_{\mathsf{io}} = \{\mathtt{in}\colon 2, \mathtt{out}^0\colon 1, \mathtt{out}^1\colon 1\}$ for I/O with a single bit. We focus on a single-bit I/O for simplicity. Evaluation of a term $\mathtt{in}(t_0, t_1)$ proceeds with t_i if the input value is i, for each $i \in \{0, 1\}$. Evaluation of a term $\mathtt{out}^i(t)$ outputs the value $i \in \{0, 1\}$ and proceeds with t.

We use the (left-to-right) call-by-value *linear substitution calculus* (LSC) [5] equipped with algebraic effects Ω and arithmetic. The LSC has been used as a cost model of various abstract machines for the λ-calculus [3]. The LSC exploits explicit substitutions $[x \leftarrow t]$ to disclose cost of the traditional β-reduction.

We present reduction semantics of the LSC, with a signature Ω, as an NA defined by Fig. 2. Transitions are labelled in a similar way as the original reduction semantics for algebraic effects [23]; labels consist of τ representing *silent* transitions, $\mathbb{N}$ representing result values, and a set $\overline{\Omega} = \{f_i \mid f \in \Omega, i \in \{0, \ldots, \mathrm{ar}(f)-1\}\}$ labelling *effect* transitions. We add a *return* transition $\underline{n} \overset{n}{\rightsquigarrow} \checkmark$

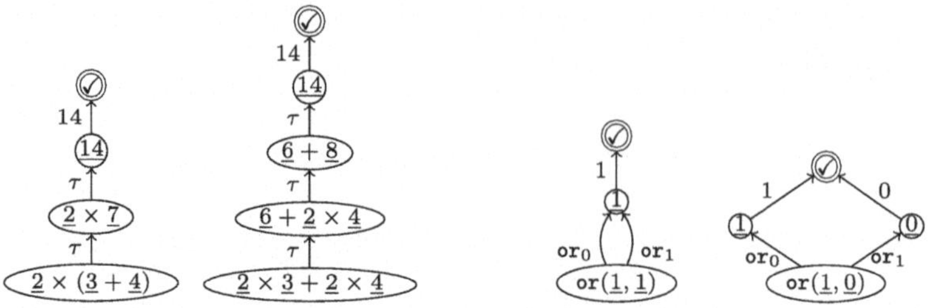

(a) $\mathcal{A}_{\Omega_{\mathsf{err}}}(\underline{2} \times (\underline{3}+\underline{4}))$ and $\mathcal{A}_{\Omega_{\mathsf{err}}}(\underline{2} \times \underline{3}+\underline{2} \times \underline{4})$

(b) $\mathcal{A}_{\Omega_{\mathsf{nd}}}(\mathtt{or}(\underline{1},\underline{1}))$ and $\mathcal{A}_{\Omega_{\mathsf{nd}}}(\mathtt{or}(\underline{1},\underline{0}))$

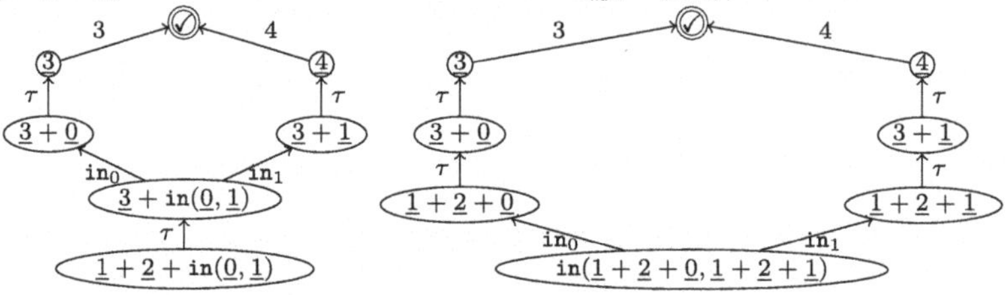

(c) $\mathcal{A}_{\Omega_{\mathsf{io}}}(\underline{1}+\underline{2}+\mathtt{in}(\underline{0},\underline{1}))$ and $\mathcal{A}_{\Omega_{\mathsf{io}}}(\mathtt{in}(\underline{1}+\underline{2}+\underline{0},\underline{1}+\underline{2}+\underline{1}))$

Fig. 3. Example pairs of NAs

from each ground value to a sole accepting state $\checkmark$. This extra transition enables us to observe an evaluation result as a transition label. For $t \in \mathbf{T}_\Omega$, an NA $\mathcal{A}_\Omega(t)$ is the restriction of $\mathcal{A}_\Omega$ to the states that are reachable from t.

A successful evaluation of a term t is given by a sequence $t \stackrel{w}{\twoheadrightarrow} L\langle\underline{n}\rangle \stackrel{n}{\rightarrow} \checkmark$ of silent or effect transitions followed by a return transition. We refer to the word wn as a *trace* of t, and to n as a *result* of the execution. By removing the silent transitions $\stackrel{\tau}{\rightarrow}$, we obtain a word w' from w. We refer to w' as an *effect trace* of t. Note that the NA $\mathcal{A}_{\Omega_{\mathsf{err}}}$ is in fact branching-free.

Example 2 (pairs of NAs).

1. In Fig. 3a, $\underline{2} \times (\underline{3}+\underline{4})$ has a trace $\tau\tau 14$, and $\underline{2} \times \underline{3} + \underline{2} \times \underline{4}$ has a trace $\tau\tau\tau 14$.
2. In Fig. 3b, terms $\mathtt{or}(\underline{1},\underline{1})$ and $\mathtt{or}(\underline{1},\underline{0})$ have the same effect traces $\{\mathtt{or}_0, \mathtt{or}_1\}$.
3. In Fig. 3c, terms $\underline{1}+\underline{2}+\mathtt{in}(\underline{0},\underline{1})$ and $\mathtt{in}(\underline{1}+\underline{2}+\underline{0},\underline{1}+\underline{2}+\underline{1})$ have the same results $\{\underline{3},\underline{4}\}$ and the same effect traces $\{\mathtt{in}_0, \mathtt{in}_1\}$, but not the same traces.

We can now formalise program refinement, using NAs $\mathcal{A}_\Omega$ for the signatures in Example 1. The notion is quantitative, in the sense that it is parameterised by a *length preorder* $Q \subseteq \mathbb{N} \times \mathbb{N}$.

Definition 1 ((quantitative) refinement). *Let Q be a preorder on $\mathbb{N}$ (dubbed* length preorder*).*

1. For Ω_{err}, $t \preceq^Q_{\mathsf{err}} u$ is defined by $\forall w.(t \stackrel{w}{\twoheadrightarrow} \checkmark \implies \exists w'.u \stackrel{w'}{\twoheadrightarrow} \checkmark \wedge |w| Q |w'|)$.

2. *For* Ω_{nd}, $t \preceq^{Q}_{\mathsf{nd}} u$ *is defined by* $\forall w.(t \overset{w}{\twoheadrightarrow} \checkmark \implies \exists w'.u \overset{w'}{\twoheadrightarrow} \checkmark \wedge |w|Q|w'| \wedge w =_{\mathsf{rem}_{\{\tau\}\cup\overline{\Omega_{\mathsf{nd}}}}} w')$.
3. *For* Ω_{io}, $t \preceq^{Q}_{\mathsf{io}} u$ *is defined by* $\forall w.(t \overset{w}{\twoheadrightarrow} \checkmark \implies \exists w'.u \overset{w'}{\twoheadrightarrow} \checkmark \wedge |w|Q|w'| \wedge w =_{\mathsf{rem}_{\{\tau\}}} w')$.

The refinement $t \preceq^{Q}_{\mathsf{err}} u$ focuses on successful termination, and additionally, compares the numbers of steps using Q. It asserts that when evaluation of t terminates in m steps, evaluation of u also terminates in n steps, and moreover, mQn. An example of Q is the total relation $\mathbb{N} \times \mathbb{N}$. This yields the asymmetric version of the basic notion of *observational equivalence*. Another example of Q is the greater-than-equal relation $\geq$, with which refinement $t \preceq^{\geq} u$ can assert that u terminates in a fewer steps.

Definition 1(2) and Definition 1(3) use the filtered equality differently to deal with different observations made for nondeterministic choice and I/O. The refinement $t \preceq_{\mathsf{nd}} u$ for nondeterministic choice asserts that when evaluation of t terminates, evaluation of u also terminates with the same result. The filtered equality $w =_{\mathsf{rem}_{\{\tau\}\cup\overline{\Omega_{\mathsf{nd}}}}} w'$ in Definition 1(2) *ignores* effect traces, and only ensures the coincidence of results. We note that Definition 1(2) corresponds to program refinement for angelic nondeterminism [29]. In contrast, the refinement $t \preceq_{\mathsf{io}} u$ for I/O asserts that when evaluation of t terminates, evaluation of u also terminates with the same effect trace and result. The filtered equality $w =_{\mathsf{rem}_{\{\tau\}}} w'$ in Definition 1(3) *observes* effect traces.

The three pairs of NAs in Fig. 3 all exhibit refinement. We have $\underline{2}\times(\underline{3}+\underline{4}) \preceq^{\leq}_{\mathsf{err}} \underline{2}\times\underline{3}+\underline{2}\times\underline{4}$, $\mathtt{or}(\underline{1},\underline{1}) \preceq^{=}_{\mathsf{nd}} \mathtt{or}(\underline{1},\underline{0})$, and $\underline{1}+\underline{2}+\mathtt{in}(\underline{0},\underline{1}) \preceq^{=}_{\mathsf{io}} \mathtt{in}(\underline{1}+\underline{2}+\underline{0},\underline{1}+\underline{2}+\underline{1})$. We do not always have the opposite: i.e. $\mathtt{or}(\underline{1},\underline{0}) \not\preceq^{=}_{\mathsf{nd}} \mathtt{or}(\underline{1},\underline{1})$.

3 Counting Simulation and Its Deficiencies

We recall the simulation notion that was introduced for an abstract machine for programs with at most one result [22]. We call it *counting simulation*, albeit with no connection to counter automata. It is parameterised by a preorder Q on natural numbers, dubbed *length preorder*, to *count* and compare the number of steps. The original presentation [22, Def. 4.4.1] of the counting simulation is for an unlabelled transition system, and it is equipped with an up-to technique. We here present its naive extension to NAs, but without any up-to technique.

In this section, we let $\mathcal{A}_i = (X_i, \Sigma, \rightsquigarrow_i, F_i)$ $(i \in \{1,2\})$ be two NAs with the same alphabet, and $Q \subseteq \mathbb{N} \times \mathbb{N}$ be an s-closed preorder.

Definition 2 (counting simulations). *A binary relation* $R \subseteq X_1 \times X_2$ *is a* Q-counting simulation *from* $\mathcal{A}_1$ *to* $\mathcal{A}_2$ *if, for any* $(x, y) \in R$, *the following* **C-Final** *and* **C-Step** *hold.*

C-Final *If* $x \in F_1$, *then* $y \in F_2$.
C-Step *For each* $x' \in X_1$ *and* $a \in \Sigma$ *such that* $x \overset{a}{\rightsquigarrow}_1 x'$, *either of the following holds.*

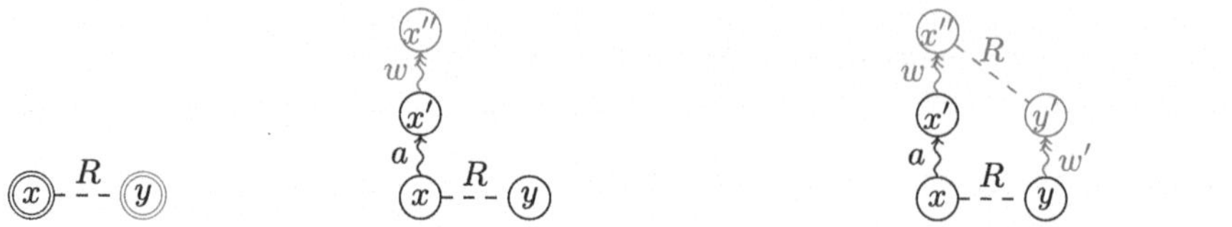

(a) ***C-Final*** (b) ***C-Step*** (1) where $x'' \not\rightsquigarrow$ (c) ***C-Step*** (2) where $|aw|Q|w'|$

Fig. 4. Conditions of Definition 2. Black parts are universally quantified, and magenta parts are existentially quantified.

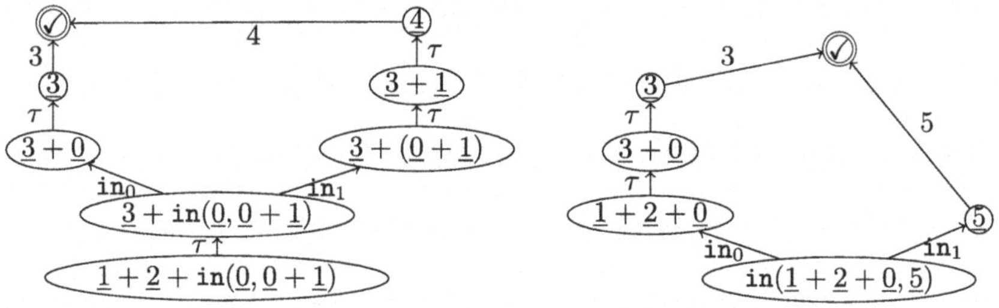

Fig. 5. Unsoundness of counting simulation

1. *There exist* $x'' \in X_1$ *and* $w \in \Sigma^*$ *such that* $x' \overset{w}{\rightsquigarrow\!\!\!\!\rightarrow}_1 x'' \not\rightsquigarrow_1$ *and* $x'' \notin F_1$.
2. *There exist* $x'' \in X_1$, $y' \in X_2$ *and* $w, w' \in \Sigma^*$, *such that* $x' \overset{w}{\rightsquigarrow\!\!\!\!\rightarrow}_1 x''$, $y \overset{w'}{\rightsquigarrow\!\!\!\!\rightarrow}_2 y'$, $|aw|Q|w'|$, *and* $x''Ry'$.

Figure 4 illustrates the conditions of Definition 2. The condition ***C-Final*** is standard. The condition ***C-Step*** (1) deals with a stuck and non-accepting state $x'' \in X_1$ (i.e. $x'' \not\rightsquigarrow$ and $x'' \notin F_1$). A term $\mathtt{err}() \in \mathbf{T}_{\Omega_{\mathsf{err}}}$ is an example of such a state in $\mathcal{A}_{\Omega_{\mathsf{err}}}$. Lastly, ***C-Step*** (2) is the key condition of Definition 2. It asserts that any transition $x \overset{a}{\rightsquigarrow} x'$ followed by *some* transitions $x' \overset{w}{\rightsquigarrow\!\!\!\!\rightarrow} x''$ in $\mathcal{A}_1$ can be simulated by some transitions $y \overset{w'}{\rightsquigarrow\!\!\!\!\rightarrow} y'$ in $\mathcal{A}_2$.

Thanks to ***C-Step***, which compares not just single steps but numbers of steps, counting simulation can witness quantitative refinement for exception.

Proposition 1 (correctness wrt. refinement). *If* R *is a* Q*-counting simulation from* $\mathcal{A}_{\Omega_{\mathsf{err}}}$ *to* $\mathcal{A}_{\Omega_{\mathsf{err}}}$, *then* $tRu \implies t \preceq^{Q}_{\mathsf{err}} u$ *holds for any* $t, u \in \mathbf{T}_{\Omega_{\mathsf{err}}}$. □

Example 3. For the pair of branching-free NAs in Fig. 3a, a relation $R_{\mathsf{distr}} = \{(\underline{2} \times (\underline{3} + \underline{4}), \underline{2} \times \underline{3} + \underline{2} \times \underline{4}), (\checkmark, \checkmark)\}$ is a $\leq$-counting simulation. The length preorder $\leq$ asserts that $\underline{2} \times (\underline{3} + \underline{4})$ has better efficiency. The relation R_{distr} represents the distributive law, without relating any intermediate states.

Nevertheless, counting simulation cannot witness refinement for effects such as nondeterminism and I/O. This is due to two challenges. The first challenge is varying observation. While we ignore effect traces for nondeterminism, we observe effect traces for I/O. However, counting simulation can neither ignore nor observe effect traces correctly. It simply compares the lengths of traces using the length preorder Q. The second challenge is branching. It is branching effects that counting simulation becomes unsound for.

Example 4 (unsoundness for I/O). For the pair of NAs in Fig. 5, refinement does not hold, i.e. $\underline{1}+\underline{2}+\mathtt{in}(\underline{0},\underline{0}+\underline{1}) \not\precsim_{\mathsf{io}} \mathtt{in}(\underline{1}+\underline{2}+\underline{0},\underline{5})$, because right branches have traces $\tau\mathtt{in}_1\tau\tau 4 \neq_{\mathsf{rem}_\tau} \mathtt{in}_1 5$. However, the relation $\{(\underline{1}+\underline{2}+\mathtt{in}(\underline{0},\underline{0}+\underline{1}), \mathtt{in}(\underline{1}+\underline{2}+\underline{0},\underline{5})), (\checkmark,\checkmark)\}$ is an $=$-counting simulation. It only asserts that left branches have "identical" traces (i.e. $\tau\mathtt{in}_0\tau 3$ and $\mathtt{in}_0\tau\tau 3$), and it does not inspect the right branches with distinct traces.

This unsoundness is because counting simulation does not necessarily inspect all possibilities of branching. Technically, this is due to the existential quantification on $x' \overset{w}{\rightsquigarrow} x''$ in ***C-Step*** (2).

4 Preorder-Constrained Simulation

4.1 Definition

We present our main contribution, the notion of *preorder-constrained simulation.* It generalises counting simulation from branching-free NAs to general NAs, and hence characterises a notion of observational refinement for a wider class of effects (Corollary 1 below). The generalisation is technically two-fold, dealing with the two challenges we discussed in Sect. 3.

Firstly, preorder-constrained simulation is parameterised by a c-closed preorder $\mathbf{Q}$ on traces Σ^*, dubbed *observation preorder*, instead of the length preorder $Q \subseteq \mathbb{N}\times\mathbb{N}$ that parameterises counting simulation. Using preorders like the filtered equality $=_{\mathsf{rem}_X}$, preorder-constrained simulation can flexibly *observe* and compare traces, adapting to varying observations.

Example 5 (observation preorders). Note that the following preorders are all c-closed.

1. The equality $=$ on words.
2. Each preorder $Q \subseteq \mathbb{N}\times\mathbb{N}$ that is s-closed induces a preorder $\dot{Q}$ on words such that $w\dot{Q}w' \iff |w|Q|w'|$, which is c-closed.
3. Each subset $\Sigma' \subseteq \Sigma$ induces the preorder $=_{\mathsf{rem}_{\Sigma'}} \subseteq \Sigma^*\times\Sigma^*$ of filtered equality.
4. The *substring* preorder $\subseteq_{\mathsf{sub}}$, where $w \subseteq_{\mathsf{sub}} w'$ means that w is a substring of w'.
5. Assume Σ is the powerset 2^{AP} of some set AP. Let $\subseteq^*$ be a preorder where $a_1 \dots a_k \subseteq^* a'_1 \dots a'_{k'}$ means $k = k'$ and $\forall i.\ a_i \subseteq a'_i$.
6. Let $\mathbb{R}$ be the set of real numbers, and $\bullet$ be a binary operation on $\mathbb{R}$ such as summation $+$ and multiplication $\times$. When $\Sigma = \mathbb{R}$, let $\leq_\bullet$ be a preorder where $a_1 \dots a_n \leq_\bullet a'_1 \dots a'_{n'}$ means $a_1 \bullet \dots \bullet a_n \leq a'_1 \bullet \dots \bullet a'_{n'}$. Its inverse $\geq_\bullet$ is also a preorder.

(a) ***Final***^{M} where $|w| < M \ \wedge\ w\mathbf{Q}w'$ (b) ***Step***^{M} where $a_1 \cdots a_k \mathbf{Q} w'$

Fig. 6. Conditions of Definition 3. Black parts are universally quantified, and magenta parts are existentially quantified.

Secondly, preorder-constrained simulation limits the existential quantification, namely that on $x' \overset{w}{\rightsquigarrow\!\!\!\!\rightarrow} x''$ in ***C-Step*** (2), to overcome the unsoundness due to incomplete inspection of branching. The idea is to put the existential quantification inside universal quantification, so that every branch gets inspected by the resultant simulation notion.

Additionally, preorder-constrained simulation is parameterised by a positive number M dubbed *look-ahead bound*. It determines the degree of awareness of branching; we leave the study of the look-ahead bound to Sect. 4.3. We let $\mathcal{A}_i = (X_i, \Sigma, \rightsquigarrow_i, F_i)$ $(i \in \{1, 2\})$ be two NAs with the same alphabet, $Q \subseteq \mathbb{N} \times \mathbb{N}$ be an s-closed preorder, and $\mathbf{Q} \subseteq \Sigma^* \times \Sigma^*$ be a c-closed preorder.

Definition 3 (**($M, \mathbf{Q}$)-simulations**). *For each $M \in \mathbb{N}_+$, a binary relation $R \subseteq X_1 \times X_2$ is an M-*bounded $\mathbf{Q}$-constrained simulation *(($M, \mathbf{Q}$)-*simulation *in short) from $\mathcal{A}_1$ to $\mathcal{A}_2$ if, for any $(x, y) \in R$, the following **Final**^{M} and **Step**^{M} hold.*

Final^M *For each $w = a_1 \dots a_n \in \Sigma^*$ and $x_1 \dots x_n \in X_1^*$ such that $n < M$, $x \overset{a_1}{\rightsquigarrow}_1 x_1 \cdots \overset{a_n}{\rightsquigarrow}_1 x_n$ and $x_n \in F_1$, there exist $w' \in \Sigma^*$ and $y' \in X_2$ such that $w\mathbf{Q}w'$, $y \overset{w'}{\rightsquigarrow\!\!\!\!\rightarrow}_2 y'$ and $y' \in F_2$.*

Step^M *For each $a_1 \dots a_M \in \Sigma^M$ and $x_1 \dots x_M \in X_1^M$ such that $x \overset{a_1}{\rightsquigarrow}_1 x_1 \cdots \overset{a_M}{\rightsquigarrow}_1 x_M$, there exist $k \in \{1, \dots, M\}$, $w' \in \Sigma^*$ and $y' \in X_2$ such that $a_1 \cdots a_k \mathbf{Q} w'$, $y \overset{w'}{\rightsquigarrow\!\!\!\!\rightarrow}_2 y'$ and $x_k R y'$.*

When an ($M, \mathbf{Q}$)-simulation relates x and y, we say x is ($M, \mathbf{Q}$)-*similar* to y and write $x \lesssim_{M,\mathbf{Q}} y$. When R is an ($M, \mathbf{Q}$)-simulation from $\mathcal{A}$ to $\mathcal{A}$, we say it is *on* $\mathcal{A}$.

Figure 6 illustrates the conditions of Definition 3. The difference between Fig. 4c and Fig. 6b is crucial to overcome the incomplete inspection of branching that counting simulation suffers from. In Fig. 6b, existential quantification is limited to x_k, which is an intermediate state of the sequence $x \rightsquigarrow\!\!\!\!\rightarrow x_M$ that is universally quantified.

4.2 Soundness

Thanks to the observation preorder $\mathbf{Q}$ and the limited existential quantification, preorder-constrained simulation can characterise notions of quantitative refine-

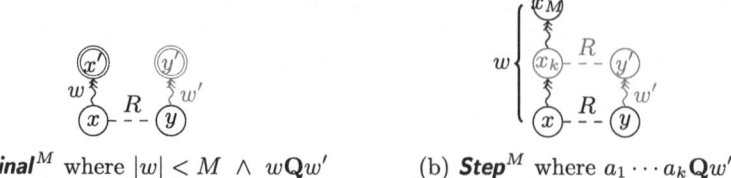

ment $\preceq^{Q}$ for all the signatures Ω in Example 1. Namely, $(M, \mathbf{Q})$-simulations provide a sufficient condition for refinement.

Corollary 1 (correctness of $(M, \mathbf{Q})$-simulations wrt. refinement).

1. *For any* $M \in \mathbb{N}_+$ *and* $t, u \in \mathbf{T}_{\Omega_{\mathsf{err}}}$, $t \lesssim_{M,\dot{Q}} u \implies t \preceq^{Q}_{\mathsf{err}} u$.
2. *For any* $M \in \mathbb{N}_+$ *and* $t, u \in \mathbf{T}_{\Omega_{\mathsf{nd}}}$, $t \lesssim_{M,\dot{Q}\cap =_{\mathsf{rem}_{\overline{\{\tau\}\cup\Omega_{\mathsf{nd}}}}}} u \implies t \preceq^{Q}_{\mathsf{nd}} u$.
3. *For any* $M \in \mathbb{N}_+$ *and* $t, u \in \mathbf{T}_{\Omega_{\mathsf{io}}}$, $t \lesssim_{M,\dot{Q}\cap =_{\mathsf{rem}_{\{\tau\}}}} u \implies t \preceq^{Q}_{\mathsf{io}} u$. □

Example 6 (Fig. 3 revisited).

1. For the NAs in Fig. 3a, $\{(\underline{2} \times (\underline{3} + \underline{4}), \underline{2} \times \underline{3} + \underline{2} \times \underline{4}), (\underline{14}, \underline{14})\}$ is a $(2, \dot{\leq})$-simulation. This simulation represents the distributive law.
2. For Fig. 3b, $\{(\mathtt{or}(\underline{1}, \underline{1}), \mathtt{or}(\underline{1}, \underline{0})), (\underline{1}, \underline{1}), (\checkmark, \checkmark)\}$ is a $(1, \dot{=}\cup =_{\mathsf{rem}_{\{\tau\}\cup\Omega_{\mathsf{nd}}}})$-simulation.
3. For Fig. 3c, let $t_1 \equiv \underline{1} + \underline{2} + \mathtt{in}(\underline{0}, \underline{1})$ and $t_2 \equiv \mathtt{in}(\underline{1} + \underline{2} + \underline{0}, \underline{1} + \underline{2} + \underline{1})$. The filtered equality $=_{\mathsf{rem}_{\{\tau\}}}$ distinguishes traces of length 1 from t_1, t_2, namely: $\tau \neq_{\mathsf{rem}_{\{\tau\}}} \mathtt{in}_i$ $(i \in \{0, 1\})$. This leads to non-existence of any $(1, \dot{=}\cap =_{\mathsf{rem}_{\{\tau\}}})$-simulation that includes the pair (t_1, t_2). In contrast, the filtered equality can identify traces of length 2 from t_1, t_2, namely: $\tau\mathtt{in}_i =_{\mathsf{rem}_{\{\tau\}}} \mathtt{in}_i\tau$ $(i \in \{0, 1\})$. This leads to existence of a $(2, \dot{=}\cap =_{\mathsf{rem}_{\{\tau\}}})$-simulation. It can be given by $\{(t_1, t_2), (\underline{3} + \underline{0}, \underline{3} + \underline{0}), (\underline{3} + \underline{1}, \underline{3} + \underline{1}), (\checkmark, \checkmark)\}$.

Corollary 1 is a consequence of a soundness property of preorder-constrained simulations. Namely, $(M, \mathbf{Q})$-simulations are sound with respect to a novel generalisation (Definition 4 below) of trace inclusion that is also parameterised by the observation preorder $\mathbf{Q}$.

Definition 4 (Q-trace inclusion). *We write* $x \sqsubseteq_{\mathbf{Q}} y$ *and say* $\mathbf{Q}$-trace inclusion *holds between* x *and* y, *if the following holds:* $\forall w \in L^*_{A_1}(x).\ \exists w' \in L^*_{A_2}(y).\ w\mathbf{Q}w'$.

Theorem 1 ((†) soundness). *Let* $M \in \mathbb{N}_+$. *For any* $(x, y) \in X_1 \times X_2$, *it holds that* $x \lesssim_{M,\mathbf{Q}} y \implies x \sqsubseteq_{\mathbf{Q}} y$.

The three refinement relations for specific NAs $\mathcal{A}_{\Omega_{\mathsf{err}}}$, $\mathcal{A}_{\Omega_{\mathsf{nd}}}$ and $\mathcal{A}_{\Omega_{\mathsf{io}}}$, defined in Sect. 2.2, are instances of the generalised trace inclusion. Corollary 1 follows from Theorem 1 above and Proposition 2 below.

Proposition 2 (refinement as trace inclusion).

1. *For* $\mathcal{A}_{\Omega_{\mathsf{err}}}$, *for any* $t, u \in \mathbf{T}_{\Omega_{\mathsf{err}}}$, $t \preceq^{Q}_{\mathsf{err}} u \iff t \sqsubseteq_{\dot{Q}} u$.
2. *For* $\mathcal{A}_{\Omega_{\mathsf{nd}}}$, *for any* $t, u \in \mathbf{T}_{\Omega_{\mathsf{nd}}}$, $t \preceq^{Q}_{\mathsf{nd}} u \iff t \sqsubseteq_{\dot{Q}\cap =_{\mathsf{rem}_{\overline{\{\tau\}\cup\Omega_{\mathsf{nd}}}}}} u$.
3. *For* $\mathcal{A}_{\Omega_{\mathsf{io}}}$, *for any* $t, u \in \mathbf{T}_{\Omega_{\mathsf{io}}}$, $t \preceq^{Q}_{\mathsf{io}} u \iff t \sqsubseteq_{\dot{Q}\cap =_{\mathsf{rem}_{\{\tau\}}}} u$. □

The most basic instance of the $\mathbf{Q}$-trace inclusion is when $\mathbf{Q}$ is the equality; it coincides with the standard notion of (finite) trace inclusion. Another basic instance is when $\mathbf{Q}$ is the filtered equality $=_{\mathsf{rem}_{\{\tau\}}}$; it corresponds to the well-known notion of weak trace inclusion.

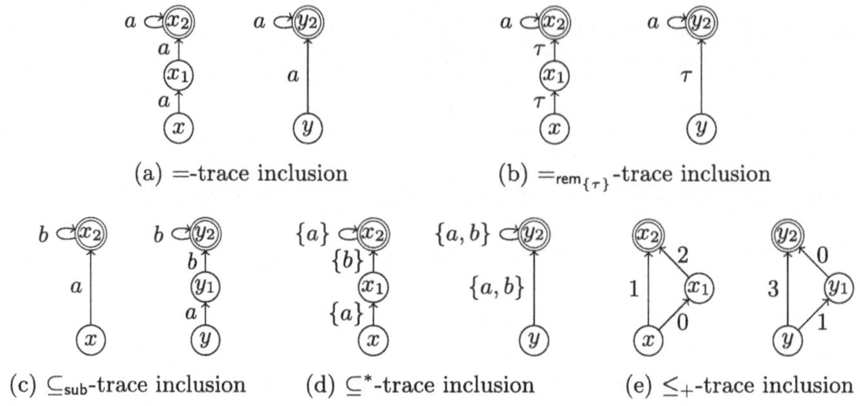

Fig. 7. Pairs of NAs that exhibit **Q**-trace inclusion

*Example 7 (**Q**-trace inclusions for various preorders).* Fig. 7 shows examples of **Q**-trace inclusions for some of the preorders from Example 5.

1. In Fig. 7a, we have $L^*(x) = aaa^* = \{a^{n+2} | n \in \mathbb{N}\}$ and $L^*(y) = aa^* = \{a^{n+1} | n \in \mathbb{N}\}$, and therefore, $L^*(x) \subseteq L^*(y)$. We have $x \sqsubseteq_= y$ for the equality $=$.
2. In Fig. 7b, we have $L^*(x) = \tau\tau a^*$ and $L^*(y) = \tau a^*$. These two sets coincide when we ignore τ. We therefore have $x \sqsubseteq_{=_{\mathsf{rem}_{\{\tau\}}}} y$ for the filtered equality $=_{\mathsf{rem}_{\{\tau\}}}$.
3. In Fig. 7c, we have $L^*(x) = ab^*$ and $L^*(y) = abb^*$. We hence have $x \sqsubseteq_{\subseteq_{\mathsf{sub}}} y$ where $\subseteq_{\mathsf{sub}}$ is the substring preorder from Example 5(4).
4. In Fig. 7d, we have $L^*(x) = \{a\}\{b\}\{a\}^*$ and $L^*(y) = \{a,b\}\{a,b\}^*$. We hence have $x \sqsubseteq_{\subseteq^*} y$ where $\subseteq^*$ is from Example 5(5).
5. In Fig. 7d, we have $L^*(x) = \{1, 02\}$ and $L^*(y) = \{3, 10\}$. It holds that $1 \leq 3$ and $0 + 2 \leq 3$. We hence have $x \sqsubseteq_{\leq_+} y$ where $\leq_+$ is from Example 5(6).

4.3 Basic Properties

Here we investigate the look-ahead bound M of preorder-constrained simulation. As observed in Example 6 (3), the look-ahead bound determines the degree of awareness of branching; as M increases, $(M, \mathbf{Q})$-simulation can inspect and identify further branches. Technically, preorder-constrained simulation has a monotonicity property with respect to M.

Lemma 1 ((†) monotonicity of *Final* and *Step*). *Let $M, N \in \mathbb{N}_+$ such that $M \leq N$.*

1. $\mathsf{Final}^N \implies \mathsf{Final}^M$
2. $\mathsf{Step}^M \implies \mathsf{Step}^N$

3 $\mathsf{Step}^M \wedge \mathsf{Final}^M \implies \mathsf{Final}^N$ □

Corollary 2 (monotonicity). *Let $M, N \in \mathbb{N}_+$ such that $M \leq N$. Each $(M, \mathbf{Q})$-simulation from $\mathcal{A}_1$ to $\mathcal{A}_2$ is also an $(N, \mathbf{Q})$-simulation from $\mathcal{A}_1$ to $\mathcal{A}_2$.*

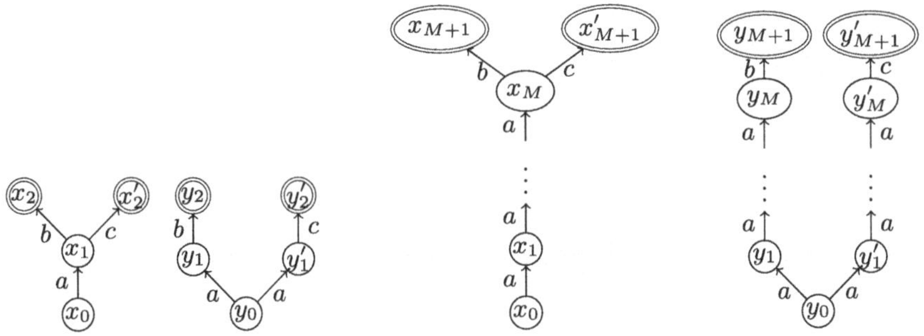

(a) For $M = 1$ (b) For arbitrary M

Fig. 8. Pairs of NAs that distinguish $(M, =)$-similarity and $(M + 1, =)$-similarity

When the observation preorder $\mathbf{Q}$ is the equality $=$, this monotonicity is strict; in other words, strict inclusion $\lesssim_{M,=} \subsetneq \lesssim_{M+1,=}$ holds. There exists a pair of NAs that does not have an $(M, =)$-simulation but has an $(M + 1, =)$-simulation. Figure 8 shows such pairs; one for $M = 1$, and the other for an arbitrary M. For Fig. 8a, we have a $(2, =)$-simulation $\{(x_0, y_0), (x_1, y_1), (x_1, y'_1), (x_2, y_2), (x'_2, y'_2)\}$. Similarly, for Fig. 8a, we have a $(M+1, =)$-simulation $\{(x_0, y_0), (x_M, y_M), (x_M, y'_M), (x_{M+1}, y_{M+1}), (x'_{M+1}, y'_{M+1})\}$.

The monotonicity is in contrast to that of *nested simulations* [8] which are also parameterised by a positive number. As M increases, $(M, \mathbf{Q})$-simulations become larger (as a relation), while M-nested simulations become smaller. When $M = 1$ (and $\mathbf{Q}$ is the equality), $(1, =)$-simulation coincides with the standard simulation, and hence with 1-nested simulation.

Corollary 2 validates the spectrum of preorder-constrained similarities in Fig. 1, which is governed by the look-ahead bound. Its "limit" can be given by the notion of $\mathbf{Q}$-trace inclusion, thanks to the soundness property (Theorem 1).

The spectrum in Fig. 1 is *generative* because, by instantiating the observation preorder $\mathbf{Q}$, we can obtain various concrete spectra, as shown in Table 1. For instance, when $\mathbf{Q}$ is the equality $=$ or the filtered equality $=_{\mathsf{rem}_{\{\tau\}}}$, the spectrum refines a part of (the asymmetric version of) van Glabbeek's spectrum [9,10]. The other spectra of Table 1 are more of our interest here, which yield simulation notions that characterise program refinement.

Finally, while $(M, \mathbf{Q})$-simulations are closed under union, they are not closed under the standard composition of relations.

Lemma 2 ((†) basic properties). *Let $M \in \mathbb{N}_+$, and I be an arbitrary set.*

1. *Given a family $\{R_i\}_{i \in I}$ of $(M, \mathbf{Q})$-simulations, $\bigcup_{i \in I} R_i$ is an $(M, \mathbf{Q})$-simulation.*
2. *The $(M, \mathbf{Q})$-similarity $\lesssim_{M,\mathbf{Q}}$ is the largest $(M, \mathbf{Q})$-simulation.*

Example 8 (no closedness under composition).

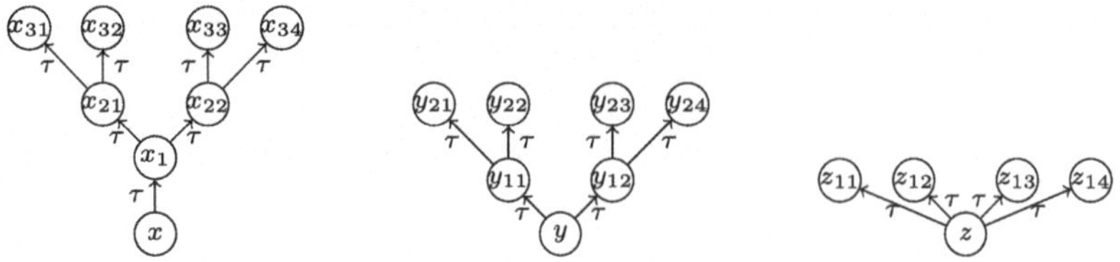

Fig. 9. A counterexample of closedness under composition

Fig. 10. More counterexamples of closedness under composition

1. Let $\top = \{\tau\}^* \times \{\tau\}^*$. For the three NAs in Fig. 9, $\{(x, y), (x21, y11), (x22, y12), (x31, y21), (x32, y22), (x33, y23), (x34, y24)\}$ and $\{(y, z), (y21, z11), (y22, z12), (y23, z13), (y24, z14)\}$ are $(2, \top)$-simulations, but their composition $\{(x, z), (x31, z11), (x32, z12), (x33, z13), (x34, z14)\}$ is not a $(2, \top)$-simulation. The composition is a $(3, \top)$-simulation instead.
2. For the three NAs in Fig. 10a, $\{(x, y), (x_1, y_1)\}$ and $\{(y, z)\}$ are $(2, =)$-simulations, but their composition $\{(x, z)\}$ is not a $(2, =)$-simulation.
3. Let $\mathbf{Q}_e$ be the reflexive and c-closed closure of $\{(a, bc), (b, d)\}$. For the three NAs in Fig. 10b, $\{(x, y), (x_1, y_2)\}$ and $\{(y, z), (y_1, z_1)\}$ are $(2, \mathbf{Q}_e)$-simulations, but their composition $\{(x, z)\}$ is not a $(2, \mathbf{Q}_e)$-simulation.

5 Game-Theoretic Characterisation

Preorder-constrained simulations can be characterised by two-player reachability games. The game is parameterised by the observation preorder $\mathbf{Q}$, the look-ahead bound M, and additionally a *catch-up bound* N. Both numerical bounds M, N are now taken from $\mathbb{N}_+ \cup \{\infty\}$.

Definition 5 **($\mathcal{G}^{M,N,\mathbf{Q}}_{\mathcal{A}_1,\mathcal{A}_2}$).** *Let $M, N \in \mathbb{N}_+$. A two-player game $\mathcal{G}^{M,N,\mathbf{Q}}_{\mathcal{A}_1,\mathcal{A}_2}$ between* Challenger *and* Simulator *is defined by Fig. 11. Simulator wins if they reach the state* sim-win*, Challenger has no possible move, or the play continues forever.*

In a game $\mathcal{G}^{M,N,\mathbf{Q}}_{\mathcal{A}_1,\mathcal{A}_2}$, Challenger is in charge of $\mathcal{A}_1$, and Simulator is in charge of $\mathcal{A}_2$. Most of the positions are of the form (w, x, y), where $w \in \Sigma^*$ represents a queue of labels that Challenger has inputted into $\mathcal{A}_1$. The two numerical

Position	Player	Move	Guard	
(w, x, y) $\in \Sigma^* \times X_1 \times X_2$	Challenger	(wa, x', y)	$x \overset{a}{\rightsquigarrow}_1 x'$	①
		$(\checkmark, w, x, y)$	$x \in F_1$	②
(w, x', y) $\in \Sigma^* \times X_1 \times X_2$	Simulator	(w, x', y)	$\lvert w \rvert < M$	③
		(ε, x', y')	$\exists w' \in \Sigma^*.$ $\lvert w' \rvert < N \wedge y \overset{w'}{\rightsquigarrow}_2 y' \wedge w\mathbf{Q}w'$	④
$(\checkmark, w, x, y)$ $\in \{\checkmark\} \times \Sigma^* \times X_1 \times X_2$	Simulator	sim-win	$\exists w' \in \Sigma^*.\exists y' \in F_2.$ $\lvert w' \rvert < N \wedge y \overset{w'}{\rightsquigarrow}_2 y' \wedge w\mathbf{Q}w'$	⑤

① Challenger chooses $x \overset{a}{\rightsquigarrow}_1 x'$ from the current state x and enqueues the label a.
② Challenger is at an accepting state $x \in F_1$. Challenger forces Simulator to check whether an accepting state is reachable from $y \in X_2$.
③ Simulator skips the turn. This move is always possible when $M = \infty$.
④ Simulator simulates Challenger's moves in the queue w.
⑤ Simulator simulates Challenger's moves in the queue w and reaches an accepting state.

Fig. 11. Two-player game $\mathcal{G}^{M,N,\mathbf{Q}}_{\mathcal{A}_1,\mathcal{A}_2}$ characterising (M-bounded) $\mathbf{Q}$-constrained simulation.

parameters M, N both constrain Simulator's ability to make a move. The look-ahead bound M limits the length of the queue w, and equivalently the number of turns that Simulator can skip consecutively. The catch-up bound N limits the number of transitions Simulator can make in $\mathcal{A}_2$ to dequeue w.

The games $\mathcal{G}^{M,N,\mathbf{Q}}_{\mathcal{A}_1,\mathcal{A}_2}$ satisfy two expected properties: monotonicity with respect to both M and N (Lem. 3 below), and correspondence with $(M, \mathbf{Q})$-similarity (Proposition 3 below). These results validate a generative (two-dimensional) spectrum of games and preorder-constrained simulations shown in Fig. 12.

Lemma 3 (monotonicity). *Let $M, M' \in \mathbb{N}_+$ such that $M \leq M'$, and let $N, N' \in \mathbb{N}_+$ such that $N \leq N'$. If Simulator is winning from a state (w, x, y) in $\mathcal{G}^{M,N,\mathbf{Q}}_{\mathcal{A}_1,\mathcal{A}_2}$, Simulator is also winning from the state in $\mathcal{G}^{M',N',\mathbf{Q}}_{\mathcal{A}_1,\mathcal{A}_2}$.* □

Proposition 3 (correctness). *Let $M \in \mathbb{N}_+$. Simulator is winning from a state (ε, x, y) in $\mathcal{G}^{M,\infty,\mathbf{Q}}_{\mathcal{A}_1,\mathcal{A}_2}$, if and only if $x \lesssim_{M,\mathbf{Q}} y$.* □

We conclude this section with complexity analysis of determining winning positions in $\mathcal{G}^{M,N,\mathbf{Q}}_{\mathcal{A}_1,\mathcal{A}_2}$, and hence of determining $(M, \mathbf{Q})$-similarity $\lesssim_{M,\mathbf{Q}}$.

Proposition 4 (complexity). *Assume that $w\mathbf{Q}w'$ can be checked in linear time to the lengths of w and w'. For $M, N \in \mathbb{N}_+$, whether Simulator is winning from a state (ε, x, y) in $\mathcal{G}^{M,N,\mathbf{Q}}_{\mathcal{A}_1,\mathcal{A}_2}$ can be checked in $\mathcal{O}(\lvert\Sigma\rvert^{M+N} \times \lvert X_1 \rvert \times \lvert X_2 \rvert^2)$ time.*

Proof. For a set X and for $k, l \in \mathbb{N}$ such that $k \leq l$, let $X^{[k,l]}$ denote the set $\{x_1 x_2 \ldots x_n \mid k \leq n \leq l; x_1, \ldots, x_n \in X\}$.

$$
\begin{array}{ccccccc}
\lesssim_{1,\mathbf{Q}} & \longrightarrow & \lesssim_{2,\mathbf{Q}} & \longrightarrow \cdots \longrightarrow & \mathbf{Q}\text{-trace inclusion } \sqsubseteq_{\mathbf{Q}} \\
\uparrow & & \uparrow & & \\
\vdots & & \vdots & & \\
\uparrow & & \uparrow & & \\
\mathcal{G}^{1,2,\mathbf{Q}}_{\mathcal{A}_1,\mathcal{A}_2} & \longrightarrow & \mathcal{G}^{2,2,\mathbf{Q}}_{\mathcal{A}_1,\mathcal{A}_2} & \longrightarrow \cdots \longrightarrow & \mathbf{Q}\text{-trace inclusion } \sqsubseteq_{\mathbf{Q}} \\
\uparrow & & \uparrow & & \\
\mathcal{G}^{1,1,\mathbf{Q}}_{\mathcal{A}_1,\mathcal{A}_2} & \longrightarrow & \mathcal{G}^{2,1,\mathbf{Q}}_{\mathcal{A}_1,\mathcal{A}_2} & \longrightarrow \cdots \longrightarrow & \mathbf{Q}\text{-trace inclusion } \sqsubseteq_{\mathbf{Q}}
\end{array}
$$

Fig. 12. A generative spectrum of games $\mathcal{G}^{M,N,\mathbf{Q}}_{\mathcal{A}_1,\mathcal{A}_2}$ and similarities $\lesssim_{M,\mathbf{Q}}$

We first approximate the complexity to construct the game $\mathcal{G}^{M,N,\mathbf{Q}}_{\mathcal{A}_1,\mathcal{A}_2}$, assuming that the membership to F_1, F_2, $\rightsquigarrow_1$ and $\rightsquigarrow_2$ can be checked in constant time. The number of positions of the game is $\mathcal{O}(|\Sigma|^M \times |X_1| \times |X_2|)$, because Challenger's positions are elements of $\Sigma^{[0,M-1]} \times X_1 \times X_2$, and Simulator's positions are elements of $\Sigma^{[0,M]} \times X_1 \times X_2$ or $\{\checkmark\} \times \Sigma^{[0,M]} \times X_1 \times X_2$.

For each pair (p, q) of positions, we approximate the complexity to determine if a move is possible from p to q, by case analysis on the possible moves (see Fig. 11).

- Case ①. It suffices to check $x \overset{a}{\rightsquigarrow}_1 x'$, which can be done in constant time.
- Case ②. It suffices to check $x \in F_1$, which can be done in constant time.
- Case ③. It suffices to check $|w| < M$, which can be done in constant time, because M is a constant here.
- Case ④. By the condition $|w'| < N$, w' can be chosen from $\Sigma^{[0,N-1]}$. Checking $y \overset{w'}{\rightsquigarrow\!\!\!\!\!\!\rightarrow}_2 y'$ is $\mathcal{O}(N-1)$ time and hence constant time, and checking $w\mathbf{Q}w'$ is also constant time, because both M and N are constants here. Overall, this move can be determined in $\mathcal{O}(|\Sigma|^N)$ time.
- Case ⑤. This case is similar to the case ④, with extra choice of y' from $F_2 \subseteq X_2$. This move can be determined in $\mathcal{O}(|\Sigma|^N \times |X_2|)$ time.

Once the game is constructed, its winning region (i.e. the positions from which Simulator is winning) can be determined in $\mathcal{O}(|\Sigma|^M \times |X_1| \times |X_2|)$ time multiplied by $\mathcal{O}(|\Sigma|^N \times |X_2|)$ time, because the game is a reachability game. Consequently, whether Simulator is winning from a state (ε, x, y) in $\mathcal{G}^{M,N,\mathbf{Q}}_{\mathcal{A}_1,\mathcal{A}_2}$ can be checked in $\mathcal{O}(|\Sigma|^{M+N} \times |X_1| \times |X_2|^2)$ time. □

6 Preorder-Constrained Simulation Up-To

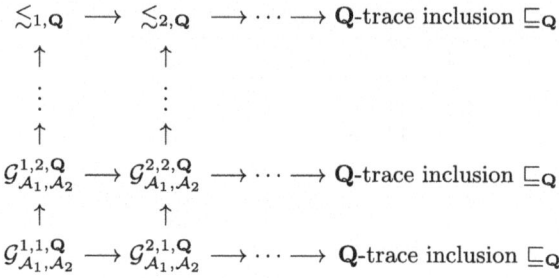

We here integrate the up-to technique [26] into preorder-constrained simulation. The technique is widely used for enhancing (bi-)simulation notions. It allows a smaller relation, which is not necessarily a simulation itself, to witness trace inclusion. The up-to version of preorder-constrained simulation is additionally parameterised by a pair of binary relations (R_1, R_2) on state spaces. The following definition is obtained by simply replacing R in Definition 3 with $R_1; R; R_2$, where ; is the composition of binary relations.

Definition 6 (**$(M, \mathbf{Q})$-simulations up to** (R_1, R_2)). *Let $R_1 \subseteq X_1 \times X_1$ and $R_2 \subseteq X_2 \times X_2$. For each $M \in \mathbb{N}_+$, a binary relation $R \subseteq X_1 \times X_2$ is an $(M, \mathbf{Q})$-simulation up to (R_1, R_2) from $\mathcal{A}_1$ to $\mathcal{A}_2$ if, for any $(x, y) \in R$, **Final**^{M} (see Definition 3) and **U-Step**^{M} below hold.*

U-Step^M *For each $w = a_1 \dots a_M \in \Sigma^M$ and $x_1 \dots x_M \in X_1^M$ such that $x \overset{a_1}{\rightsquigarrow}_1 x_1 \overset{a_2}{\rightsquigarrow}_1 \cdots \overset{a_M}{\rightsquigarrow}_1 x_M$, there exist $k \in \{1, \dots, M\}$, $w' \in \Sigma^*$ and $y' \in X_2$ such that $a_1 \cdots a_k \mathbf{Q} w'$, $y \overset{w'}{\rightsquigarrow\!\!\!\!\rightarrow}_2 y'$ and $x_k(R_1; R; R_2)y'$.*

The question now is when a preorder-constrained simulation up to (R_1, R_2) is sound with respect to $\mathbf{Q}$-trace inclusion. The relations R_1 and R_2 cannot be arbitrary, and they should be *consistent* to the $\mathbf{Q}$-trace inclusion $\sqsubseteq_{\mathbf{Q}}$. Formally, they should satisfy $R_1 \subseteq \sqsubseteq_{\mathbf{Q}_1}$ and $R_2 \subseteq \sqsubseteq_{\mathbf{Q}_2}$ for some preorders $\mathbf{Q}_1, \mathbf{Q}_2 \subseteq \Sigma^* \times \Sigma^*$ such that $(\mathbf{Q}_1; \mathbf{Q}; \mathbf{Q}_2) \subseteq \mathbf{Q}$.

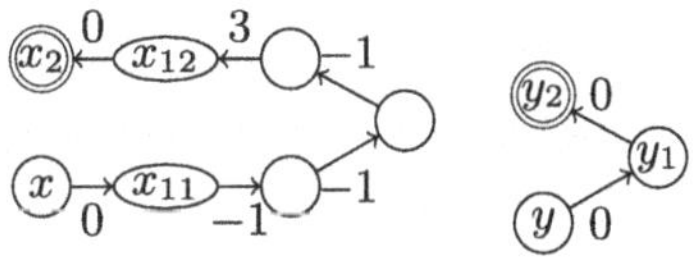

Fig. 13. A $(1, \geq_+)$-simulation up-to

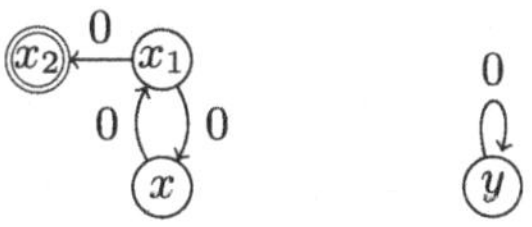

Fig. 14. An unsound $(1, \geq_+)$-simulation up-to

Example 9 (*$(M, \geq_+)$-constrained simulation up to* $(\sqsubseteq_{\geq_+}, \sqsubseteq_{\geq_+})$). Recall the preorder $\geq_+$ from Example 5(6). It is feasible to use a $\geq_+$-constrained simulation up to $(\sqsubseteq_{\geq_+}, \sqsubseteq_{\geq_+})$. The first reason is that the consistence property is satisfied, thanks to transitivity of the preorder $\geq_+$. The second reason is that the $\geq_+$-trace inclusion $\sqsubseteq_{\geq_+}$ between two states of the same NA can be checked efficiently with a graph algorithm. Specifically, for two states x, x' of the same NA, $x \sqsubseteq_{\geq_+} x'$ holds if the summation of weights on each path from x to x' is non-negative.

Example 10 (*$(1, \geq_+)$-simulation up-to*). For a pair of NAs shown in Fig. 13, $x \sqsubseteq_{\geq_+} y$ holds, but there exists no $(1, \geq_+)$-simulation that relates x with y. In contrast, a relation $R = \{(x, y), (x_{12}, y_1), (x_2, y_2)\}$ is a $(1, \geq_+)$-simulation up to $(\sqsubseteq_{\geq_+}, \sqsubseteq_{\geq_+})$. The pair (x, y) satisfies the condition **U-Step**^{1} of Definition 6, because we have $x_{11} \sqsubseteq_{\geq_+} x_{12}\ R\ y_1 \sqsubseteq_{\geq_+} y_1$. The up-to allows us to *move* **U-Step**^{1} *towards the accepting state* x_2, that is, it allows us to deal with **U-Step**^{1} of (x_{12}, y_1) instead of (x_{11}, y_1).

However, the consistence property $(\mathbf{Q}_1; \mathbf{Q}; \mathbf{Q}_2) \subseteq \mathbf{Q}$ is not enough to achieve soundness. A naive combination of the weak simulation notion, which coincides with our similarity $\lesssim_{1,=_{\mathsf{rem}_{\{\tau\}}}}$, and an up-to technique is known to be unsound, and requires special care [24,25]. Fig 14 The following counterexample is inspired by the one for weak simulation from the literature [24,25].

Example 11 (unsound $(1, \geq_+)$-simulation up-to). For a pair of NAs shown in Fig. 13, $x \sqsubseteq_{\geq_+} y$ does not hold, because no accepting state is reachable from y. However, a relation $\{(x, y)\}$ is a $(1, \geq_+)$-simulation up to $(\sqsubseteq_{\geq_+}, \sqsubseteq_{\geq_+})$. The pair (x, y) indeed satisfies ***U-Step***^{1}, because $x_1 \sqsubseteq_{\geq_+} x \; R \; y \sqsubseteq_{\geq_+} y$. Here, the up-to allows us to *move* ***U-Step***^{1} *away from the accepting state* x_2: namely, it allows us to deal with ***U-Step***^{1} of (x, y) instead of (x_1, y).

From the two examples above, we can extract an additional condition on the preorders $\mathbf{Q}, \mathbf{Q}_1, \mathbf{Q}_2$, namely on $\mathbf{Q}_1$. We can observe that, while it is safe to move ***U-Step***^{M} *toward* an accepting state (Example 10), it is unsafe to move it *away from* an accepting state (Example 11), for the sake of soundness. To prohibit the unsafe move, the additional condition is required, namely $w \mathbf{Q}_1 w' \implies |w| \geq |w'|$. Examples of such $\mathbf{Q}_1$ are the equality $=$, the preorder $\dot{\geq}$ induced by $\geq \subseteq \mathbb{N} \times \mathbb{N}$, the inverse $(\subseteq_{\mathsf{sub}})^{-1}$, and $\subseteq^*$ (see Example 5).

The condition is inspired by a soundness criterion for counting simulation up-to [22, Def. 4.3.13]. A similar idea can be found in Pous' work [24,25] on weak simulation up-to.

Theorem 2 (($\dagger$) soundness). *Let $M \in \mathbb{N}_+$. Let $R \subseteq X_1 \times X_2$, $R_1 \subseteq \sqsubseteq_{\mathbf{Q}_1}$ and $R_2 \subseteq \sqsubseteq_{\mathbf{Q}_2}$ be binary relations, for preorders $\mathbf{Q}_1, \mathbf{Q}_2 \subseteq \Sigma^* \times \Sigma^*$ such that (i) $(\mathbf{Q}_1; \mathbf{Q}; \mathbf{Q}_2) \subseteq \mathbf{Q}$ and (ii) $w \mathbf{Q}_1 w' \implies |w| \geq |w'|$. If R is an $(M, \mathbf{Q})$-simulation up to (R_1, R_2), it holds that $xRy \implies x \sqsubseteq_{\mathbf{Q}} y$ for any $(x, y) \in X_1 \times X_2$.*

Example 12 (weak simulation up to "expansion/contraction"). Let $\mathbf{Q}_e$ be a preorder $=_{\mathsf{rem}_{\{\tau\}}} \cap \dot{\geq}$. The $(1, \mathbf{Q}_e)$-similarity $\lesssim_{1,\mathbf{Q}_e}$ is akin to *expansion* [28] and *contraction* [27] that have been used with an up-to technique. It is indeed valid to have the weak simulation up to the similarity $\lesssim_{1,\mathbf{Q}_e}$. More precisely, we can have a $(1, =_{\mathsf{rem}_{\{\tau\}}})$-simulation, which is the weak simulation, up to $(\lesssim_{1,\mathbf{Q}_e}, =)$. The three preorders $=_{\mathsf{rem}_{\{\tau\}}}$, $\mathbf{Q}_e$ and $=$ satisfy the two conditions in Theorem 2; we have $(\mathbf{Q}_e; =_{\mathsf{rem}_{\{\tau\}}}; =) \subseteq \; =_{\mathsf{rem}_{\{\tau\}}}$, and the preorder $\mathbf{Q}_e$ satisfies $\mathbf{Q}_e \subseteq \dot{\geq}$. We also have $\lesssim_{1,\mathbf{Q}_e} \subseteq \sqsubseteq_{\mathbf{Q}_e}$ by soundness (Theorem 1), and $= \subseteq \sqsubseteq_=$.

7 Related Work

Our two-player game $\mathcal{G}^{M,N,\mathbf{Q}}_{\mathcal{A}_1,\mathcal{A}_2}$ is similar to *buffered simulation game* [12,13]. While ours is for NAs with finite languages, the latter is for Büchi automata. We contribute to coinductively defining the simulations and investigating the generative spectrum.

Quantitative simulation notions are known for *weighted automata*: many are for probabilistic systems [11,15,20]; a general simulation notion for automata weighted with semirings (e.g. $\mathbb{R}$ with $+$ and $\times$, and $\mathbb{R}$ with max and $+$ a.k.a. tropical semiring) was introduced as a matrix over real numbers [31].

Preorder-constrained simulations are defined for NAs, but can be quantitative, in the sense that they can compare lengths of accepted runs (e.g. using the preorder $\dot{\geq}$). Known (bi-)simulation notions such as *expansion* [28] and *contraction* [27] are also capable of such quantitative comparison. As mentioned in Example 12, the $(1, =_{\mathsf{rem}_{\{\tau\}}} \cap \dot{\geq})$-similarity is akin to these (bi-)simulation notions seen as simulation notions.

Proposition 4 suggests that we can reduce the problem of checking refinement between programs to determining a winning region of the reachability game $\mathcal{G}^{M,N,\mathbf{Q}}_{\mathcal{A}_1,\mathcal{A}_2}$. *Algorithmic game semantics* [2,16] is another approach to reduce program refinement to solving games. A notable difference is that algorithmic game semantics restricts *types* of programs, while our approach would require a program t to induce a *finite* automaton $\mathcal{A}_\Omega(t)$.

Our work is not the first to characterise or refine the LT–BT spectrum using a simulation notion or a game. It would be interesting to compare existing work, e.g. [6,7], to ours in details. Ordered words are not a new research topic either, but the literature seems to focus on decidability of their theory, e.g. [18,19].

8 Conclusion and Future Work

We proposed a notion of preorder-constrained simulation. Being parameterised by the observation preorder on traces, it can uniformly characterise quantitative notions of observational refinement for different algebraic effects: exception, nondeterminism and I/O. We demonstrated this using reduction semantics for the LSC.

Being additionally parameterised by the look-ahead bound, preorder-constrained simulations form a generative spectrum. Its "limit" is given by a novel generalisation of trace inclusion. The spectrum is generative in the sense that it can be instantiated variously according to the observation preorder.

We additionally presented a characterisation of preorder-constrained simulation as a two-player reachability game, and showed that preorder-constrained similarity can be determined in polynomial time in the finite setting. Finally, we studied enhancement of preorder-constrained simulation, showing how to integrate an up-to technique, in terms of observation preorders.

One direction of future work is to extend the characterisation of program refinement to probabilistic choice. This would require preorder-constrained simulation to work on weighted automata instead of NAs. One can try to accommodate probabilistic choice into the current work, using the preorder $\leq_+$, but a naive approach does not work. It results in a false refinement such as $\mathtt{or}_{0.5}(\underline{1},\underline{1}) \sqsubseteq_{\leq_+} \mathtt{or}_{0.5}(\underline{0},\underline{1})$, where $\mathtt{or}_{0.5}$ is an operation that chooses either argument with probability 0.5. It would also be interesting to connect preorder-constrained simulation to a generic metatheory of algebraic effects [14,29].

Another future work is to develop a methodology to constructing a preorder-constrained simulation. For the automaton $\mathcal{A}_\Omega$ induced by a signature Ω, it would be particularly important to construct a preorder-constrained simulation that is closed under term construction. Such a simulation would characterise *contextual refinement*, which asserts refinement between two terms in an arbitrary context. Counting simulation was originally used in this way [22].

Acknowledgments. We are grateful to Ichiro Hasuo and Shigeru Chiba for insightful comments. The first and second authors are supported by JST, ACT-X Grant No. JPMJAX190U, Japan, and JSPS, KAKENHI Project No. 22K17850, Japan. The third author is supported by JST ERATO HASUO Metamathematics for Systems Design Project (No. JPMJER1603).

References

1. Abramsky, S.: The lazy lambda-calculus, pp. 65–117. Addison Wesley (1990)
2. Abramsky, S.: Algorithmic game semantics. In: Schwichtenberg, H., Steinbrüggen, R. (eds.) Proof and System-Reliability. NATO Science Series, vol. 62, pp. 21–47. Springer, Dordrecht (2002). https://doi.org/10.1007/978-94-010-0413-8_2
3. Accattoli, B., Barenbaum, P., Mazza, D.: Distilling abstract machines. In: ICFP 2014, pp. 363–376. ACM (2014). https://doi.org/10.1145/2628136.2628154
4. Accattoli, B., Dal Lago, U., Vanoni, G.: The machinery of interaction. In: PPDP 2020: 22nd International Symposium on Principles and Practice of Declarative Programming, Bologna, Italy, 9–10 September 2020, pp. 4:1–4:15. ACM (2020). https://doi.org/10.1145/3414080.3414108
5. Accattoli, B., Kesner, D.: The structural λ-calculus. In: Dawar, A., Veith, H. (eds.) CSL 2010. LNCS, vol. 6247, pp. 381–395. Springer, Heidelberg (2010). https://doi.org/10.1007/978-3-642-15205-4_30
6. Bisping, B., Jansen, D.N., Nestmann, U.: Deciding all behavioral equivalences at once: a game for linear-time-branching-time spectroscopy. Log. Methods Comput. Sci. **18**(3) (2022). https://doi.org/10.46298/lmcs-18(3:19)2022
7. de Frutos-Escrig, D., Gregorio-Rodríguez, C., Palomino, M., Romero-Hernández, D.: Unifying the linear time-branching time spectrum of process semantics. Log. Methods Comput. Sci. **9**(2) (2013). https://doi.org/10.2168/LMCS-9(2:11)2013
8. de Frutos-Escrig, D., Gregorio-Rodríguez, C.: Constrained simulations, nested simulation semantics and counting bisimulations. In: Pimentel, E. (ed.) Proceedings of the Seventh Spanish Conference on Programming and Computer Languages, PROLE 2007, Zaragoza, Spain, 12–14 September 2007. Electronic Notes in Theoretical Computer Science, vol. 206, pp. 41–58. Elsevier (2007). https://doi.org/10.1016/j.entcs.2008.03.074
9. Glabbeek, R.J.: The linear time - branching time spectrum. In: Baeten, J.C.M., Klop, J.W. (eds.) CONCUR 1990. LNCS, vol. 458, pp. 278–297. Springer, Heidelberg (1990). https://doi.org/10.1007/BFb0039066
10. Glabbeek, R.J.: The linear time — branching time spectrum II. In: Best, E. (ed.) CONCUR 1993. LNCS, vol. 715, pp. 66–81. Springer, Heidelberg (1993). https://doi.org/10.1007/3-540-57208-2_6
11. Hughes, J., Jacobs, B.: Simulations in coalgebra. Theor. Comput. Sci. **327**(1–2), 71–108 (2004). https://doi.org/10.1016/j.tcs.2004.07.022
12. Hutagalung, M.: Buffered Simulation for Büchi Automata. Ph.D. thesis, University of Kassel, Germany (2019). https://kobra.uni-kassel.de/bitstream/handle/123456789/11329/DissertationMilkaHutagalung.pdf?sequence=7&isAllowed=y

13. Hutagalung, M., Lange, M., Lozes, É.: Buffered simulation games for büchi automata. In: Ésik, Z., Fülöp, Z. (eds.) Proceedings 14th International Conference on Automata and Formal Languages, AFL 2014, Szeged, Hungary, 27–29 May 2014. EPTCS, vol. 151, pp. 286–300 (2014). https://doi.org/10.4204/EPTCS.151.20
14. Johann, P., Simpson, A., Voigtländer, J.: A generic operational metatheory for algebraic effects. In: Proceedings of the 25th Annual IEEE Symposium on Logic in Computer Science, LICS 2010, 11–14 July 2010, Edinburgh, United Kingdom, pp. 209–218. IEEE Computer Society (2010). https://doi.org/10.1109/LICS.2010.29
15. Jonsson, B., Larsen, K.G.: Specification and refinement of probabilistic processes. In: Proceedings of the Sixth Annual Symposium on Logic in Computer Science (LICS 1991), Amsterdam, The Netherlands, 15–18 July 1991, pp. 266–277. IEEE Computer Society (1991). https://doi.org/10.1109/LICS.1991.151651
16. Kiefer, S., Murawski, A.S., Ouaknine, J., Wachter, B., Worrell, J.: Algorithmic probabilistic game semantics - playing games with automata. Formal Methods Syst. Des. **43**(2), 285–312 (2013). https://doi.org/10.1007/s10703-012-0173-1
17. Koutavas, V., Levy, P.B., Sumii, E.: From applicative to environmental bisimulation. In: Mislove, M.W., Ouaknine, J. (eds.) Twenty-seventh Conference on the Mathematical Foundations of Programming Semantics, MFPS 2011, Pittsburgh, PA, USA, 25–28 May 2011. Electronic Notes in Theoretical Computer Science, vol. 276, pp. 215–235. Elsevier (2011). https://doi.org/10.1016/j.entcs.2011.09.023
18. Kuske, D.: Theories of orders on the set of words. RAIRO Theor. Inform. Appl. **40**(1), 53–74 (2006). https://doi.org/10.1051/ita:2005039
19. Kuske, D., Zetzsche, G.: Languages ordered by the subword order. In: Bojańczyk, M., Simpson, A. (eds.) FoSSaCS 2019. LNCS, vol. 11425, pp. 348–364. Springer, Cham (2019). https://doi.org/10.1007/978-3-030-17127-8_20
20. Larsen, K.G., Skou, A.: Bisimulation through probabilistic testing. Inf. Comput. **94**(1), 1–28 (1991). https://doi.org/10.1016/0890-5401(91)90030-6
21. Morris Jr, J.H.: Lambda-calculus models of programming languages. Ph.D. thesis, Massachusetts Institute of Technology (1969). https://dspace.mit.edu/handle/1721.1/64850
22. Muroya, K.: Hypernet semantics of programming languages. Ph.D. thesis, University of Birmingham, UK (2020). http://ethos.bl.uk/OrderDetails.do?uin=uk.bl.ethos.817915
23. Plotkin, G., Power, J.: Adequacy for algebraic effects. In: Honsell, F., Miculan, M. (eds.) FoSSaCS 2001. LNCS, vol. 2030, pp. 1–24. Springer, Heidelberg (2001). https://doi.org/10.1007/3-540-45315-6_1
24. Pous, D.: Up-to techniques for weak bisimulation. In: Caires, L., Italiano, G.F., Monteiro, L., Palamidessi, C., Yung, M. (eds.) ICALP 2005. LNCS, vol. 3580, pp. 730–741. Springer, Heidelberg (2005). https://doi.org/10.1007/11523468_59
25. Pous, D.: New up-to techniques for weak bisimulation. Theor. Comput. Sci. **380**(1–2), 164–180 (2007). https://doi.org/10.1016/j.tcs.2007.02.060
26. Sangiorgi, D.: On the bisimulation proof method. Math. Struct. Comput. Sci. **8**(5), 447–479 (1998). https://doi.org/10.1017/S0960129598002527
27. Sangiorgi, D.: Equations, contractions, and unique solutions. ACM Trans. Comput. Log. **18**(1), 4:1–4:30 (2017). https://doi.org/10.1145/2971339
28. Sangiorgi, D., Milner, R.: The problem of "weak bisimulation up to". In: Cleaveland, W.R. (ed.) CONCUR 1992. LNCS, vol. 630, pp. 32–46. Springer, Heidelberg (1992). https://doi.org/10.1007/BFb0084781

29. Simpson, A., Voorneveld, N.: Behavioural equivalence via modalities for algebraic effects. In: Ahmed, A. (ed.) ESOP 2018. LNCS, vol. 10801, pp. 300–326. Springer, Cham (2018). https://doi.org/10.1007/978-3-319-89884-1_11
30. Tiuryn, J., Wand, M.: Untyped lambda-calculus with input-output. In: Kirchner, H. (ed.) CAAP 1996. LNCS, vol. 1059, pp. 317–329. Springer, Heidelberg (1996). https://doi.org/10.1007/3-540-61064-2_46
31. Urabe, N., Hasuo, I.: Generic forward and backward simulations III: quantitative simulations by matrices. In: Baldan, P., Gorla, D. (eds.) CONCUR 2014. LNCS, vol. 8704, pp. 451–466. Springer, Heidelberg (2014). https://doi.org/10.1007/978-3-662-44584-6_31
32. Yachi, T., Sumii, E.: A sound and complete bisimulation for contextual equivalence in λ-calculus with call/cc. In: Igarashi, A. (ed.) APLAS 2016. LNCS, vol. 10017, pp. 171–186. Springer, Cham (2016). https://doi.org/10.1007/978-3-319-47958-3_10

Automata and Coalgebras in Categories of Species

Fosco Loregian(✉)

Tallinn University of Technology, Tallinn, Estonia
fosco.loregian@gmail.com

Abstract. We study generalized automata (in the sense of Adámek-Trnková) in Joyal's category of (set-valued) combinatorial species, and as an important preliminary step, we study coalgebras for its derivative endofunctor ∂ and for the 'Euler homogeneity operator' $L \circ \partial$ arising from the adjunction $L \dashv \partial \dashv R$.

1 Introduction

The theory of combinatorial species arose in the work of André Joyal [71, 72] as categorification of the theory of generating functions [130]; crafting a bijective proof [96] to grok numerical identities in terms of bijections between finite sets is acknowledged as *the* fundamental problem in modern combinatorics. For Joyal, a 'species of structure' is a functor having domain the category of finite sets and bijections; properties of the category of all such functors can then be given combinatorial meaning, and combinatorial identities acquire meaning as bijective proofs (=isomorphisms of functors). Joyal's insighftul proof of Cayley's counting of trees [19], paved the way to a booming development of techniques (propelled by the support of an insider of enumerative combinatorics, and genius, as C.G. Rota) in domains such as representation theory of groups [20, 83, 111, 133], the study of set partitions [17, 68, 95], Möbius functions [96, 114, 115], graph theory [100], up to the exciting field of *combinatorial differential equations* [12, 82, 83, 97]. This wealth of applications is by no means limited to the field of enumerative combinatorics; the operation of *plethystic substitution* [10, 104, 105] is recognized as the fundamental building block in the theory of *operads* envisioned by J.P. May [93, 94] and finds applications to algebraic topology and algebraic geometry [30, 36, 84, 106], logic and computer science [32, 33, 135], theoretical physics [37, 38], and more.

At about the same time, another application of category theory gained momentum: the idea of interpreting *abstract state machines* inside general categories. The line of research initiated by Arbib–Manes [5, 109], Goguen [43–45],

F. Loregian was supported by the Estonian Research Council grant PRG1210.
Some of these computations were suggested by Todd Trimble, who pointed out the existence of the 'Euler' derivation and proposed Example 3, which in turn suggested a simple description of $\mathsf{Spc}^{\mathfrak{L}}$ and more examples, by analogy. The author is extremely grateful to Todd for his invaluable contribution and his mathematical generosity.

Published by Springer Nature Switzerland AG 2024
B. König and H. Urbat (Eds.): CMCS 2024, LNCS 14617, pp. 65–92, 2024.
https://doi.org/10.1007/978-3-031-66438-0_4

Naudé [101,102], and others [39,58,63,126] culminated into Ehrig's monograph [25] on automata 'valued' in an abstract monoidal category $\mathcal{K}$. This provides a systematic, category–theoretic insight into the transition from determinism to non-determinism, that can be seen as the passage from automata in a monoidal category [98], to automata in the Kleisli category of an opmonoidal monad [51,65] (such as for example the probability distribution monads for convex spaces, [24,31,64,66,92] or one of its companions –the subdistribution or unnormalized distribution monad).

The category-theoretic content of such an approach to 'machines' goes a long way: a tentative chronology follows, but it can only scratch the surface of an immense, often submerged, body of research.

- [1,3] introduced the notion of an F-*automaton* in order to abstract even further from the monoidal case the 'dynamics' igniting the behaviour of an abstract machine; the progression in abstraction is as follows: from Cartesian machines, i.e. spans $E \leftarrow A \times B \to B$, one goes to monoidal ones, i.e. spans $E \leftarrow A \otimes B \to B$; these are the objects of categories $\mathsf{Mly}_{(\mathcal{K},\otimes)}(A, B)$. Subsequently, one abstracts the action of $A \otimes \_$ on E even further, using a generic endofunctor $F : \mathcal{K} \to \mathcal{K}$; this is the category $\mathsf{Mly}_{\mathcal{K}}(F, B)$.
- Only few years prior, extensive work of Betti-Kasangian [13,14] and Kasangian-Rosebrugh [74] pushed for the adoption of 'profunctorial' models for automata, capable to pinpoint their behaviour, and their minimization, as a universal property [42,45].
- An insightful idea of Katis, Sabadini and Walters [75,76] recognized that categories of automata organize themselves as the hom-categories of a bicategory $\mathsf{KSW}(\mathcal{K})$.[1]
- in [48,51] René Guitart introduces the bicategory Mac as a refinement of a bicategory of spans.[2] In [56], Guitart proves Mac is simply the Kleisli bicategory of the 2-monad of cocompletion under lax colimits. This theme is reprised in [50] where Guitart introduces the notion of *lax coend* [60,85] as a technical preliminary to expand on the theme of [56].

Pushing further these ideas intersects the most prolific branches of modern category theory.

Building on [25], R. Paré proposed in [107] the notion of a *Mealy morphism* as a proxy between strong functors and profunctors in any $\mathcal{V}$-enriched category $\mathcal{C}$. The paper culminates in the impressively general and elegant[3] result

[1] Interestingly enough, KSW category can be seen as a lax analogue of the category of 'categories with endofunctor' upon which one builds the *Spanier-Whitehead stabilization* of the category of (pointed) CW-complexes, a staple construction in stable homotopy theory [123], [87, Chapter I].

[2] Note in passing that this is related to Betti, Kasangian, and Rosebrugh idea as two-sided fibrations and profunctors are well-known equivalent ways to present the same bicategory.

[3] The reader suspecting that this is an overstatement shall rest with the thought that this straightforward statement bestows the bicategory $\mathcal{V}$-Mly with a clear-cut

that the bicategory of $\mathcal{V}$-Mealy maps is simply the Kleisli bicategory of the lax idempotent 2-monad of $\mathcal{V}$-copower completion.[4]

In a joint work [16] we explore how KSW's 'circuits' and Guitart's Mac connect via a *local adjunction* [67,73], and can be used to enhance categorical automata into widgets 'typed' over a bicategory with possibly more than one object; in short, it allows the passage from a bicategory of automata to *automata in a bicategory*, drawing some ideas from Bainbridge's [6,7]. Despite its relative obscurity, likely due to its cutting-edge nature, Bainbridge's work recognized the importance of bicategory theory as a foundational language for the theory of abstract automata and, in particular, proposed the idea of left/right Kan extensions along an 'input scheme' to analyze behaviour and minimization.

To sum up, we find ourselves in the following situation today: a forgotten school of category theorists hid an exciting claim behind a curtain of 2-dimensional algebra:

> A piece of *formal category theory* as envisioned by [47,49,52,120,127,128, 131] serves as the mathematical foundation of abstract state machines.

This intriguing hypothesis is scattered across various sources, often unaware of each other; it has been hinted at multiple times and continues to leave traces of its presence for those willing to follow it. We are left with a conjecture and a clear work plan: can this fundamental guiding principle be taken seriously and formalized? Whoever is willing to take up the challenge of verifying this claim is now tasked with lifting the curtain and exploring a rich fauna of categorical widgets.

The present work grafts on top of the wide branches of this overarching project, studying categorical automata theory specialized to the *differential 2-rig* (a notion introduced by the author in [86]) of Joyal's combinatorial species. The category Spc of species is a presheaf topos equipped with a plethora of tightly-knit monoidal structures interacting with a differential structure; this richness implies that when used as an ambient category for monoidal/functorial automata, it gives rise to an interesting theory that, when stated at the correct level of abstraction, is 'stable under small perturbations', which means that similar results to the ones presented here export without much effort to presheaf categories equipped with a plethystic substitution operation, such as *coloured* species [99], *linear* species (both in the sense of [82] and in the sense of k-Mod-enriched, [4,36]), *Möbius* species [96], nominal sets [108],... and it allows to predict what happens when abstract automata are interpreted in a differential 2-rig other than Spc, generalizing Theorem 4.

---

universal property generalizing, in one fell swoop, KSW and Guitart's approach to every base of enrichment.

[4] The reader will have noticed a repeating theme: categories that naturally arise organizing computational machines share a universal property of Kleisli type (they are initial in some sense, for ways of factoring a certain monad), and the monad is 'of property type', i.e. it is a 2-monad of cocompletion under certain shapes [80,136].

1.1 Outline of the Paper

The basic terminology about the category of species that we need is classical, drawing upon various sources such as [11,32,132,135]; we rework an equally 'classical' construction of the categories $\mathsf{Mly}_{\mathcal{K}}(F, B)$ and $\mathsf{Mre}_{\mathcal{K}}(F, B)$, drawing from [25,51]. In Proposition 5, we introduce the concept of 'ω-differential limit', as an intuition for what the terminal object in $\mathsf{Mly}_{\mathcal{K}}(F, B)/\mathsf{Mre}_{\mathcal{K}}(F, B)$ should represent; the terminology is somewhat borrowed from ergodic theory (specifically, the notion of ω-limit, see [40, Def. 1.12]). Later, in Sect. 3.1, we thoroughly explore the fibrational properties of the $\mathsf{Mly}_{\mathcal{K}}$ construction, yielding the 2-fibration of the *total Mealy 2-category* **Mly**, along with two-sided fibrations [119] $\mathscr{Mly}_{\mathcal{K}}/\mathscr{Mre}_{\mathcal{K}}$ allowing to consider all dynamics and all outputs at the same time, coherently. In (13) we define the *monoidal Mealy fibration* as a particular instance of this construction. The fundamental result of [75], defining the KSW category of a monoidal category $(\mathcal{K}, \otimes)$ arises (Theorem 3) when the profunctor associated to the monoidal Mealy two-sided fibration carries the structure of a promonad, of which KSW$(\mathcal{K}, \otimes)$ is the Kleisli object. In Proposition 7 we address the issue of lifting accessibility from $\mathcal{K}$ to $\mathsf{Mly}_{\mathcal{K}}/\mathsf{Mre}_{\mathcal{K}}$, consolidating the idea that nice properties of the ambient category lift easily to its category of automata. Interestingly, assuming $\mathcal{K}$ is a differential 2-rig in the sense of [86], $\mathscr{Mly}_{\mathcal{K}}$ and $\mathscr{Mre}_{\mathcal{K}}$ are differential 2-rigs: an upshot of [86] is that differential structures are 'difficult to create', and yet categories of $\mathcal{K}$-valued automata exhibit an additional differential 2-rig structure, simply but not trivially related to $\mathcal{K}$.

Finally we turn to the task of studying (Mealy) automata in species, focusing on the particular case where $\mathcal{K}$ is the category of Sect. 2; given its structure of differential 2-rig, we are particularly interested in studying *differential dynamics*, i.e. in studying categories $\mathsf{Mly}_{\mathsf{Spc}}(F, B)$ where the generator F of dynamics is induced by the derivative functor. Given the results in [110], recalled in Theorem 1, there is plenty of choice for such F's: the triple of adjoints $L \dashv \partial \dashv R$ generates four functors, a comonad-monad adjunction $L\partial \dashv R\partial$ and a monad-comonad adjunction $\partial L \dashv \partial R$ (paying tribute to the 'twelvefold way' of [116], we dub the study of this quadruple of pairwise adjoint functors the 'fourfold way'); each of these adjunctions generate monads or comonads $R\partial L\partial$, $L\partial R\partial$, $\partial R\partial L$, $\partial L\partial R$ and all these are finitely accessible functors because R is.

2 The Category of Species

Issues of page count force us to condense a wealth of material in a small space; the reader will find excellent introductory texts and surveys on the category of species in [11,32,132,135] or in the original papers by Joyal [71,72].

After defining the category Spc of combinatorial species, we recall its various monoidal structures and outline how they relate, with particular attention to the *Day convolution* monoidal structure, and its *differential 2-rig* structure in the sense of [86], with particular attention to the fact that the derivative functor $\partial : \mathsf{Spc} \to \mathsf{Spc}$ has both a left and a right adjoint. We study co/monoids in

$(\mathsf{Spc}, \otimes_{\mathrm{Day}})$, since monoidal automata theory in a category with countable sums forces us to understand the structure of the subcategory of $\otimes_{\mathrm{Day}}$-monoids at a fundamental level.

Definition 1 (Species and $\mathscr{V}$-species). *Let S be a set, and $\mathscr{V}$ a symmetric monoidal closed category admitting all limits and colimits. The category $(S, \mathscr{V})$-Spc of (S-colored) $\mathscr{V}$-species is defined as the category of functors $\mathfrak{F} : \mathsf{P}[S] \to \mathscr{V}$, where $\mathsf{P}[S]$ is the free symmetric monoidal category on S, regarded as a discrete category.*

We will particularly be interested in the category $(1, \mathsf{Set})$-Spc (that we dub simply Spc in the following), where 1 is a singleton.[5] The category $\mathsf{P} = \mathsf{P}[1]$ is called the *groupoid of natural numbers*, having as objects the nonnegative integers $[0], [1], [2], \dots$ and where morphisms $n \to m$ are the symmetric group S_n of permutations of the set $[n] = \{1, \dots, n\}$ (so in particular $[0]$ is the empty set and S_0 is the trivial group) if $n = m$, and the empty set otherwise. As such, P is the skeleton of the *groupoid of finite sets* Bij, the category having objects the finite sets $A, B, \dots$ and morphisms $A \to B$ the set of all bijections between A and B (so, in particular, if A and B do not have the same cardinality, $\mathsf{Bij}(A, B)$ is empty). Note that Bij is the core (=larger subcategory that is a groupoid) of the category of finite sets. The (commutative, i.e. strictly symmetric) monoidal structure on P is given by sum of natural numbers, i.e. $[n] \oplus [m] = [n + m]$, the unit is $[0]$ and permutations act by juxtaposition. In the following we denote a species as $c : \mathsf{P} \to \mathsf{Set}$ and call an element $s \in \mathfrak{F}[n]$ a *species of $\mathfrak{F}$-structure.*

Corollary 1. *The universal property of P entails that there is an isomorphism of categories $\mathsf{P} \cong \sum_n S_n$ where the right-hand side is the coproduct in the category of groupoids and as a consequence $\mathsf{Spc} \cong \prod_n \mathsf{Set}^{S_n}$ where each Set^{S_n} is the category of left S_n-set.*

As a consequence, species can equivalently be presented as a symmetric sequence $\{X_n \mid n \geq 0\}$ *of sets, each of which is equipped with a (left) S_n-action $S_n \times X_n \to X_n$.*

Definition 2 (Change of base for species). *Let $\mathscr{V}$ be a monoidal category monadic over Set such that the functor $K : \mathscr{V} \to \mathsf{Set}$ is lax monoidal (for example the forgetful functor $U : \mathsf{Mod}_R \to \mathsf{Set}$); then there is a* base change *adjunction $F_* : \mathsf{Spc} \rightleftarrows \mathscr{V}\text{-}\mathsf{Spc} : U_*$ induced through the free-forgetful $F \dashv U$. For example, if $F : \mathsf{Set} \to \mathsf{Mod}_k$ is the free k-vector space functor, we denote $k\langle L \rangle$ the k-linear species $F_* L$ induced by a Set-species L.*

Example 1 (Some important species). Many more examples of species can be found in [11, Ch. 1].

---

[5] Other possible choices for $\mathscr{V}$ are the category Mod_R of modules over a ring R (if R is a field, we call a Mod_k-species just a k-linear species, see [84] for a comprehensive introduction) or the category Top_* of pointed topological spaces equipped with the smash product [121, 3.6.2] (for applications to algebraic topology, see e.g. [91]; for a broader notion of operad cf. the excellent readings [22,124]).

ES1) Given an object V of $\mathcal{V}$, there is a unique symmetric monoidal $\mathcal{V}$-species c_V sending $[n]$ to $V^{\otimes n}$. If $V = I$ is the monoidal unit, c_I is called the 'exponential species' $\mathfrak{E}$. The exponential Set-species is just the constant functor at the terminal object.[6]

ES2) The species $\wp$ of subsets sends an n-set A to the 2^n-set of all its subsets; a permutation acts in an obvious way, since a bijection $\sigma : A \to A$ induces a bijection $\sigma^* : 2^A \to 2^A$ by functoriality.

ES3) The species $\mathfrak{L}$ of total orders[7] sends $[n]$ to the set of total orders on $[n]$, identified with the *set* $|S_n|$ of bijections of $[n]$, over which S_n acts by left multiplication.

ES4) The species $\mathfrak{S}$ of permutations sends each finite set $[n]$ into the (carrier of the) symmetric group on n letters, S_n. The symmetric group acts on itself by conjugation: if $\tau \in S_n$, $\sigma : S_n \to S_n$ is the map sending $\tau \mapsto \sigma\tau\sigma^{-1}$.

ES5) The species $\mathfrak{Cyc}$ of *oriented cycles* sends a finite set $[n]$ to the set of inequivalent (i.e. not related by a cyclic permutation) ways to sit n people at a round table, or more formally, in the set of *cylic orderings* of $\{x_1, \dots, x_n\}$. As $\mathfrak{Cyc}[n]$ identifies with the set of cosets S_n/C_n (C_n the cyclic group), one derived that $|\mathfrak{Cyc}[n]| = (n-1)!$.

The category of species exhibits a fairly rich structure that we now review.[8]

Proposition 1. Spc *is the free cocompletion under small colimits [125, Remark 2.29] of* P*; as such, for every cocomplete category* $\mathcal{D}$ *there is an equivalence of categories*

$$\mathsf{Cat}(\mathsf{P}, \mathcal{D}) \cong \{\textit{colimit preserving functors } \mathsf{Spc} \to \mathcal{D}\} \tag{1}$$

given by 'Yoneda extension' [85, Ch. 2].

Proposition 2. *Following [2, 57, 71]* Spc *is the (nonfull) subcategory of* analytic endofunctors *of* Set, *i.e. those endofunctors* $F : \mathsf{Set} \to \mathsf{Set}$ *such that, if* $J : \mathsf{Bij} \to \mathsf{Set}$ *is the tautological functor* $[n] \mapsto [n]$*, the left Kan extension of* FJ *along* J *coincides with* J*. The usual coend formula [88] to express* $\mathrm{Lan}_J FJ$ *entails that* F *is analytic if and only if it acts on a set* X *as*

$$FX \cong \int^n F[n] \times X^n \tag{2}$$

i.e. if and only if F *admits a 'Taylor expansion'* $\sum_{n=0}^{\infty} F[n]\frac{X^n}{n!}$*; hence the name. The series* $g_F(X) = \sum_{n=0}^{\infty} |F[n]|\frac{X^n}{n!} \in \mathbb{Q}[\![X]\!]$*, where* $|S|$ *denotes the cardinality of a set, is called the (exponential)* generating series *[11, §1.2] of the species* F*.*

[6] In a serendipitous choice, the notation $\mathfrak{E}$ for this species hints at the same time that $\mathfrak{E}$ is the species of sets, or *éspèce des ensembles*, and that it's an analogue of the exponential function, as $\partial\mathfrak{E}[n] = \mathfrak{E}[n]$, for the derivative functor of Remark 4.

[7] Elsewhere customarily called the species of *linear* orders, but this might conflict with linear as in 'k-linear' if k is a ring.

[8] An important additional universal property we do not need in our analysis is that Spc is a Grothendieck topos, precisely the classifying topos [89, Ch. VIII] for $\mathcal{P}$-torsors, where $\mathcal{P}$ is the category P regarded as a groupoid.

Proposition 3 ([86, §5]). *Spc is the free cocomplete 2-rig on a singleton; as such, given a cocomplete 2-rig $\mathcal{R}$ there is an equivalence of categories*

$$\mathcal{R} \cong \{\textit{colimit preserving 2-rig functors } \mathsf{Spc} \to \mathcal{R}\} \tag{3}$$

*In [86, §5] we observe how to construct the free cocomplete (symmetric) 2-rig on a given category $\mathcal{A}$ it suffices to take the free (symmetric) monoidal category on $\mathcal{A}$, call it $\mathsf{P}[\mathcal{A}]$ and subsequently, its free cocompletion $\mathsf{Cat}(\mathsf{P}[\mathcal{A}]^{\mathrm{op}}, \mathsf{Set})$. The notion of morphism of 2-rigs is given indirectly as pseudomorphism for the particular 'doctrine of **D**-rigs' in study.*

This last characterization requires a more fine-grained analysis of the various monoidal structures Spc can be equipped with.

Remark 1. The category Spc of species carries

MS1) the *Cartesian* (or *Hadamard* [4, 8.1.2]) monoidal structure, the product of species being taken pointwise; the monoidal unit for the Hadamard product is the species that is constant at the singleton. Dually, the *coCartesian* monoidal structure, the coproduct of species being taken pointwise (together with the structure above, Spc is $\times$-distributive and forms a 'biCartesian closed' category in the sense of [122]); however, its biCartesian structure is not very interesting, compared to

MS2) the *Day convolution* (or *Cauchy* [4, 8.1.2]) monoidal structure, given by the universal property of Spc as the free monoidally cocomplete category on P [62] as the coend

$$(F \otimes_{\mathsf{Day}} G)[p] := \int^{mn} F[m] \times G[n] \times \mathsf{P}(m+n, p) \tag{4}$$

(Note in passing that the $\otimes_{\mathsf{Day}}$-monoidal structure is symmetric and closed with an internal hom $\{-,-\}_{\mathsf{Day}}$.) In particular, P is monoidally equivalent to the subcategory of Spc spanned by representables, and thus the $\otimes_{\mathsf{Day}}$-monoidal unit is $y[0]$.

MS3) the *substitution* (or *plethystic*, cf. [95,103]) monoidal structure, defined for $F, G : \mathsf{P} \to \mathsf{Set}$ as $(F \circ G)[p] = \int^n Fk \times G^{\otimes_{\mathsf{Day}} k}[p]$, where $G^{\otimes_{\mathsf{Day}} k} := G \otimes_{\mathsf{Day}} G \otimes_{\mathsf{Day}} \dots \otimes_{\mathsf{Day}} G$ (k times). The $\circ$-monoidal unit is the representable $y[1]$. Note in passing that the $\circ$-monoidal structure is *not* symmetric, and only *right* closed, i.e. only $\_ \circ G$ has a right adjoint.

All these monoidal structures are tightly related:

Remark 2. The Hadamard and Day convolution product give Spc the structure of a *duoidal category* in the sense of [34]: $(\mathsf{Spc}, \times, \otimes_{\mathsf{Day}})$ and $(\mathsf{Spc}, \otimes_{\mathsf{Day}}, \times)$ [4, 8.13.5] are both duoidal; *positive* species, i.e. those for which $F[\varnothing] = \varnothing$ form a duoidal category under substitution and Hadamard product, [B.6.1, *ibi*]. All these results extend to $\mathcal{V}$-species. The plethystic structure makes Spc monoidally equivalent to the category of analytic functors under composition [2,71].

Remark 3. As an additional demonstration of how tightly the Hadamard, Cauchy and plethystic structures are related, observe how all these identification between combinatorial species hold [11]:

CI1) the species of subsets $\wp : A \mapsto 2^A$ is isomorphic to $\mathfrak{E} \otimes_{\text{Day}} \mathfrak{E}$;
CI2) the species $\mathfrak{S}$ of permutations of ES4 is isomorphic to the substitution $\mathfrak{E} \circ \mathfrak{Cyc}$;
CI3) more generally, for every species $\mathfrak{F}$ the substitution $\mathfrak{E} \circ \mathfrak{F}$ sends A to a r-partition $(U_1, \dots, U_r)$ of A and picks a $\mathfrak{F}$-structure on each U_i.

An important structure enjoyed by Spc that we will analyze in this paper is that of a *differential 2-rig*: the notion was introduced in [86] as a unifying language for instances of a monoidal category $(\mathcal{C}, \otimes, I)$ where each $A \otimes \_, \_ \otimes B$ is cocontinuous and an endofunctor ∂ is 'linear and Leibniz' in the following sense.

As it is true in all presheaf categories, the tensor product $\otimes_{\text{Day}}$ preserves colimits separately in each variable (i.e., each $A \otimes_{\text{Day}} \_$ and $\_ \otimes_{\text{Day}} B$ is cocontinuous); moreover, the following is true.

Remark 4 (The differential structure of Spc*).* The category Spc of species is equipped with a 'derivative' endofunctor $\partial : \mathsf{Spc} \to \mathsf{Spc}$ (cf. [11, §1.4 and *passim*]; ∂F is the species sending $[n]$ to $F[n \oplus [1]]$) such that

D1) ∂ is 'linear', i.e. it preserves all colimits (in particular, coproducts);
D2) ∂ is 'Leibniz', i.e. it is equipped with tensorial strengths $\tau' : \partial A \otimes B \to \partial(A \otimes B)$ and $\tau'' : A \otimes \partial B \to \partial(A \otimes B)$ such that the unique map induced by τ', τ'' from the coproduct of their domains is invertible, to the effect that ∂ 'satisfies the Leibniz rule'

$$\partial A \otimes B + A \otimes \partial B \cong \partial(A \otimes B). \tag{5}$$

Definition 3. *Every monoidal category* $(\mathcal{K}, \otimes)$ *equipped with an endofunctor* ∂ *that satisfies the same three properties is called a* differential 2-rig *(for the doctrine of colimits) in [86].*

In the case of species, the proof that $\partial(F \otimes_{\text{Day}} G) \cong \partial F \otimes_{\text{Day}} G + F \otimes_{\text{Day}} \partial G$ appears in Joyal's original papers introducing combinatorial species. Moreover, it was known to Joyal that ∂ satisfies the 'chain rule' in that $\partial(F \circ G) \cong (\partial F \circ G) \otimes_{\text{Day}} \partial G$; cf. [86, Theorem 5.18] for a conceptual proof of this latter result. To a very large extent, the combinatorial differential calculus of species agrees with the classical differential calculus of formal power series. In particular, observe that $g_{\partial F}(X)$ is the formal derivative $\frac{d}{dX} g_F(X)$ of the series in Proposition 2.

Remark 5. Part of the fairly rich structure enjoyed by the differential 2-rig $(\mathsf{Spc}, \otimes, \partial)$ can be explained with the fact that ∂ also preserves all *limits*: ∂F is precomposition with the $\_ \oplus [1]$ functor; but then, call $\Delta = \_ \oplus [1]$, the left (resp., right) adjoint to ∂ is the left (resp., right) Kan extension along Δ, which exists since Spc is a presheaf category.[9]

We just proved the following result:

Theorem 1. *The derivative functor* $\partial : \mathsf{Spc} \to \mathsf{Spc}$ *sits in a triple of adjoints* $L \dashv \partial \dashv R$, *and* L, R *are obtained as Kan extensions.*

This fact was first observed in [110], where the explicit descriptions

$$LF : A \mapsto \sum_{a \in A} F[A \setminus \{a\}] \qquad RF : A \mapsto \prod_{a \in A} F[A \setminus \{a\}] \tag{6}$$

are given in terms of F as a functor $\mathsf{Bij} \to \mathsf{Set}$, and some useful combinatorial identities expressing $L\partial, R\partial, \partial L, \partial R$ in simpler terms are also analyzed.

Notation 2 (Scopic 2-rig). *We introduce the terminology* scopic[10] *2-rig to refer to a differential 2-rig* $(\mathcal{R}, \otimes, D)$ *whose derivative functor* D *has both a left and a right adjoint.*

Algebraic Structures and Co/algebras in Spc. We end the section reviewing the characterization of monoids, comonoids and Hopf monoids in Spc. First of all, Hadamard co/monoids are simply co/monoid-valued species, i.e. functors $F : \mathsf{P} \to \mathsf{Mon}$ or $\mathsf{P} \to \mathsf{Comon}$ into the categories of monoids and comonoids in Set (and this result extends to $\mathcal{V}$-species, a Hadamard monoid in $\mathcal{V}\text{-}\mathsf{Spc}$ being just a functor $\mathsf{P} \to \mathsf{Mon}(\mathcal{V})$).

Cauchy co/monoids (i.e. co/monoids for the Day convolution, whence our preference for calling them Day co/monoids) are far more interesting, as well as substitution co/monoids (the latter are called *co/operads* and have an extremely long history, excellent surveys geared towards the different areas of Mathematics using them are [22,32,77,91]). The first remark on $\otimes_{\mathrm{Day}}$-co/monoids is simply that there aren't any among representables.

Remark 6. There are no nontrivial representable $\otimes_{\mathrm{Day}}$-magmas, for the simple reason that the subcategory spanned by representables is monoidally equivalent to $(\mathsf{P}, \oplus)$, and in the latter a binary operation $[n] \oplus [n] = [2n] \to [n]$ can exist only if $2n = n$. For a similar reason, there are no nontrivial k-ary cooperations $[n] \to [n]^{\oplus k}$.

Remark 7. It is worth to explicitly spell out what a $\otimes_{\mathrm{Day}}$-monoid (M, μ, η) in Spc must be made of:

---

9 An alternative proof of the same fact is in terms of the Day convolution structure: one sees that there is a natural isomorphism $\partial F \cong \{y[1], F\}_{\mathrm{Day}}$ where $\{-,-\}_{\mathrm{Day}}$ is the internal hom, and $y[1] = \mathsf{P}(1, -)$ the corepresentable functor on $[1]$; now, certainly $\{y[1], -\}_{\mathrm{Day}}$ must have $y[1] \otimes_{\mathrm{Day}} -$ as left adjoint, but since in every presheaf category representables are tiny objects, ∂ must also be cocontinuous, hence a left adjoint by the special adjoint functor theorem.

10 From the Proto-Indo-European root **speḱ-*, derived the Latin word *speciēs* and the Greek verb *skopéo*, related to the verb 'to see'.

- the unit consists of a species morphism $\eta : y[0] \to M$ which by Yoneda is just an element $e \in M[0]$.
- the multiplication splits into a cowedge $\mu_{pq} : M[p] \times M[q] \to M[n]$ for each pair of integers p, q such that $p + q = n$, natural for the action of symmetric groups, under the shuffling maps $S_p \times S_q \to S_{p+q}$ sending a pair of permutations (σ, τ) to the one acting as σ on $\{1, \dots, p\}$ and as τ on $\{p+1, \dots, p+q\}$.

The following is implied joining [2, Example 2.3] and adapting [4, 8.16]: in particular, the species $\mathfrak{L}$ of Example 1 has a convenient universal property.

Proposition 4. *([11, p. 7], [4, §8.1]) The species $\mathfrak{L}$ of total orders is the free monoid on $y[1]$. The species $\mathfrak{L}_+$ of* nonempty *linear orders is the free semigroup on $y[1]$. Thus,*

$$\mathfrak{L} \cong \textstyle\sum_{n \geq 0} y[n] \qquad\qquad \mathfrak{L}_+ \cong \textstyle\sum_{n \geq 1} y[n]. \tag{7}$$

(A terminological note. [4] calls 'positive' what we tend to dub 'nonempty', considering species as monadic over graded vector spaces.) In fact, in a k-linear setting (k a field) the structure of $\mathfrak{L}$ is way richer: $k\langle\mathfrak{L}\rangle$ (cf. Definition 2; it's the species assigning to $[n]$ the k-vector space having the set $\mathfrak{L}[n]$ as a basis) carries the structure of a *Hopf monoid.* Following Remark 7, the monoid structure of $\mathfrak{L}$ arises as a cowedge $\mathfrak{L}[p] \times \mathfrak{L}[q] \to \mathfrak{L}[n]$ for every $p+q = n$, defined as $(l, l') \mapsto l \cdot l'$ where the later is the *ordinal sum* or concatenation of the linear orders l on $[p]$ and l' on $[q]$; ordinal sum is an associative operation, equivariant under the shuffling maps. The unit is the only element of $\mathfrak{L}[1]$.

The Hopf monoid structure of $k\langle\mathfrak{L}\rangle$ is extensively studied and described in [4, §8.5].

Co/algebras for Endofunctors of Spc. This subsection studies algebras and coalgebras for a few interesting endofunctors M defined over Spc. Despite its naturality, this idea is seemingly unexplored thus far.

It becomes particularly intriguing to explore the interactions between M and the structures on Spc mentioned in Remark 1, Remark 4; clearly, this is essential to study (M, B)-automata, defined in Definition 5 as a pullback along M-algebras.

Definition 4 (The category $\mathsf{Spc}^{\mathfrak{L}}$). *The category $\mathsf{Spc}^{\mathfrak{L}}$ is, up to equivalence, described as any of the following:*

L1) *the category of endofunctor algebras for $y[1] \otimes_{\mathrm{Day}} -$;*
L2) *the category of endofunctor coalgebras for ∂;*
L3) *the Eilenberg–Moore category of the monad $\mathfrak{L} \otimes_{\mathrm{Day}} -$;*
L4) *the coEilenberg–Moore category of the comonad $\{\mathfrak{L}, -\}_{\mathrm{Day}}$.*

These identifications follow from the freeness of $\mathfrak{L}$ and the general fact that whenever $F \dashv G$ is an adjunction between endofunctors, $\mathsf{Alg}(F) \cong \mathsf{coAlg}(G)$.

Representing objects of $\mathsf{Spc}^{\mathfrak{L}}$ as Eilenberg–Moore algebras is particularly convenient, as a $\mathfrak{L}$-module is the same thing as a $\otimes_{\mathrm{Day}}$-monoid homomorphism $\mathfrak{L} \to \{F, F\}_{\mathrm{Day}}$, which since $\mathfrak{L}$ is the free monoid generated on $y[1]$, amounts to a single element of $\{F, F\}_{\text{Day}}[1]$; equivalently, if one uses characterization 4 above, a structure of type $y[1] \otimes_{\text{Day}}$ on $[n]$ consists of a choice of point in $[n]$, together with an F-structure on the complement of that point.[11]

Remark 8. Limits and colimits in $\mathsf{Spc}^{\mathfrak{L}}$ are computed exactly as in Spc, i.e. pointwise (since Spc is monadic over $\mathsf{Set}^{\mathbb{N}} = \prod_{n\geq 1} \mathsf{Set}$), given that $\mathsf{Spc}^{\mathfrak{L}}$ is at the same time a category of algebras (for $\mathfrak{L} \otimes_{\text{Day}} -$, hence limits are created in Spc) and of coalgebras (for the right adjoint comonad $\{\mathfrak{L}, -\}_{\text{Day}}$, hence colimits are created in Spc). We just proved that

Lemma 1. *The terminal object of* $\mathsf{Spc}^{\mathfrak{L}}$ *is the exponential species of Example 1, whence the isomorphism* $\partial\mathfrak{E} \cong \mathfrak{E}$ *characterizing* $\mathfrak{E}$ *as a 'Napier object' of the differential 2-rig of species.*[12]

Armed with these explicit computations, we can attempt to unveil the structure of the category $\mathsf{Spc}^{\mathfrak{L}}$ in any of the equivalent forms given in Definition 4 as a building block of $\mathsf{Mly}_{\mathsf{Spc}}(\mathfrak{L}, \_)$.

We now collect some examples of: a species that has only a few structures of $\mathfrak{L}$-algebra (=structures of ∂-coalgebra); a species that has at least uncountably many; a species with *no* such structure as a Set-species, that however becomes interesting when 'changing base' (cf. Definition 2).

Example 2. Structures of ∂-coalgebra on the species of subsets of ES2 correspond to S_n-equivariant maps $\theta : \wp \to \partial\wp$ and using the Leibniz rule over the isomorphism $\wp \cong E \otimes_{\text{Day}} E$ of [11, §1.3, Eq. (33)] one gets that $\theta : \wp \to \wp + \wp$. Using elementary group theory on the components θ_A one sees that there are only four such θ: embedding a subset $U \subseteq A$ in the first summand, embedding a subset $U \subseteq A$ in the second summand, embedding $U^c = A \setminus U$ in the first summand, embedding $U^c = A \setminus U$ in the second summand.

Example 3. [11, Example 9, (37)] yields $\partial\mathfrak{L} \cong \mathfrak{L} \otimes_{\text{Day}} \mathfrak{L}$, whence a natural choice for a coalgebra structure $s : \mathfrak{L} \to \partial\mathfrak{L}$, given a finite set A, is specified on components s_A in terms of a choice of decomposition $A = I \sqcup J$ and a splitting of the total order on A as a total order on I and a total order on J. This choice is made independently for every finite set A, so this argument shows that there is an uncountable infinity of coalgebra structures on $\mathfrak{L}$.

Example 4. Let $\mathfrak{Cyc}$ be the species of cyclic orders, Example 1.ES5; then, we immediately get $\partial\mathfrak{Cyc} \cong \mathfrak{L}$ from manipulating generating series. A ∂-coalgebra structure on $\mathfrak{Cyc}$ now would be a natural transformation $\vartheta : \mathfrak{Cyc} \to \mathfrak{L}$, and no such map can exist by cardinality reasons: since $\mathfrak{Cyc}[n]$ identifies with the coset space $S_n/\mathbb{Z}_n$, over which S_n acts transitively, an S_n-equivariant map $\vartheta_n : \mathfrak{Cyc}[n] \to S_n$ must be surjective (the translation action $S_n \times \mathfrak{Cyc}[n] \to \mathfrak{Cyc}[n] : (\sigma, \tau) \mapsto \sigma\tau$ is also transitive). Yet, $|S_n| = n! > (n-1)! = |\mathfrak{Cyc}[n]|$.

[11] One can read off the fact that these descriptions are equivalent from the end defining $\{F, F\}_{\text{Day}}[n]$, cf. [77, Equation (2.6)].

[12] The rationale behind the terminology is that, evidently, 'exponential object' already has a different, conflicting meaning.

Example 5. Let $\mathfrak{S}$ be the species of permutations of Example 1.ES4; from Remark 3 it follows that $\partial\mathfrak{S} \cong \mathfrak{S} \otimes_{\mathrm{Day}} \mathfrak{L}$, so that ∂-coalgebra structures (i.e. Eilenberg–Moore algebras for $\mathfrak{L} \otimes_{\mathrm{Day}} \_$) correspond under adjunction to monoid homomorphisms $\mathfrak{L} \to \{\mathfrak{S}, \mathfrak{S}\}_{\mathrm{Day}}$.

3 Abstract Automata in Spc

Let $\mathcal{K}$ be a category and $F : \mathcal{K} \to \mathcal{K}$ be an endofunctor that we think of as a categorification of a dynamical system and its iterates $F, F^2, F^3, \dots, F^n : \mathcal{K} \to \mathcal{K}$, cf. [3]. We also fix an object $B \in \mathcal{K}$ (an 'output' object, cf. [25,51]).

Definition 5. *We define the category* $\mathsf{Mly}_{\mathcal{K}}(F, B)$ *and* $\mathsf{Mre}_{\mathcal{K}}(F, B)$ *as the following strict 2-pullbacks in* Cat *respectively:*

$$\begin{array}{ccc} \mathsf{Mly}_{\mathcal{K}}(F,B) & \longrightarrow & F/B \\ \downarrow & \lrcorner & \downarrow{\scriptstyle U} \\ \mathsf{Alg}(F) & \xrightarrow[V]{} & \mathcal{K} \end{array} \qquad \begin{array}{ccc} \mathsf{Mre}_{\mathcal{K}}(F,B) & \longrightarrow & \mathcal{K}/B \\ \downarrow & \lrcorner & \downarrow{\scriptstyle U'} \\ \mathsf{Alg}(F) & \xrightarrow[V]{} & \mathcal{K} \end{array} \tag{8}$$

where $\mathsf{Alg}(F)$ *is the category of* endofunctor algebras *of* F, F/B *the comma category of arrows* $FX \to B$, *and* $\mathcal{K}/B$ *the comma category of arrows* $X \to B$ *(and* U, V, U', V' *are the most obvious forgetful functors).*

Remark 9 (Limits and colimits in categories of automata). If F admits a right adjoint R, and $\mathcal{K}$ is complete and cocomplete, so are $\mathsf{Mly}_{\mathcal{K}}(F, B)$ and $\mathsf{Mre}_{\mathcal{K}}(F, B)$; this can be easily argued using an argument in [88, V.6, Ex. 3] and the fact that U, U' create colimits and connected limits, together with the fact that $F/B \cong \mathcal{K}/RB$; then, the terminal object of $\mathsf{Mly}_{\mathcal{K}}(F, B)$ is $\prod_{n\geq 1} R^n B$ and the terminal object of $\mathsf{Mre}_{\mathcal{K}}(F, B)$ is $\prod_{n\geq 0} R^n B$.

Remark 10 (Accessibility of categories of automata). Repeatedly applying the completeness theorem of the 2-category Acc of accessible categories [90, Ch. 5] one can prove that if $\mathcal{K}$ is locally presentable (say for a regular cardinal κ) and F is κ-accessible, then $\mathsf{Mly}_{\mathcal{K}}(F, B)$, $\mathsf{Mre}_{\mathcal{K}}(F, B)$ are both locally presentable (but in general, for a much higher cardinal κ).

Remark 11. A particular instance of Remark 9 is when $\mathcal{K}$ is monoidal and $F : \mathcal{K} \to \mathcal{K}$ is the tensor product $A \otimes -$ for a fixed object of $\mathcal{K}$. Then, we shorten $\mathsf{Mly}_{\mathcal{K}}(F, B)$ and $\mathsf{Mre}_{\mathcal{K}}(F, B)$ to $\mathsf{Mly}_{\mathcal{K}}(A, B)$ and $\mathsf{Mre}_{\mathcal{K}}(A, B)$ and we observe that

- if $\mathcal{K}$ has countable sums, $\mathsf{Alg}(F) = \mathsf{Alg}(A \otimes -)$ is the Eilenberg-Moore category of the monad $A^* \otimes -$ where $A^* := \sum_{n=0}^{\infty} A^{\otimes n}$ is the free monoid on A;

- if $\mathcal{K}$ is monoidal closed, complete and cocomplete, then $\mathsf{Mly}_{\mathcal{K}}(A,B)$ and $\mathsf{Mre}_{\mathcal{K}}(A,B)$ are complete and cocomplete; if $\mathcal{K}$ is locally κ-presentable, so are $\mathsf{Mly}_{\mathcal{K}}(A,B)$ and $\mathsf{Mre}_{\mathcal{K}}(A,B)$ (generally, for a larger cardinal $\kappa' \gg \kappa$). The terminal object in $\mathsf{Mly}_{\mathcal{K}}(A,B)$ is $[A^+,B]$, A^+ being the free semigroup on A (resp., in $\mathsf{Mre}_{\mathcal{K}}(A,B)$ it's $[A^*,B]$, A^* being the free monoid).

Unwinding Definition 5 in this particular case, the typical object $\Vert\frac{E}{d,s}\Vert$ of $\mathsf{Mly}_{\mathcal{K}}(A,B)$ is a span as in the left of the following diagram, and the typical object $\Vert\frac{E}{d,s}|$ of $\mathsf{Mre}_{\mathcal{K}}(A,B)$ a (disconnected) diagram as in the right

$$\Vert\tfrac{E}{d,s}\Vert\colon E \xleftarrow{d} A\otimes E \xrightarrow{s} B \qquad \Vert\tfrac{E}{d,s}|\colon E \xleftarrow{d} A\otimes E, E \xrightarrow{s} B. \tag{9}$$

Remark 12. Remarks 9, 10, 11 all apply to $\mathcal{K} = \mathsf{Spc}$ considered with the Day convolution structure (and in fact to all $\mathcal{V}$-Spc when $\mathcal{V}$ is complete, cocomplete and monoidal closed). In particular, for every fixed combinatorial species $B : \mathsf{P} \to \mathsf{Set}$ we can easily study $\mathsf{Mly}_{\mathsf{Spc}}(L,B) = \mathsf{Mly}_{\mathsf{Spc}}(y[1],B)$ as the category having objects the diagrams $E \xleftarrow{d} y[1] \otimes_{\mathrm{Day}} E \xrightarrow{s} B$, or more concisely as the category obtained as the pullback $\mathsf{Spc}^{\mathfrak{L}} \times_{\mathsf{Spc}} (\mathsf{Spc}/B)$ where $\mathsf{Spc}^{\mathfrak{L}}$ is as in Definition 4.

Note that this is equivalent to the category of coalgebras for the functor $E \mapsto \partial B \times \partial E$. From this coalgebraic characterization, we deduce that

Proposition 5. *The terminal object of $\mathsf{Mly}_{\mathsf{Spc}}(L,B)$ is the 'ω-differential limit'*[13] *of B defined as*

$$\textstyle\prod_{n\ge 1}\partial^n B \cong \prod_{n\ge 1}\{y[1]^{\otimes_{\mathrm{Day}} n}, B\}_{\mathrm{Day}} \cong \Big\{\sum_{n\ge 1} y[n], B\Big\}_{\mathrm{Day}} = \{y[1]^+, B\}_{\mathrm{Day}} \tag{10}$$

where again $y[1]^+$ is the free semigroup on $y[1]$: given Proposition 4, $y[1]^+ \cong \mathfrak{L}_+$.

Remark 13. Consider two endofunctors $F : \mathcal{K} \to \mathcal{K}$, $G : \mathcal{H} \to \mathcal{H}$. If $P : \mathcal{K} \to \mathcal{H}$ is a functor equipped with an intertwiner $\pi : GP \Rightarrow PF$ we can define a functor $\pi^* : \mathsf{Alg}(F) \to \mathsf{Alg}(G)$ by application of P and precomposition with π, a functor $\mathcal{K}/B \to \mathcal{H}/PB$ in the obvious way, and in turn a unique functor

$$\varpi^* : \mathsf{Mly}_{\mathcal{K}}(F,B) \longrightarrow \mathsf{Mly}_{\mathcal{H}}(G,PB) \tag{11}$$

[13] The name is chosen in analogy with the notion of ω-limit set of a dynamical system $f : X \to X$ defined over a metric space, see e.g. [40, Def. 1.12], where the (ω-)limit set of x under f is defined as

$$\omega(x,f) = \textstyle\bigcap_{n\in\mathbb{N}} \overline{\{f^k(x) : k > n\}},$$

the topological closure of the 'eventual f-orbits' of x.

3.1 Fibrational Properties of the Mly Construction

The fact that Definition 5 is functorial in (F, B) motivates us to examine the fibrational properties of such associations $(F, B) \mapsto \mathsf{Mly}_{\mathcal{K}}(F, B)$ and $(F, B) \mapsto \mathsf{Mre}_{\mathcal{K}}(F, B)$. This yields total categories where all dynamics and all outputs can be considered simultaneously and coherently. The entire section takes place under the assumption that $\mathcal{K}$ is locally presentable.

Definition 6. *The* total Mealy 2-category **Mly** *is defined as follows:*

- *the objects are triples* $(\mathcal{K}; F, B)$ *where* $F : \mathcal{K} \to \mathcal{K}$ *is an endofunctor of a category* $\mathcal{K}$*, and* B *an object of* $\mathcal{K}$*;*
- *the morphisms* $(P, \pi, u) : (\mathcal{K}; F, B) \to (\mathcal{H}; G, B')$ *are triples where* $P : \mathcal{K} \to \mathcal{H}$ *is a functor,* $\pi : GP \Rightarrow PF$ *is an* intertwiner *natural transformation between* F *and* G *and* $u : PB \to B'$ *is a morphism;*
- *2-cells* $\gamma : (P, \pi, u) \Rightarrow (Q, \theta, v)$ *consist of natural transformations* $\gamma : P \Rightarrow Q$ *compatible with the intertwiners* π, θ *in the obvious sense, and such that* $v \circ \gamma_B = u$.

From such a domain **Mly**, *sending* $(\mathcal{K}, F, B)$ *to* $\mathsf{Mly}_{\mathcal{K}}(F, B)$ *results in a strict 2-functor* **Mly** $\to$ **Cat** *(***Cat** *is the 2-category of categories, functors, natural transformations).*

It is, however, rarely needed to vary the domain $\mathcal{K}$ of the automata in study (but cf. Remark 18 for an instance of when this 'change of scalars' might be required). A simpler (=lower-dimensional) approach is convenient if we are content with keeping $\mathcal{K}$ fixed.

Definition 7 (The total categories of automata). *Definition 5 entails at once that the correspondence* $(F, B) \mapsto \mathsf{Mly}_{\mathcal{K}}(F, B)$ *is a (pseudo)functor of type* $\mathsf{Mly}_{\mathcal{K}} : \mathsf{Cat}(\mathcal{K}, \mathcal{K})^{\mathrm{op}} \times \mathcal{K} \to \mathsf{Cat}$, *i.e. a pseudo-profunctor* $\mathsf{Cat}(\mathcal{K}, \mathcal{K}) \nrightarrow \mathcal{K}$ *from which we can extract a span*

$$\mathsf{Cat}(\mathcal{K}, \mathcal{K}) \xleftarrow{p} \mathcal{M}ly_{\mathcal{K}} \xrightarrow{q} \mathcal{K} \tag{12}$$

such that p *is a fibration,* q *is an opfibration,* p*-Cartesian lifts are* q*-vertical and* q*-opCartesian lifts are* p*-vertical, whose tip* $\mathcal{M}ly_{\mathcal{K}}$ *we call the* total Mealy category.

Similar considerations allow to construct the total Moore category $\mathcal{M}re_{\mathcal{K}}$ *from the pseudo-profunctor* $(F, B) \mapsto \mathsf{Mre}_{\mathcal{K}}(F, B)$, *and obtain a two-sided fibration* $\mathsf{Cat}(\mathcal{K}, \mathcal{K}) \leftarrow \mathcal{M}re_{\mathcal{K}} \to \mathcal{K}$, *the* total Moore category.

Remark 14. Unwinding the definition, it is easy to establish how reindexings of the total Mealy and Moore fibration act. In particular, given $\alpha : F \Rightarrow G$ a natural transformation between left adjoints $F \dashv R$ and $G \dashv Q$, and a morphism $f : B \to B'$, the reindexing functor $\mathcal{M}ly_{\mathcal{K}}(\alpha, f) : \mathcal{M}ly_{\mathcal{K}}(G, B) \to \mathcal{M}ly_{\mathcal{K}}(F, B')$

preserves all colimits –and thus, in the blanket assumption of presentability of $\mathcal{K}$, is a left adjoint; however, it fails to preserve limits (even terminal objects).[14]

If $\mathcal{K}$ is monoidal its tensor functor $\_ \otimes - : \mathcal{K} \times \mathcal{K} \to \mathcal{K}$ now curries to the 'left regular representation' $\lambda : \mathcal{K} \to \mathsf{Cat}(\mathcal{K}, \mathcal{K}) : A \mapsto A \otimes -$ of $\mathcal{K}$ on itself, and as a consequence, we can pullback the total Mealy fibration and the total Moore fibration to obtain the left leg of the diagram

$$\begin{array}{ccc} \mathcal{Mly}^{\otimes}_{\mathcal{K}} & \longrightarrow & \mathcal{Mly}_{\mathcal{K}} \\ \downarrow & \lrcorner & \downarrow \\ \mathcal{K}^{\mathrm{op}} \times \mathcal{K} & \xrightarrow[\lambda^{\mathrm{op}} \times \mathcal{K}]{} & \mathsf{Cat}(\mathcal{K}, \mathcal{K})^{\mathrm{op}} \times \mathcal{K} \end{array} \tag{13}$$

which gives rise to the *monoidal Mealy* (two-sided) *fibration*

$$\mathcal{K} \xleftarrow{p^{\otimes}} \mathcal{Mly}^{\otimes}_{\mathcal{K}} \xrightarrow{q^{\otimes}} \mathcal{K} \tag{14}$$

(Similar considerations define $\mathcal{Mre}^{\otimes}_{\mathcal{K}}$, but we refrain from doing so for some technical reasons that make $\mathcal{Mly}^{\otimes}_{\mathcal{K}}$ a better-behaved object, cf. [15].) In fact, the terminology is chosen to inspire the fact that we have restricted the total Mealy category to the case where F-actions are monoidal and hint at the following result.

Proposition 6. *The monoidal Mealy fibration is a* monoidal two-sided fibration, *in the sense of [118, 134], and the monoidal product interfiber is given by componentwise tensor product,*

$$(A, B; \|\overset{E}{\underset{d,s}{\longmapsto}}\|) \otimes (A', B', \|\overset{E'}{\underset{d',s'}{\longmapsto}}\|) = (A \otimes A', B \otimes B'; \|\overset{E \otimes E'}{\underset{d \otimes d', s \otimes s'}{\longmapsto}}\|) \tag{15}$$

Theorem 3. ([76], [112, **Def. 1**] **rephrased**). *If $\mathcal{K}$ is* Cartesian *monoidal, the profunctor $\mathcal{K}^{\mathrm{op}} \times \mathcal{K} \to \mathsf{Cat}$ obtained from* (14) *carries the structure of a (pseudo)promonad, and it gives rise to a bicategory $\mathsf{Mly}_{\mathcal{K}}$ whose hom-categories are precisely the $\mathsf{Mly}_{\mathcal{K}}(A, B)$.*

The terminology introduced so far gives us enough leeway to introduce our first main theorem:

Theorem 4. *Let $(\mathcal{K}, \otimes, \partial)$ be a differential 2-rig; then the total categories of the monoidal Mealy and Moore fibrations are themselves differential 2-rigs for a universal choice of derivative functor $\bar{\partial} : \mathcal{Mly}^{\otimes}_{\mathcal{K}} \to \mathcal{Mly}^{\otimes}_{\mathcal{K}}$ such that the projection functors $p^{\otimes}, q^{\otimes}$ in* (14) *are (strict) morphisms of differential 2-rigs.*

---

[14] It is probably interesting to devise under which conditions the canonical map $\mathcal{Mly}_{\mathcal{K}}(\alpha, f)\left(\prod_{n \geq 1} Q^n B\right) \to \prod_{n \geq 1} R^n B'$, is well behaved in some sens (for example, under mild conditions that there exist at least one 'point' in its domain, the map is a split epi).

Corollary 2. *The total category* $\mathcal{M}\ell y^{\otimes}_{\mathsf{Spc}}$ *obtained coupling Definition 7,* (13), *and Proposition 6 is a differential 2-rig such that* $\bar\partial$ *commutes with all colimits that are preserved by* ∂.

Proposition 7. *The category* $\mathcal{M}\ell y^{\otimes}_{\mathsf{Spc}}$ *is locally presentable, so by the special adjoint functor theorem* $\bar\partial$ *has a left adjoint; in fact, more is true:*

- *the* fibration *of* (14) *is accessible (and cocomplete, hence locally presentable) in the sense of [90, 5.3.1], i.e. the total category* $\mathcal{M}\ell y^{\otimes}_{\mathsf{Spc}}$ *is locally presentable, the projection* $\langle p, q\rangle$*, all reindexing functors are accessible, and the pseudo-functor associated to the fibration preserves filtered colimits.*
- *the* $\bar\partial$ *functor is also continuous,* $(\mathcal{M}\ell y^{\otimes}_{\mathsf{Spc}}, \otimes_{\mathrm{Day}}, \bar\partial)$ *is a* scopic *differential 2-rig.*

The following lemma splits the verification that $\mathsf{Mly}_{\mathcal{K}}$, defined in Definition 7, preserves filtered colimits in both components into two parts. The first part is a straightforward consequence of the fact that $\mathsf{Alg}(\_)$ preserves filtered colimits, in the sense that if $\mathcal{J}$ is a λ-filtered category, $\mathsf{Alg}(\mathrm{colim}_{\mathcal{J}} F_i) \cong \lim_{\mathcal{J}} \mathsf{Alg}(F_i)$. The key result allowing us to prove the second part is the fact that R as described in Eq. 6 also preserves filtered colimits, hence for every filtered diagram, one has $T/\mathrm{colim}_{\mathcal{J}} B_i \cong \mathrm{colim}_{\mathcal{J}}(T/B_i)$.

Lemma 2. *For every fixed output object* $B \in \mathcal{K}$*, the functor* $\mathsf{Mly}_{\mathcal{K}}(-, B)$ *preserves filtered colimits. For every fixed dynamics* $F : \mathcal{K} \to \mathcal{K}$*, the functor* $\mathsf{Mly}_{\mathcal{K}}(F, -)$ *preserves filtered colimits.*

4 Differential and Co/monadic Dynamics

Besides usual monoidal automata, which have a distinguished differential flavour by the above remarks, one can exploit the other adjunction $\partial \dashv R$ where ∂ sits and look at categories $\mathsf{Mly}_{\mathsf{Spc}}(\partial, B)$ of *differential automata*, where dynamics are induced by the subsequent derivatives of a state object $E, \partial E, \dots, \partial^n E = E^{(n)}, \dots$.

Then, from every triple of adjoints $L \dashv \partial \dashv R$, 'monad-comonad' and 'comonad-monad' adjunctions $L\partial \dashv R\partial$ and $\partial L \dashv \partial R$ arise. One can then put the categories $\mathsf{Mly}_{\mathsf{Spc}}(L\partial, B)$ and $\mathsf{Mly}_{\mathsf{Spc}}(\partial L, B)$ under the spotlight using the language of Sect. 3.

This is, respectively, what we do in Sect. 5 below after we address the problem in more generality.

We want to study categories $\mathsf{Mly}_{\mathcal{K}}(T, B)$ of $(T \dashv S)$-automata where T is a left adjoint monad, and dually, categories $\mathsf{Mly}_{\mathcal{K}}(Q, B)$ of $(Q \dashv R)$-automata where Q is a left adjoint comonad.

In the case of a left adjoint monad, several technical results can be used to make the description of the categories $\mathsf{Mly}_{\mathcal{K}}(T, B)$ easier:

- [18, 4.3.2] if T is a left adjoint monad, with S as right adjoint comonad, its Eilenberg–Moore category $\mathcal{K}^T$ is cocomplete, with colimits preserved by the forgetful functor; in fact more is true:
- [18, 4.4.6] if T is a left adjoint monad, with S as right adjoint comonad, colimits in $\mathcal{K}^T$ are *created* by U, which in fact is comonadic and $\mathcal{K}^T$ identifies with the category of coEilenberg–Moore S-coalgebras.

The first general observation is completely elementary but already useful: considering that co/monads admit co/unit natural transformations to/from the identity functor, and given the functoriality of $\mathsf{Mly}_{\mathcal{K}}(-, B)$, we get canonical choices of functors

$$\mathsf{Mly}_{\mathcal{K}}(\mathrm{id}_{\mathcal{K}}, B) \longrightarrow \mathsf{Mly}_{\mathcal{K}}(Q, B) \qquad \mathsf{Mly}_{\mathcal{K}}(T, B) \longrightarrow \mathsf{Mly}_{\mathcal{K}}(\mathrm{id}_{\mathcal{K}}, B) \tag{16}$$

One can immediately prove by inspection that

Remark 15. The category $\mathsf{Mly}_{\mathcal{K}}(\mathrm{id}_{\mathcal{K}}, B)$ is the category of coalgebras for the functor $\_ \times B$.

Arguing again by functoriality, the monad structure on the functor T specifying the dynamics yields an augmented simplicial object [41], [129, 8.6]:

$$\mathsf{Mly}_{\mathcal{K}}(\boldsymbol{T}, B)_\bullet = \left(\mathsf{Mly}_{\mathcal{K}}(\mathrm{id}_{\mathcal{K}}, B) \xleftarrow{\eta^*} \mathsf{Mly}_{\mathcal{K}}(T, B) \underset{(T\eta)^*}{\overset{(\eta T)^*}{\substack{\longleftarrow \\ \xrightarrow{\mu^*} \\ \longleftarrow}}} \mathsf{Mly}_{\mathcal{K}}(T^2, B) \substack{\longleftarrow \\ \longrightarrow \\ \longleftarrow \\ \longrightarrow \\ \longleftarrow} \dots \right) \tag{17}$$

obtained feeding the bar resolution of T to the functor $\mathsf{Mly}_{\mathcal{K}}(-, B)$.

Dually, the cobar resolution of a left adjoint comonad Q yields an augmented cosimplicial object

$$\mathsf{Mly}_{\mathcal{K}}(\boldsymbol{Q}, B)_\bullet = \left(\dots \substack{\longleftarrow \\ \longrightarrow \\ \longleftarrow \\ \longrightarrow \\ \longleftarrow} \mathsf{Mly}_{\mathcal{K}}(Q^2, B) \underset{(\epsilon Q)^*}{\overset{(Q\epsilon)^*}{\substack{\longleftarrow \\ \xrightarrow{\sigma^*} \\ \longleftarrow}}} \mathsf{Mly}_{\mathcal{K}}(Q, B) \xleftarrow[\epsilon^*]{} \mathsf{Mly}_{\mathcal{K}}(\mathrm{id}_{\mathcal{K}}, B) \right) \tag{18}$$

Definition 8 (Bar and cobar Mealy complexes). *Both constructions are natural in the output object B, hence the above construction sets up functors $\mathsf{Mly}_{\mathcal{K}}(\boldsymbol{Q})_\bullet : \mathcal{K} \times \Delta \to \mathsf{Cat}$ and $\mathsf{Mly}_{\mathcal{K}}(\boldsymbol{T})_\bullet : \mathcal{K} \times \Delta^{\mathrm{op}} \to \mathsf{Cat}$. We refer to these as the* bar complex *of T-automata and the* cobar complex *of Q-automata.*

Remark 16. Let $\mathcal{K}$ be locally presentable. Given that $\mu^* : \mathsf{Mly}_{\mathcal{K}}(T, B) \to \mathsf{Mly}_{\mathcal{K}}(T^2, B)$ acts by precomposition with μ, sending $\left\|\frac{E}{d,s}\right\|$ to $\left\|\frac{E}{d\mu_E, s\mu_E}\right\|$ a swift application of the adjoint functor theorem yields a right adjoint μ_*.

Remark 17 (On monadic automata). It is reasonable to describe *Eilenberg–Moore*[15] Mealy automata, refining the pullbacks in Definition 5 by using the

[15] The 'Moore' of 'Moore automaton' and the Moore of 'Eilenberg–Moore' are two different people; the notion of 'Eilenberg–Moore Moore automaton' makes perfect sense as a category $\mathsf{Mre}_{\mathcal{K}}(T, B)$ arising as a pullback $\mathcal{K}^T \times_{\mathcal{K}} \mathcal{K}/B$. However, we leave Eilenberg–Moore Moore automata out of this note.

forgetful from $\mathcal{K}^T$ (the Eilenberg–Moore category of T) instead of $\mathsf{Alg}(T)$, and obtaining categories $\mu\mathsf{Mly}_{\mathcal{K}}(T, B)$ and $\mu\mathsf{Mre}_{\mathcal{K}}(T, B)$; in this case, some of the observations listed here carry over:

- $\mu\mathsf{Mly}_{\mathcal{K}}(\mathrm{id}_{\mathcal{K}}, B)$ is just the slice $\mathcal{K}/B$, so the free-forgetful adjunction $F^T : \mathcal{K} \rightleftarrows \mathcal{K}^T : U^T$ induces a 'pulled-back' adjunction $\mu\mathsf{Mly}_{\mathcal{K}}(T, B) \rightleftarrows T/B$.
- Let S, T be monads on $\mathcal{K}$. Whenever a morphism of monads $\lambda : T \Rightarrow S$ in the sense of [8, §6.1] is given, the induced (colimit-preserving) functor $\mathcal{K}^S \to \mathcal{K}^T$ (cf. [*ibi*, Thm. 6.3]) induces in turn a (colimit-preserving) functor $\mu\mathsf{Mly}_{\mathcal{K}}(S, B) \to \mu\mathsf{Mly}_{\mathcal{K}}(T, B)$.

Remark 18. Working in the more restrictive case of Eilenberg–Moore automata is, however, rather unrewarding for a variety of reasons: first of all, there is the trivial remark that as soon as a carrier E has a structure $a : TE \to E$ of T-algebra, its 'dynamics' is pretty trivial, as a must be a split epi with a privileged right inverse η_E; thus, the composition $s \circ \eta_E$ 'knows everything' about the evolution of $\left\|\frac{E}{d,s}\right\|$. Second, the conditions for a natural transformation to induce functors between Eilenberg–Moore categories are fairly more imposing, and third, the morphisms inducing an analogue of (17),(18) are simply not available.

Something can be said, however, if we work 'interfiber' using Definition 6. A monad morphism in the sense of [117] induces a monad $\hat{S}$ on $\mathcal{K}^T$ so that the forgetful $U^T : \mathcal{K}^T \to \mathcal{K}$ is an intertwiner, hence leveraging on Definition 6 we can induce a functor

$$\mathsf{Mly}_{\mathcal{K}^T}(\hat{S}, (B, b)) \longrightarrow \mathsf{Mly}_{\mathcal{K}}(S, B). \tag{19}$$

Dually, one can try to render the free functor $F_T : \mathcal{K} \to \mathcal{K}_T$ into the Kleisli category of T strong monoidal for a monoidal structure on $\mathcal{K}_T$; this will yield functors $\mathsf{Mly}_{\mathcal{K}}(S, B) \to \mathsf{Mly}_{\mathcal{K}_T}(\check{S}, F_T B)$. The matter is investigated in the second part of [51] when $F = A \otimes \_$. For example, consider $\mathcal{K}$ monoidal and with countable sums preserved by the tensor; then, every oplax monoidal monad $T : \mathcal{K} \to \mathcal{K}$ lifts a monoidal structure on $\mathcal{K}_T$ and one can then consider $\mathcal{K}_T$-valued F-machines, cf. [51, Prop. 30].

Remark 19 (On the proper choice of output objects). The construction of Definition 5 depends not only on F, but also on an output object B, usually thought as a 'space of responses' the machine $\left\|\frac{E}{d,s}\right\|$ can give as output. The choice of what B best models a given problem has to be made each time according to the nature of the problem itself. However, one is almost always led to consider choices of B that are 'spaces of truth values', like a Heyting or Boole algebra, or spaces of probabilities, like the closed unit interval $[0, 1]$. The co/completeness of $\mathsf{Mly}_{\mathcal{K}}(F, B)$ and $\mathsf{Mre}_{\mathcal{K}}(F, B)$ established in Remark 9 entails that all algebraic structures (=all essentially algebraic theories) can be interpreted in such categories, and the nature of Spc as a presheaf topos entails that the construction of an object of internal real numbers is more or less straightforward. In particular,

- Hadamard Heyting/Boole algebra objects are just species $B : \mathsf{P} \to \mathsf{Set}$ which factor through the subcategory Heyt or Bool, the simplest case being the constant species $\mathfrak{B}$ at the booleans $\mathbf{B} = \{0 < 1\}$, with trivial action of each S_n ($\mathfrak{B}$ is the subobject classifier of Spc; another example of a Boolean algebra object in Spc is the species $\wp$ of subsets of Example 1.ES2);
- regarding Spc as a presheaf topos, it is easy to determine that the NNO, the object of integers, and of rationals, and of internal Dedekind reals [89, §VI.1] can be constructed as constant functors $c_{\mathbb{N}}, c_{\mathbb{Z}}, c_{\mathbb{Q}}, c_{\mathbb{R}}$ at natural, integers, rationals and reals in Set.

5 $L\partial$- and ∂L-Algebras, the Fourfold Way

Remark 20 (On the structure of $L\partial$ and ∂L). Rajan [110] provides explicit formulas for the monad and comonad associated to $L \dashv \partial \dashv R$. Let $\mathfrak{F} : \mathsf{P} \to \mathsf{Set}$ be a species. Then,

- $L\partial\mathfrak{F}$ acts as $y[1] \otimes_{\mathrm{Day}} \partial\mathfrak{F}$; a structure of type $L\partial\mathfrak{F}$ on a finite set A chooses a point of A, and an $\mathfrak{F}$-structure on the complement of that point.
- $R\partial\mathfrak{F}$ acts as $A \mapsto \prod_{a \in A} \mathfrak{F}[(A \setminus \{a\}) \sqcup \{\bullet\}]$, i.e. as $A \mapsto \mathfrak{F}A^A$; a structure of type $R\partial\mathfrak{F}$ on a finite set A chooses an $\mathfrak{F}$-structure on A for every $a \in A$. With a similar reasoning,
- $\partial L\mathfrak{F} = \partial(y[1] \otimes_{\mathrm{Day}} \mathfrak{F})$ is the functor $\mathfrak{F} + L\partial\mathfrak{F}$;[16] Note in particular that the unit of the monad ∂L is the first coproduct injection.
- $\partial R\mathfrak{F}$ acts as $A \mapsto \mathfrak{F}[A]^A \times \mathfrak{F}[A] = R\partial\mathfrak{F}[A] \times \mathfrak{F}[A]$.

Both the following claims follow at once from the definition (and leave the question of when a generic scopic 2-rig admits a Euler derivation open).[17]

Remark 21 (The Euler derivation on Spc). The functor $L\partial = y[1] \otimes_{\mathrm{Day}} \partial$ is a derivation in the sense of [86], and furthermore a left adjoint (with right adjoint $R\partial$), hence a colimit-preserving derivative functor.

Armed with these explicit descriptions, we can attempt to unveil the structure of the categories $\mathsf{Alg}(L\partial)$, $\mathsf{Alg}(\partial L)$, as building blocks for the category $\mathsf{Mly}_{\mathsf{Spc}}(L\partial, B)$, $\mathsf{Mly}_{\mathsf{Spc}}(\partial L, B)$. A thorough analysis of co/algebra structures for

[16] This gives rise to the evocative formula: $[\partial, L] = \partial L - L\partial = 1$, i.e. to the canonical commutation relation between position and momentum (up to a sign); in the language of *virtual species* [69,70,132,133] and [11, §2.5] such an equation can be made completely formal. As for its meaning, *hanc marginis exiguitas non caperet*, but see Problem 1 below.

[17] The differential operator $\Upsilon = \sum_{i=1}^{n} x_i \frac{\partial}{\partial x_i}$ in $\mathbb{R}^n$ is called 'Euler homogeneity operator', cf. [35, p. 296]; another name for the same operation, 'numbering derivation', comes from Physics where if X^n represents something like a state of n bosons, like photons in a laser, then the differential operator $X \cdot D$ takes X^n to nX^n, where the coefficient 'counts' or 'numbers' the of bosons.

such interesting endofunctors of Spc seems to be missing from the existing literature. Rajan [110] goes as close as determining in painstaking detail the monad and comonad structures on $\partial L, \partial R, L\partial, R\partial$, but doesn't seem to provide a characterization for their endofunctor or Eilenberg–Moore algebras, or even for the (much easier, and somewhat more inspiring) bare endofunctor algebras. As one would expect from the adjunction relations $L\partial \dashv R\partial$ and $\partial L \dashv \partial R$ the structures of $L\partial$-algebras (=$R\partial$-coalgebras) and ∂L-algebras (=∂R-coalgebras) are tightly related. The following computations all follow a general argument, given Remark 20 a ∂L-algebra structure on a species F consists of a pair $\left[\begin{smallmatrix} u \\ v \end{smallmatrix}\right] : F + L\partial F \to F$ of maps $u : F \to F$ and $v : \partial F \to \partial F$ of endomorphisms, one for F and one for ∂F.

Example 6. A ∂L-algebra structure on the exponential species E reduces to a pair $u : E \to E$ and $v : LE \to E$, which in turn reduces to another endomap of E, given how E is a Napier object. Then, ∂L-algebra structures on E are representations of the free monoid $\mathbb{N}\langle d, c\rangle$ (cf. [53–55]) on 2 generators d, c over the set $E[1]$ (because endomaps of E are in bijection with elements of $E[1]$, by Yoneda). For set species, this must be trivial, for linear species this amounts to a 'character' for the monoid representation $\mathbb{N}\langle d, c\rangle$.

Example 7. For the species $\mathfrak{L}$ of linear orders, a ∂L-algebra map is a map $L \otimes L \to L$, since

$$\mathfrak{L} + L\partial(\mathfrak{L}) = \mathfrak{L} + y[1] \otimes_{\text{Day}} \mathfrak{L} \otimes_{\text{Day}} \mathfrak{L} \tag{20}$$

but then $\mathfrak{L} + y[1] \otimes_{\text{Day}} \mathfrak{L} \otimes_{\text{Day}} \mathfrak{L} = \mathfrak{L} \otimes_{\text{Day}} (1 + y[1] \otimes_{\text{Day}} \mathfrak{L})$, and the fact that $1 + y[1] \otimes_{\text{Day}} \mathfrak{L} \cong \mathfrak{L}$ is exactly the universal property satisfied by $\mathfrak{L}$ as initial algebra of $1 + y[1] \otimes_{\text{Day}} -$.

Example 8. A similar line of reasoning leads to the characterization of ∂L-algebra structures on the species of cycles, Example 1.ES5: since $\partial\mathfrak{Cyc} \cong \mathfrak{L}$, structures of ∂L-algebras are pairs, $\mathfrak{Cyc} \to \mathfrak{Cyc}$ and $\mathfrak{L} \to \mathfrak{L}$ of endomorphisms.

Example 9. For the species $\mathfrak{S}$ of permutations of Example 1, a ∂L-algebra structure consists of a pair $\left[\begin{smallmatrix} u \\ v \end{smallmatrix}\right] : \mathfrak{S} + L\partial\mathfrak{S} \to \mathfrak{S}$, where v can in turn be simplified into $\mathfrak{S} \otimes_{\text{Day}} (1 + y[1] \otimes_{\text{Day}} \mathfrak{L}) \cong \mathfrak{S} \otimes_{\text{Day}} \mathfrak{L}$ using Example 5.

6 Conclusions and Future Work

Problem 1. Let $\mathbf{K}$ be a strict 2-category with all finite weighted limits. Consider objects $X, B \in \mathbf{K}$ in a diagram of the following form:

$$X \xrightarrow[\mathrm{id}_X]{} X \xleftarrow[f]{} X \xrightarrow[f]{} X \xleftarrow[b]{} B \tag{21}$$

The *Vaucanson limit* [59][18] obtained from (21) consists of the limit obtained (cf. [28, 78])

[18] Jacques de Vaucanson (∗1709–†1782) was, besides the inventor of the modern lathe and of automatic loom, the creator of sophisticated and almost lifelike mechanical toys such as the '*flûteur automate*' and the '*canard défécateur*'. The mechanical duck appeared to have the ability to eat kernels of grain, and to metabolize and defecate them.

- the *inserter* $X \xleftarrow{u} \mathcal{I}(f, \mathrm{id}_X) \xrightarrow{u} X$ of the left cospan;
- the *comma object* $X \xleftarrow{v} f/b \xrightarrow{q} B$ of the right cospan;
- the strict pullback $\mathcal{I}(f, \mathrm{id}_X) \times_X (f/b)$ of u, v.

If $\mathbf{K}$ is the 2-category of categories, functors, and natural transformations, Vaucanson limits recover the categories $\mathsf{Mly}_{\mathcal{K}}(A, B)$ when $B = 1$ is the terminal category and b is an object therein.

Formal theory of Mealy automata is then the study of Vaucanson objects in $\mathbf{K}$. One can define analogues for $\mathsf{Mly}_{\mathcal{K}}(A, B)$, $\mathsf{Mre}_{\mathcal{K}}(A, B)$ enriched over a generic monoidal base $\mathcal{W}$ in the sense of [18, Ch. 6], [79], for example a quantale [26,113] like $[0, \infty]^{\mathrm{op}}$, so that there is a *metric space* $\mathsf{Mly}_{(X,d)}(f, b)$ [21,61,81] associated to every nonexpansive map $f : X \to X$ and point $b \in X$. This begs the question: what is this theory, and how can it profit from being studied via discrete dynamical methods? Can it be related with fixpoint theory as classically intended in [46]?

Problem 2. The canonical commutation $[\partial, L] = \partial L - L\partial = 1$ valid in Joyal's virtual species suggests how L acts as a 'conjugate operator' to ∂. Compare this with the analogue relation $[x \cdot \_, \frac{d}{dx}] = 1$ valid in the ring $C^{\omega}(\mathbb{R})$ of analytic functions on, say, the real line [27, Ch. 5], [29]. Is it the case that there is a still undiscovered 'categorified Greenfunctionology' introducing a 'Heaviside distribution' Θ with the property that the colimit of F weighted by Θ is a solution of the differential equation $\partial G = F$ on species, i.e. $\partial\left(\int^X \Theta(X, \_) \times F[X]\right) \cong F$? Compare this with the well-known integral equation $\frac{d}{dx}\left(\int \Theta(x-t) f(t) dt\right) = f(x)$ for the Heaviside function, and cf. [23] where Day sketched a categorified theory of Fourier transforms (upper and lower transforms, Parseval relations, etc.) for categories enriched over a $*$-autonomous base $\mathcal{V}$ [9], generalizing Joyal's categories of analytic functors. We intend to pursue the matter, captivated by its compelling aesthetic beauty.

References

1. Adámek, J.: Free algebras and automata realizations in the language of categories. Commentationes Mathematicae Universitatis Carolinae **015**(4), 589–602 (1974). http://eudml.org/doc/16649
2. Adámek, J., Velebil, J.: Analytic functors and weak pullbacks. Theory Appl. Categ. **21**(11), 191–209 (2008)
3. Adámek, J., Trnková, V.: Automata and Algebras in Categories. Kluwer, Boston (1990)
4. Aguiar, M., Mahajan, S.: Monoidal Functors, Species and Hopf Algebras. CRM Monograph Series. American Mathematical Society, Providence (2010)
5. Arbib, M.A., Manes, E.G.: A Categorist's view of automata and systems. In: Manes, E.G. (ed.) Category Theory Applied to Computation and Control. LNCS, vol. 25, pp. 51–64. Springer, Heidelberg (1975). https://doi.org/10.1007/3-540-07142-3_61

6. Bainbridge, E.S.: Addressed machines and duality. In: Manes, E.G. (ed.) Category Theory Applied to Computation and Control. LNCS, vol. 25, pp. 93–98. Springer, Heidelberg (1975). https://doi.org/10.1007/3-540-07142-3_66
7. Bainbridge, E.: A unified minimal realization theory, with duality, for machines in a hyperdoctrine. Technical report. Computer and Communication Sciences Department, University of Michigan
8. Barr, M., Wells, C.: Toposes, Triples and Theories. Springer, New York (1985). http://www.tac.mta.ca/tac/reprints/articles/12/tr12abs.html
9. Barr, M.: ∗-Autonomous Categories. Springer, Heidelberg (1979). https://doi.org/10.1007/bfb0064579
10. Bergeron, F.: Une combinatoire du pléthysme **46**(2), 291–305. https://doi.org/10.1016/0097-3165(87)90007-0
11. Bergeron, F., Labelle, G., Leroux, P.: Combinatorial Species and Tree-Like Structures. No. 67 in Encyclopedia of Mathematics and Its Applications. Cambridge University Press, Cambridge (1998). https://doi.org/10.1017/CBO9781107325913
12. Bergeron, F., Reutenauer, C.: Combinatorial resolution of systems of differential equations III: a special class of differentially algebraic series. Eur. J. Comb. **11**, 501–512 (1990). https://doi.org/10.1016/S0195-6698(13)80035-2
13. Betti, R., Kasangian, S.: A quasi-universal realization of automata. Università degli Studi di Trieste. Dipartimento di Scienze Matematiche
14. Betti, R., Kasangian, S.: Una proprietà del comportamento degli automi completi. Università degli Studi di Trieste. Dipartimento di Scienze Matematiche
15. Boccali, G., Femić, B., Laretto, A., Loregian, F., Luneia, S.: The semibicategory of Moore automata (2023)
16. Boccali, G., Laretto, A., Loregian, F., Luneia, S.: Bicategories of automata, automata in bicategories. In: Staton, S., Vasilakopoulou, C. (eds.) Proceedings of the Sixth International Conference on Applied Category Theory 2023, University of Maryland, 31 July - 4 August 2023. Electronic Proceedings in Theoretical Computer Science, vol. 397, pp. 1–19. Open Publishing Association (2023). https://doi.org/10.4204/EPTCS.397.1
17. Bonetti, F., Rota, G., Senato, D., Venezia, A.M.: On the foundation of combinatorial theory X. A categorical setting for symmetric functions. Stud. Appl. Math. **86**(1), 1–29 (1992). https://doi.org/10.1002/sapm19928611
18. Borceux, F.: Handbook of Categorical Algebra 2. (Categories and Structures), Encyclopedia of Mathematics and Its Applications, vol. 51. Cambridge University Press, Cambridge (1994). https://doi.org/10.1017/cbo9780511525858
19. Cayley, A.: A Theorem on Trees, vol. 23, pp. 376–378. Cambridge University Press (1889). https://doi.org/10.1017/CBO9780511703799.010
20. Chen, W.: The theory of compositionals. Discret. Math. **122**(1–3), 59–87 (1993). https://doi.org/10.1016/0012-365x(93)90287-4
21. Clementino, M.M., Tholen, W.: Metric, topology and multicategory: a common approach **179**(1), 13–47. http://www.math.yorku.ca/Who/Faculty/Tholen/MatV.pdf
22. Curien, P.: Operads, clones, and distributive laws. In: Operads and Universal Algebra, pp. 25–49. World Scientific (2012). https://doi.org/10.1142/9789814365123_0002
23. Day, B.J.: Monoidal functor categories and graphic Fourier transforms. Theory Appl. Categ. **25**(5), 118–141 (2011)
24. Doberkat, E.E.: Stochastic Relations: Foundations for Markov Transition Systems. Chapman and Hall/CRC (2007). https://doi.org/10.1201/9781584889427

25. Ehrig, H., Kiermeier, K.D., Kreowski, H.J., Kühnel, W.: Universal theory of automata. A categorical approach. https://doi.org/10.1007/978-3-322-96644-5
26. Eklund, P., Garciá, J.G., Höhle, U., Kortelainen, J.: Semigroups in Complete Lattices. Springer, Cham (2018). https://doi.org/10.1007/978-3-319-78948-4
27. Eyges, L.: The Classical Electromagnetic Field. Dover Books on Physics. Dover Publications, Mineola (1980)
28. Fiore, T.M.: Pseudo limits, biadjoints, and pseudo algebras: categorical foundations of conformal field theory. Mem. Amer. Math. Soc. **182**(860), x+171 (2006)
29. Folland, G.B.: Fourier Analysis and Its Applications, vol. 4. American Mathematical Society (2009)
30. Fresse, B.: Modules Over Operads and Functors. Lecture Notes in Mathematics, vol. 1967. Springer, Heidelberg (2009). https://doi.org/10.1007/978-3-540-89056-0
31. Fritz, T.: Convex spaces I: definition and examples (2015)
32. Gambino, N., Joyal, A.: On operads, bimodules and analytic functors (2017). https://doi.org/10.1090/memo/1184
33. Gambino, N., Garner, R., Vasilakopoulou, C.: Monoidal Kleisli bicategories and the arithmetic product of coloured symmetric sequences (2022)
34. Garner, R., Franco, I.L.: Commutativity. J. Pure Appl. Algebra **220**(5), 1707–1751 (2016). https://doi.org/10.1016/j.jpaa.2015.09.003
35. Gel'fand, I.M., Shilov, G.E.: Generalized Functions, vol. I. Academic Press Inc, New York (1968)
36. Getzler, E., Kapranov, M.M.: Modular operads. Compos. Math. **110**(1), 65–125 (1998). https://doi.org/10.1023/a:1000245600345
37. Getzler, E.: Operads revisited. In: Algebra, Arithmetic, and Geometry, Progr. Math., vol. 269, pp. 675–698. Birkhäuser Boston, Inc., Boston (2009). https://doi.org/10.1007/978-0-8176-4745-2_16. http://dx.doi.org/10.1007/978-0-8176-4745-2_16
38. Getzler, E., Jones, J.: Operads, homotopy algebra and iterated integrals for double loop spaces. arXiv preprint hep-th/9403055, pp. 1–70 (1994)
39. Glabbeek, R.J.: Petri Nets, configuration structures and higher dimensional automata. In: Baeten, J.C.M., Mauw, S. (eds.) CONCUR 1999. LNCS, vol. 1664, pp. 21–27. Springer, Heidelberg (1999). https://doi.org/10.1007/3-540-48320-9_3
40. Glendinning, P.: Stability, Instability and Chaos. An Introduction to the Theory of Nonlinear Differential Equations. Cambridge Texts in Applied Mathematics. CUP (1994). https://doi.org/10.1017/CBO9780511626296
41. Goerss, P., Jardine, J.: Simplicial Homotopy Theory, Modern Birkhäuser Classics, vol. 174. Birkhäuser Verlag, Basel (2009). https://doi.org/10.1007/978-3-0346-0189-4, reprint of the 1999 edition [MR1711612]
42. Goguen, J.: Minimal realization of machines in closed categories **78**, 777–783. https://doi.org/10.1090/S0002-9904-1972-13032-5
43. Goguen, J.: Discrete-time machines in closed monoidal categories. I **10**(1), 1–43. https://doi.org/10.1016/s0022-0000(75)80012-2
44. Goguen, J.: Realisation is universal. Math. Syst. Theory **6**(4). https://doi.org/10.1007/bf01843493
45. Goguen, J.A., Thatcher, J.W., Wagner, E.G., Wright, J.B.: Factorizations, congruences, and the decomposition of automata and systems. In: Blikle, A. (ed.) MFCS 1974. LNCS, vol. 28, pp. 33–45. Springer, Heidelberg (1975). https://doi.org/10.1007/3-540-07162-8_665

46. Granas, A., Dugundji, J.: Fixed Point Theory. Springer Monographs in Mathematics, 2003 edn. Springer, New York (2003). https://doi.org/10.1007/978-0-387-21593-8
47. Gray, J.W.: Formal Category Theory: Adjointness for 2-Categories. Springer, Heidelberg (1974). https://doi.org/10.1007/bfb0061280
48. Guitart, R.: Remarques sur les machines et les structures. Cahiers Topologie Géom. Différentielle Catég. **15**, 113–144 (1974)
49. Guitart, R.: Monades involutives complémentées. Cahiers de topologie et géométrie différentielle **16**(1), 17–101 (1975). http://www.numdam.org/item/CTGDC_1975__16_1_17_0/
50. Guitart, R.: Des machines aux bimodules, April 1978. http://rene.guitart.pagesperso-orange.fr/textespublications/rg30.pdf. Univ. Paris 7
51. Guitart, R.: Tenseurs et machines. Cahiers de topologie et géométrie différentielle **21**(1), 5–62 (1980). http://www.numdam.org/item/CTGDC_1980__21_1_5_0/
52. Guitart, R.: Qu'est-ce que la logique dans une catégorie? Cahiers Topologie Géom. Différentielle **23**(2), 115–148 (1982)
53. Guitart, R.: Autocategories: I. A common setting for knots and 2-categories. Cahiers Top. Geo. Diff. Cat. **LV-1**, 66–80 (2014)
54. Guitart, R.: Autocategories: II. Autographic algebras. Cahiers Top. Geo. Diff. Cat. **LV-2**, 151–160 (2014)
55. Guitart, R.: Autocategories: III. Representations, and expansions of previous examples. Cahiers Top. Geo. Diff. Cat. **LVIII-1**, 67–80 (2017)
56. Guitart, R., Van den Bril, L.: Décompositions et Lax-complétions **18**(4), 333–407. http://www.numdam.org/item/CTGDC_1977__18_4_333_0/
57. Hasegawa, R.: Two applications of analytic functors. Theoret. Comput. Sci. **272**(1–2), 113–175 (2002). https://doi.org/10.1016/s0304-3975(00)00349-2
58. Hermida, C., Mateus, P.: Paracategories II: adjunctions, fibrations and examples from probabilistic automata theory **311**(1-3), 71–103. https://doi.org/10.1016/S0304-3975(03)00317-7
59. Heudin, J.C.: Les Créatures artificielles: Des automates aux mondes virtuels. Odile Jacob, Paris (2008)
60. Hirata, K.: Notes on lax ends (2022). arXiv:2210.01522 preprint
61. Hofmann, D., Seal, G., Tholen, W.: Monoidal Topology: A Categorical Approach to Order, Metric, and Topology, vol. 153. Cambridge University Press (2014). https://doi.org/10.1017/CBO9781107517288
62. Im, G., Kelly, G.: A universal property of the convolution monoidal structure. J. Pure Appl. Algebra **43**(1), 75–88 (1986). http://dx.doi.org/10.1016/0022-4049(86)90005-8
63. Jacobs, B.: A bialgebraic review of deterministic automata, regular expressions and languages. In: Futatsugi, K., Jouannaud, JP., Meseguer, J. (eds.) Algebra, Meaning, and Computation, pp. 375–404. Springer, Heidelberg (2006). https://doi.org/10.1007/11780274_20
64. Jacobs, B.: From probability monads to commutative effectuses **94**, 200–237
65. Jacobs, B.: Introduction to Coalgebra: Towards Mathematics of States and Observation. Cambridge Tracts in Theoretical Computer Science. Cambridge University Press. https://doi.org/10.1017/CBO9781316823187
66. Jacobs, B.: Convexity, Duality and Effects. In: Calude, C.S., Sassone, V. (eds.) Theoretical Computer Science, vol. 323, pp. 1–19. Springer, Heidelberg. https://doi.org/10.1007/978-3-642-15240-5_1
67. Jay, C.: Local adjunctions. J. Pure Appl. Algebra **53**(3), 227–238 (1988). https://doi.org/10.1016/0022-4049(88)90124-7

68. Joni, S.A., Rota, G.C.: Coalgebras and bialgebras in combinatorics. Stud. Appl. Math. (1979). https://doi.org/10.1002/sapm197961293
69. Joyal, A.: Calcul intégral combinatoire et homologie du groupe symétrique. Comptes rendus mathématiques de l'Académie des sciences, La société royale du Canada **7**, 337–342 (1985)
70. Joyal, A.: Règle des signes en algèbre combinatoire. CR Math. Rep. Acad. Sci. Canada **7**(5), 285–290 (1985)
71. Joyal, A.: Foncteurs analytiques et espèces de structures. In: Labelle, G., Leroux, P. (eds.) Combinatoire énumérative. LNM, vol. 1234, pp. 126–159. Springer, Heidelberg (1986). https://doi.org/10.1007/BFb0072514. http://dx.doi.org/10.1007/BFb0072514
72. Joyal, A.: Une théorie combinatoire des séries formelles. Adv. Math. **42**(1), 1–82 (1981). https://doi.org/10.1016/0001-8708(81)90052-9. https://www.sciencedirect.com/science/article/pii/0001870881900529
73. Kasangian, S., Kelly, G., Rossi, F.: Cofibrations and the realization of non-deterministic automata. Cahiers Topologie Géom. Différentielle Catég. **24**(1), 23–46 (1983)
74. Kasangian, S., Rosebrugh, R.: Glueing enriched modules and composition of automata **31**(4), 283–290
75. Katis, P., Sabadini, N., Walters, R.: Bicategories of processes. J. Pure Appl. Algebra **115**(2), 141–178 (1997). https://doi.org/10.1016/s0022-4049(96)00012-6
76. Katis, P., Sabadini, N., Walters, R.: Feedback, trace and fixed-point semantics. RAIRO - Theoret. Inform. Appl. **36**(2), 181–194 (2010). http://eudml.org/doc/92696
77. Kelly, G.M.: On the operads of. J. P. May. Repr. Theory Appl. Categ. **13**(13), 1–13 (2005)
78. Kelly, G.: Elementary observations on 2-categorical limits. Bull. Aust. Math. Soc. **39**, 301–317 (1989). https://doi.org/10.1017/s0004972700002781
79. Kelly, G.: Basic concepts of enriched category theory. Repr. Theory Appl. Categ. **64**(10), vi+137 (2005)
80. Kock, A.: Monads for which structures are adjoint to units. J. Pure Appl. Algebra **104**(1), 41–59 (1995). https://doi.org/10.1016/0022-4049(94)00111-U
81. Lawvere, F.W.: Metric spaces, generalized logic, and closed categories. Rendiconti del Seminario Matematico e Fisico di Milano **43**(1), 135–166. https://doi.org/10.1007/BF02924844
82. Leroux, P., Viennot, G.X.: Combinatorial resolution of systems of differential equations, I. Ordinary differential equations. In: Labelle, G., Leroux, P. (eds.) Combinatoire énumérative. LNM, vol. 1234, pp. 210–245. Springer, Heidelberg (1986). https://doi.org/10.1007/BFb0072518
83. Leroux, P., Viennot, G.X.: Combinatorial resolution of systems of differential equations IV: separation of variables. Discret. Math. **72**(1), 237–250 (1988). https://doi.org/10.1016/0012-365X(88)90213-0. https://www.sciencedirect.com/science/article/pii/0012365X88902130
84. Loday, J.L., Vallette, B.: Algebraic operads, Grundlehren der Math. Wissenschaften, vol. 346. Springer, Heidelberg (2012). https://doi.org/10.1007/978-3-642-30362-3
85. Loregian, F.: Coend Calculus. London Mathematical Society Lecture Note Series, 1st edn., vol. 468. Cambridge University Press, Cambridge (2021). ISBN 9781108746120

86. Loregian, F., Trimble, T.: Differential 2-rigs. Electron. Proc. Theoret. Comput. Sci. **380**, 159–182 (2023). https://doi.org/10.4204/eptcs.380.10. http://dx.doi.org/10.4204/EPTCS.380.10
87. Lurie, J.: Higher algebra. http://www.math.harvard.edu/~lurie/ (online version 18 September 2017). http://www.math.harvard.edu/~lurie/papers/higheralgebra.pdf, preprint
88. Mac Lane, S.: Categories for the Working Mathematician, Graduate Texts in Mathematics, 2nd edn., vol. 5. Springer, New York (1998). https://doi.org/10.1007/978-1-4757-4721-8. http://link.springer.com/10.1007/978-1-4757-4721-8
89. Mac Lane, S., Moerdijk, I.: Sheaves in Geometry and Logic: A First Introduction to Topos Theory, Universitext, vol. 13. Springer, New York (1992). https://doi.org/10.1007/978-1-4612-0927-0
90. Makkai, M., Paré, R.: Accessible Categories: The Foundations of Categorical Model Theory, Contemporary Mathematics, vol. 104. American Mathematical Society, Providence (1989). https://doi.org/10.1090/conm/104
91. Markl, M., Shnider, S., Stasheff, J.: Operads in Algebra, Topology and Physics, Mathematical Surveys and Monographs, vol. 96. American Mathematical Society, Providence (2002)
92. Mateus, P., Sernadas, A., Sernadas, C.: Realization of probabilistic automata: categorial approach. https://doi.org/10.1007/978-3-540-44616-3_14
93. May, J.: The Geometry of Iterated Loop Spaces. Lectures Notes in Mathematics, vol. 271. Springer, New York (1972). https://doi.org/10.1007/bfb0067491
94. May, J.: Operads, algebras and modules. In: Operads: Proceedings of Renaissance Conferences (Hartford, CT/Luminy, 1995), Contemp. Math., vol. 202, pp. 15–31. Amer. Math. Soc., Providence (1997). https://doi.org/10.1090/conm/202/02588. http://dx.doi.org/10.1090/conm/202/02588
95. Méndez, M., Nava, O.: Colored species, c-monoids, and plethysm, I. J. Comb. Theory Ser. A. **64**(1), 102–129 (1993)
96. Méndez, M., Yang, J.: Möbius species. Adv. Math. **85**(1), 83–128 (1991)
97. Menni, M.: Combinatorial functional and differential equations applied to differential posets. Discret. Math. **308**(10), 1864–1888 (2008). https://doi.org/10.1016/j.disc.2007.04.035
98. Meseguer, J., Sols, I.: Automata in semimodule categories. In: Manes, E.G. (ed.) Category Theory Applied to Computation and Control. LNCS, vol. 25, pp. 193–198. Springer, Heidelberg (1975). https://doi.org/10.1007/3-540-07142-3_81
99. Méndez, M., Nava, O.: Colored species, c-monoids, and plethysm, I. J. Comb. Theory Ser. A (1993). https://doi.org/10.1016/0097-3165(93)90090-U
100. Méndez, M.: Species on digraphs. Adv. Math. **123**(2), 243–275 (1996). https://doi.org/10.1006/aima.1996.0073. http://dx.doi.org/10.1006/AIMA.1996.0073
101. Naudé, G.: On the adjoint situations between behaviour and realization **2**, 245–267. https://doi.org/10.1080/16073606.1977.9632546
102. Naudé, G.: Universal realization **19**(3), 277–289. https://doi.org/10.1016/0022-0000(79)90005-9
103. Nava, O., Rota, G.C.: Plethysm, categories, and combinatorics. Adv. Math. (1985). https://doi.org/10.1016/0001-8708(85)90049-0
104. Nava, O.: On the combinatorics of plethysm. J. Comb. Theory Ser. A **46**, 212–251 (1987). https://doi.org/10.1016/0097-3165(87)90004-5. https://www.semanticscholar.org/paper/ddbd4291a067388ac9ba6b092f4e901d9fd249f9
105. Nava, O., Rota, G.: Plethysm, categories, and combinatorics. Adv. Math. **58**, 61–88 (1985). https://doi.org/10.1016/0001-8708(85)90049-0

106. Obradović, J.: Cyclic operads: syntactic, algebraic and categorified aspects. Ph.D. thesis, École doctorale Sciences mathématiques de Paris centre (2017). http://www.theses.fr/2017USPCC191
107. Paré, R.: Mealy morphisms of enriched categories. Appl. Categ. Struct. **20**(3), 251–273 (2010). https://doi.org/10.1007/s10485-010-9238-8
108. Pitts, A.M.: Nominal Sets: Names and Symmetry in Computer Science. Cambridge University Press (2013). https://doi.org/10.1017/cbo9781139084673
109. Pohl, I., Arbib, M.: Theories of abstract automata **24**(111), 760. https://doi.org/10.2307/2004866
110. Rajan, D.S.: The adjoints to the derivative functor on species. J. Comb. Theory Ser. A. **62**(1), 93–106 (1993). https://doi.org/10.1016/0097-3165(93)90073-H
111. Rajan, D.S.: The equations $D^k Y = X^n$ in combinatorial species. Discret. Math. **118**(1), 197–206 (1993). https://doi.org/10.1016/0012-365X(93)90061-W
112. Rosebrugh, R., Sabadini, N., Walters, R.: Minimal realization in bicategories of automata. Math. Struct. Comput. Sci. **8**(2), 93–116 (1998). https://doi.org/10.1017/S0960129597002454
113. Rosenthal, K.: Quantales and Their Applications. No. 234 in Pitman Research Notes in Mathematics Series. Longman Scientific & Technical, Harlow (1990)
114. Rota, G.C.: On the Foundations of Combinatorial Theory I. Theory of Möbius Functions. Probability Theory and Related Fields (1964). https://doi.org/10.1007/bf00531932
115. Senato, D., Venezia, A., Yang, J.: Möbius polynomial species. Discret. Math. **173**(1–3), 229–256 (1997). https://doi.org/10.1016/s0012-365x(96)00133-1
116. Stanley, R.: Enumerative Combinatorics. Cambridge University Press (2023). https://doi.org/10.1017/9781009262538
117. Street, R.: The formal theory of monads. J. Pure Appl. Algebra **2**(2), 149–168 (1972). https://doi.org/10.1016/0022-4049(72)90019-9
118. Street, R.: Fibrations and Yoneda lemma in a 2-category. In: Kelly, G. (ed.) Proceedings Sydney Category Theory Seminar 1972/1973. Lecture Notes in Mathematics, vol. 420, pp. 104–133. Springer, Heidelberg (1974). https://doi.org/10.1007/BFb0063096
119. Street, R.: Fibrations in bicategories. Cahiers Topologie Géom. Différentielle Catég. **21**(2), 111–160 (1980)
120. Street, R., Walters, R.: Yoneda structures on 2-categories. J. Algebra **50**(2), 350–379 (1978). https://doi.org/10.1016/0021-8693(78)90160-6
121. Strom, J.: Modern Classical Homotopy Theory, vol. 127. American Mathematical Society, Providence (2011). https://doi.org/10.1090/gsm/127
122. Szabo, M.: Bicartesian closed categories. In: Algebra of Proofs, Studies in Logic and the Foundations of Mathematics, vol. 88, chap. 10, pp. 145–162. Elsevier (1978)
123. Tierney, M.: Categorical Constructions in Stable Homotopy Theory. A Seminar Given at the ETH, Zürich, in 1967. Lecture Notes in Mathematics, No. 87, Springer, New York (1969). https://doi.org/10.1007/bfb0101425
124. Trimble, T.: Towards a doctrine of operads. nLab page
125. Ulmer, F.: Properties of dense and relative adjoint functors. J. Algebra **8**(1), 77–95 (1968). https://doi.org/10.1016/0021-8693(68)90036-7
126. Venema, Y.: Automata and fixed point logic: a coalgebraic perspective **204**(4), 637–678. https://doi.org/10.1016/j.ic.2005.06.003
127. Weber, M.: Yoneda structures from 2-toposes. Appl. Categ. Structures **15**(3), 259–323 (2007). https://doi.org/10.1007/s10485-007-9079-2

128. Weber, M.: Algebraic Kan extensions along morphisms of internal algebra classifiers. Tbilisi Math. J. **9**(1), 65–142 (2016). https://doi.org/10.1515/tmj-2016-0006
129. Weibel, C.: An Introduction to Homological Algebra, Cambridge Studies in Advanced Mathematics, vol. 38. Cambridge University Press, Cambridge (1994). https://doi.org/10.1017/CBO9781139644136
130. Wilf, H.: Generating functionology, 3rd edn. Academic Press, Wellesley (1990). http://www.math.upenn.edu/%7Ewilf/DownldGF.html
131. Wood, R.: Abstract proarrows I. Cahiers de topologie et géometrie différentielle categoriques **23**(3), 279–290 (1982)
132. Yeh, Y.: On the combinatorial species of Joyal. Ph.D. thesis (1985)
133. Yeh, Y.: The calculus of virtual species and $\mathbb{K}$-species. In: Labelle, G., Leroux, P. (eds.) Combinatoire énumérative, pp. 351–369. Springer, Heidelberg (1986). https://doi.org/10.1007/BFb0072525
134. Yoneda, N.: On Ext and exact sequences. J. Fac. Sci. Univ. Tokyo Sect. I **8**, 507–576 (1960)
135. Yorgey, B.: Combinatorial Species and Labelled Structures. Ph.D. thesis, Computer and Information Science, University of Pennsylvania, September 2014. Feely available online at github
136. Zöberlein, V.: Doctrines on 2-categories. Mathematische Zeitschrift **148**, 267–280 (1976). http://eudml.org/doc/172368

Automata in W-Toposes, and General Myhill-Nerode Theorems

Victor Iwaniack(✉)

Laboratoire J. A. Dieudonné, Université Côte d'Azur, Nice, France
victor.iwaniack@unice.fr

Abstract. We extend the functorial approach to automata by Colcombet and Petrişan [7] from the category of sets to any W-topos and establish general Myhill-Nerode theorems in our setting, including an explicit relationship between the syntactic monoid and the transition monoid of the minimal automaton. As a special case we recover the result of Bojańczyk, Klin and Lasota [5] for orbit-finite nominal automata by considering automata in the Myhill-Schanuel topos of nominal sets.

Keywords: Categories of machines · Topoi · Enriched Categories

Introduction

Automata theory appeared in the second half of the XXth century. Automata are simple formal machines meant to recognise languages, that is, sets of finite words over a finite alphabet. The fundamental theorem, first proven by Kleene [11] in 1956, characterises those languages that are recognised by *finite* automata. Two years later, Nerode [15] published another characterisation of this class of languages using an equivalence relation on words, closely related to a notion of a "minimal" automaton. More recently, Colcombet and Petrişan [7] gave an entirely categorical definition of automata, in particular making transparent the construction of the minimal automaton (based on Arbib and Manes [3]) by means of purely categorical tools such as Kan extensions and (orthogonal) factorisation systems.

Our purpose here is to extend the categorical approach to automata theory by Colcombet and Petrişan [7] to more general contexts than those considered by the authors, namely to automata in an arbitrary W-topos. Topos theory is a far-reaching categorical generalisation of set theory with a strong topological flavour; the historic examples of toposes are those categories of sheaves over a topological space, but also categories of continuous actions of a topological group on discrete spaces. One of the notions crucial to the definition of an automaton in such a general context is "finiteness" and we will consider two different notions

This project has been partially funded by the European Research Council (ERC) under the European Union's Horizon 2020 research and innovation program (grant agreement No. 670624).

Published by Springer Nature Switzerland AG 2024
B. König and H. Urbat (Eds.): CMCS 2024, LNCS 14617, pp. 93–113, 2024.
https://doi.org/10.1007/978-3-031-66438-0_5

of finiteness which are well-established in topos theory: dK-finiteness (decidable Kuratowski finiteness) and decomposition-finiteness, both reducing to the classical notion of a finite set in the topos of sets. For both notions of finiteness we get a corresponding Myhill-Nerode Theorem characterising languages with a Nerode congruence of "finite type" as those recognised by "finite type" automata. The key property beneath these general Myhill-Nerode Theorems is the stability of "finite" objects under taking subquotients.

In Sects. 1 and 2 we give definitions and proprieties about toposes and finiteness conditions we will consider. In Sect. 3 we enrich the functorial approach of Colcombet and Petrişan and use it to deduce Myhill-Nerode type theorems, and in the last sections we explore automata theory in specific toposes: toposes of G-sets for a discrete group G (Sect. 4) and automata in the Myhill-Schanuel topos of nominal sets (Sect. 5).

The main contribution here is the generalisation of the Colcombet and Petrişan framework, and the Myhill-Nerode theorems in other toposes than Set. Each theorem depends on a notion of "finiteness", and while Kuratowski finiteness is one of the most important notion of finiteness (for it is definable in any elementary topos and covers finite sets, finite coverings of topological spaces, and finite discrete group actions), it is not in general stable under subquotients, which is the key ingredient for those theorems.

Notations

We use the *diagrammatical order* for composition: if $f : A \to B$ and $g : B \to C$ are morphisms of some category, $fg : A \to C$ is their composition. By "factorisation system" we will always mean "orthogonal factorisation system" unless stated otherwise. If $\mathscr{D} \xleftarrow{L} \mathscr{C} \xleftarrow{R} \mathscr{D}$ is a diagram of functors, then we write $L \dashv R$ the fact that L is left adjoint to R and will usually denote the unit by $\eta : \mathrm{id}_{\mathscr{C}} \Rightarrow LR$ and the counit by $\varepsilon : RL \Rightarrow \mathrm{id}_{\mathscr{D}}$. For any morphism $a : Lc \to d$ of $\mathscr{D}$ we denote by $a^{\dashv} : c \to Rd$ of $\mathscr{C}$ its adjunct with respect to this adjunction, and for any morphism $b : c \to Rd$ of $\mathscr{C}$, $b^{\vdash} : Lc \to d$ of $\mathscr{D}$. Finally, if it exists, we denote by $\emptyset$ an initial object and $\mathbb{1}$ a terminal object.

1 Toposes

Definition 1. *An* (elementary) topos *is a category $\mathcal{E}$ with finite limits, exponentials (i.e. for each object B, the product by B endofunctor $B \times (-)$ has a right adjoint denoted by $(-)^B$ with counit* $\mathrm{ev}^B : B \times (-)^B \Rightarrow \mathrm{id}_{\mathcal{E}}$*) and a* subobject classifier, *that is a pointed object $\top : \mathbb{1} \to \Omega$ such that for each object A and subobject $S \hookrightarrow A$, there exists a unique morphism $\chi_S : A \to \Omega$ called the* characteristic map *such that the following diagram*

$$\begin{array}{ccc} S & \xrightarrow{\;!\;} & \mathbb{1} \\ \big\downarrow & \lrcorner & \big\downarrow{\scriptstyle \top} \\ A & \xrightarrow[\chi_S]{} & \Omega \end{array} \quad \textit{is a pullback.}$$

Example 1. Examples of toposes are categories $\mathbb{B}G := [G^{\mathrm{op}}, \mathrm{Set}]$ where G is a discrete group seen as a single-object groupoid, and $\mathbb{B}G$ is (equivalent to) the category of G-sets and equivariant functions: the subobject classifier is any two element set with trivial G action, exponentials Y^X are sets of mere functions $f : X \to Y$ endowed with the action $(f \cdot g)(x) = f(x \cdot g^{-1}) \cdot g$. For topological G, the category of continuous G-sets is also a topos, also denoted by $\mathbb{B}G$, cf. Mac Lane and Moerdijk [13, Section III.9]. Most famously, categories of sheaves $\mathrm{Sh}\, B$ over a topological space B (equivalent to the category of *étalé spaces over* B i.e. local homeomorphisms $p : E \to B$) are also toposes (cf. Corollary III.7.4, ibidem).

Remark 1. A topos is enriched over itself: the hom-objects are the exponentials, and we shall denote by $\mathrm{comp}_{A,B,C} : B^A \times C^B \to C^A$ the composition morphisms defined as adjuncts of $A \times B^A \times C^B \xrightarrow{\mathrm{ev}_B^A \times C^B} B \times C^B \xrightarrow{\mathrm{ev}_C^B} C$, and the identities by $\mathrm{id}_A^{\mathcal{E}} : \mathbb{1} \to A^A$ defined as the adjuncts of $\mathrm{id}_A : A \to A$. Recall that exponentials define an internal hom functor $\mathcal{E}^{\mathrm{op}} \times \mathcal{E} \to \mathcal{E}, (A, B) \mapsto B^A$ and for all $f : A \leftarrow A'$, $g : X \leftarrow X'$, $h : B \to B'$ and $a : A \times X \to B$ in $\mathcal{E}$:

$$((f \times g)ah)^{\dashv} = g(a^{\dashv})h^f.$$

We recall useful definitions on objects we will use:

Definition 2. *Let $\mathcal{E}$ be a topos and A any of its objects.*

1. *A subobject $S \leq A$ is* complemented *if there exists a subobject $C \leq A$ such that $S \cup C = A$ and $S \cap C = \emptyset$.*
2. *A is* decidable *if the diagonal $\Delta_A : A \xhookrightarrow{(\mathrm{id}_A, \mathrm{id}_A)} A \times A$ is complemented.*
3. *A is* connected *if it admits exactly two complemented subobjects $\emptyset$ and A.*

Amongst the examples cited, a space étalé over a base space B is connected in the topos $\mathrm{Sh}\, B$ iff it is connected as a topological space, and a G-set is connected iff non-empty transitive. Any G-set is decidable because $\mathbb{B}G$ is Boolean:

Definition 3. *A topos is*

1. Boolean *if every object is decidable (cf. Acuña-Ortega and Linton [1, Observation 2.6]);*
2. locally connected *if each object is a sum of connected objects;*
3. atomic *if Boolean and locally connected.*

A W-topos *is a topos admitting a free monoid $(\Sigma^*, m_\Sigma, \varepsilon_\Sigma)$ for every object Σ.*

Remark 2. The terminology "W-topos" comes from Moerdijk and Palmgren [14]. A topos is a W-topos iff it has a so-called *natural number object* (i.e. the free monoid generated by $\mathbb{1}$, cf. ibidem). A topos with countable coproducts is a W-topos because then $\Sigma^* := \sum_{n \in \mathbb{N}} \Sigma^n$.

Toposes $\mathbb{B}G$ are always atomic. A topos $\mathrm{Sh}\, B$ is locally connected iff B is as a topological space. Both types of toposes are W-toposes, because cocomplete.

2 Notions of Finiteness

Definition 4. *Let A be an object of an elementary topos $\mathcal{E}$. The submonoid of $(\Omega^A, \vee, \emptyset)$ generated by the singleton subobject $\{\cdot\}_A : A \hookrightarrow \Omega^A$ (adjunct of the characteristic morphism of the diagonal $\Delta_A : A \xhookrightarrow{(\mathrm{id}_A, \mathrm{id}_A)} A \times A$) is denoted by $K(A)$ and called the object of* Kuratowski-finite subobjects of A.

In toposes $\mathbb{B}G$ for a discrete group G, $K(A)$ is the set of finite subsets of A, endowed with the subset action $S \cdot g := \{s \cdot g | s \in S \subset A\}$ (cf. Example 1).

Definition 5. *Let $\mathcal{E}$ be a topos.*

1. *An object is called* decomposition-finite *if it is a finite coproduct of connected subobjects.*
2. *An object A is* Kuratowski-finite *or* K-finite *if the global element $\mathbb{1} \to \Omega^A$ corresponding to $A \leq A$ factors through $K(A) \leq \Omega^A$ (cf. Johnstone [8, Subsection D5.4]). It is* decidable Kuratowski finite *(*dK-finite*) if it is also decidable.*
3. *Let P be a non-empty class of* points *of $\mathcal{E}$ (left-adjoint left exact functors $x^* : \mathcal{E} \to \mathrm{Set}$), an object A is* P-stalkwise finite *if for all points x^* in P, $x^*(A)$ is a finite set.*

The dK-finite sheaves of $\mathrm{Sh}\,B$ are exactly finite coverings. Finiteness conditions in toposes $\mathbb{B}G$ will be discussed in Sects. 4 and 5.

Proposition 1. *1. In an atomic topos, decomposition-finiteness is stable under taking subquotients.*
2. In a Boolean topos, dK-finiteness is stable under taking subquotients.
3. In any topos, stalkwise finiteness is stable under taking subquotients.

Proof. 1. If C and D are connected subobjects of a same object X, then either $C \cap D = \emptyset$, or $C \cap D \neq \emptyset$ is a complemented (because $\mathcal{E}$ is Boolean) non-empty subobject of connected C, therefore $C \cap D = C$ and by the same argument, $C \cap D = D$ so that $C = D$. This shows that if $\sum_{i \in I} A_i \leq \sum_{j \in J} B_j$ with connected A_i and B_j then $I \subset J$ so that I is finite when J is. In any topos, homomorphic images of connected objects are connected, and this fact can be used to show that decomposition-finiteness is stable under taking quotients.
2. According to Johnstone [8, Lemma 5.4.4.ii], in any topos, K-finite objects are stable under taking quotients, and if $\mathcal{E}$ is Boolean, they are moreover closed under taking subobjects (cf. Remark 5.4.20.ii⇒iii, ibidem).
3. Immediate because any point x^* preserves epimorphisms (as a left adjoint) and monomorphisms (as a left exact functor).

3 Automata in Toposes and Myhill-Nerode Theorems

Before enriching the approach of Colcombet and Petrişan [7], we recall their point of view. Consider a (complete deterministic) automaton (Q, i, F, δ) on an alphabet Σ (any set), meaning Q is a set of *states*, $i \in Q$ the *initial state*, $F \subset Q$ the set of *final states* and $\delta : Q \times \Sigma \to Q$ the *transition* function. The transition function gives, by iteration, a right action of the free monoid Σ^* generated by Σ on the set Q. In particular, we can interpret the action in a functorial (classical) way as a functor $\Sigma^* \to \mathrm{Set}$ where the monoid Σ^* is seen as a category with a single object st (for "states"). Now the initial state $i \in Q$ can be seen as a global element $\mathbb{1} \to Q$, and the subset F of final states can be represented by its characteristic function $\chi_F : Q \to \Omega$ where Ω is the subobject classifier of Set, namely any two-element set of "truth values". All this data can be expressed by a functor $\mathcal{I}_\Sigma \to \mathrm{Set}$ with source freely generated by the quiver

$$\mathrm{in} \xrightarrow{\ \triangleright\ } \mathrm{st} \circlearrowleft^{s\in\Sigma} \xrightarrow{\ \triangleleft\ } \mathrm{out}$$

and the functor corresponding to the automaton (Q, i, F, δ)

$$\mathbb{1} \xrightarrow{\ i\ } Q \circlearrowleft^{\delta(-,s),\, s\in\Sigma} \xrightarrow{\ \chi_F\ } \Omega$$

sends $(\mathrm{in}, \mathrm{st}, \mathrm{out})$ to $(\mathbb{1}, Q, \Omega)$, $\triangleright$ to the global element corresponding to i, $\triangleleft$ to the characteristic function corresponding to F, and extends the previous functor $\Sigma^* \to \mathrm{Set}$. This correspondence is in fact a bijection. The automata then organise as a (non-full) subcategory of a functor category. This approach merges the algebraic and coalgebraic point of view: an automaton seen as an algebra lacks terminal states, while an automaton seen as a coalgebra lacks an initial state.

Definition 6. *Let $\mathcal{E}$ be a W-topos, and fix Σ any object of this topos, we call an* alphabet.

- *A* language *on Σ is any subobject of Σ^*.*
- *A* (deterministic complete) automaton *on the alphabet Σ is a quadruple $\mathcal{A} = (Q, i, F, \delta)$ where*
  - *Q is the* states *object,*
  - *$i : \mathbb{1} \to Q$ is a global element, the* initial state, *of Q,*
  - *F is the subobject of Q of* final states, *which we identify with its characteristic morphism $\chi_F : Q \longrightarrow \Omega$, and*
  - *$\delta : Q \times \Sigma \to Q$ is the* transition morphism.

There is a notion of a language recognised by an automaton. To define it, observe that the adjunct $\delta^{\dashv} : \Sigma \to Q^Q$ of $\delta : Q \times \Sigma \to Q$, with respect to $Q \times (-) \dashv (-)^Q$, takes values in an internal monoid $(Q^Q, \mathrm{comp}_{Q,Q,Q}, \mathrm{id}_Q^{\mathcal{E}})$, cf. Remark 1. Because Σ^* is the free internal monoid, $\delta^{\dashv}$ extends uniquely to Σ^* into an internal monoid morphism we denote by $\delta^* : \Sigma^* \to Q^Q$.

Definition 7. *The language recognised by the automaton $\mathcal{A} = (Q, i, F, \delta)$ defined over Σ is the subobject $L(\mathcal{A})$ of Σ^* with characteristic morphism*

$$\Sigma^* \xrightarrow{\delta^*} Q^Q \xrightarrow{\chi_F^i} \Omega^{\mathbb{1}} \cong \Omega.$$

Automata $\mathcal{A}$ recognising a language L are called L-automata.

3.1 Languages and Automata as Enriched Functors

The definitions we will give in the following subsections are immediate generalisations of the definitions of Colcombet and Petrişan [7], so that we will keep essentially the same terminology.

Definition 8. *Let Σ be an alphabet of a W-topos $\mathcal{E}$. The $\mathcal{E}$-category $\mathcal{I}_\Sigma$, called the* $\mathcal{E}$-category of internal behaviours over the alphabet Σ, *is the $\mathcal{E}$-category freely generated by the $\mathcal{E}$-quiver Q_Σ with three vertices* in, st *and* out, *and objects of edges $Q_\Sigma(\mathrm{in}, \mathrm{st}) = \mathbb{1} = Q_\Sigma(\mathrm{st}, \mathrm{out})$, $Q_\Sigma(\mathrm{st}, \mathrm{st}) = \Sigma$ and $Q_\Sigma(X, Y) = \emptyset$ for every other cases. Spelt out, $\mathcal{I}_\Sigma$ is defined by:*

Objects in, st *and* out.
Objects of morphisms *given by the table:*

$\mathcal{I}_\Sigma(\downarrow, \rightarrow)$	in	st	out
in	$\mathbb{1}$	Σ^*	Σ^*
st	$\emptyset$	Σ^*	Σ^*
out	$\emptyset$	$\emptyset$	$\mathbb{1}$

Composition morphisms *being of the following form, depending on the domain and codomain of definition:*
- $\Sigma^* \times \Sigma^* \xrightarrow{m_\Sigma} \Sigma^*$,
- $\mathbb{1} \times \Sigma^* \cong \Sigma^*$,
- $\Sigma^* \times \mathbb{1} \cong \Sigma^*$ *and*
- *if the source is $\emptyset$ or the target is $\mathbb{1}$, then the composition is trivial.*

Proposition 2. *Automata $\mathcal{A} = (Q, i, F, \delta)$ over Σ are in bijective correspondence with $\mathcal{E}$-functors $\underline{\mathcal{A}} : \mathcal{I}_\Sigma \to \mathcal{E}$ sending* $(\mathrm{in}, \mathrm{out})$ *to* $(\mathbb{1}, \Omega)$.

Proof. We use the fact that $\mathcal{I}_\Sigma$ is a free $\mathcal{E}$-category: $\mathcal{E}$-functors $\underline{\mathcal{A}}$ sending $(\mathrm{in}, \mathrm{out})$ to $(\mathbb{1}, \Omega)$ correspond to $\mathcal{E}$-quiver morphisms $\alpha : Q_\Sigma \to \mathcal{E}$ sending $(\mathrm{in}, \mathrm{out})$ to $(\mathbb{1}, \Omega)$, which correspond to automata (Q, i, F, δ) over Σ:

- α takes $(\mathrm{in}, \mathrm{st}, \mathrm{out})$ to $(\mathbb{1}, Q, \Omega)$
- $\alpha_{\mathrm{in},\mathrm{st}} = i : \mathbb{1} \to Q$
- $\alpha_{\mathrm{st},\mathrm{out}} = \ulcorner F \urcorner : \mathbb{1} \to \Omega^Q$ adjunct of $Q \times \mathbb{1} \cong Q \xrightarrow{\chi_F} \Omega$
- $\alpha_{\mathrm{st},\mathrm{st}} = \delta^\dashv : \Sigma \to Q^Q$ adjunct of $\delta : Q \times \Sigma \to Q$.

Definition 9. *The full sub-$\mathcal{E}$-category of $\mathcal{I}_\Sigma$ spanned by the objects* in *and* out *is denoted by $\mathcal{O}_\Sigma \overset{\iota_\Sigma}{\hookrightarrow} \mathcal{I}_\Sigma$ and called the* $\mathcal{E}$-category of observable behaviours over the alphabet Σ.

Remark 3. All the non-trivial data of an $\mathcal{E}$-functor $F : \mathcal{O}_\Sigma \to \mathrm{Set}$ sending $(\mathrm{in}, \mathrm{out})$ to $(\mathbb{1}, \Omega)$ is contained in $F_{\mathrm{in,out}} : \Sigma^* \to \Omega$, therefore such functors bijectively correspond to languages L over Σ and we denote by $\underline{L}$ the corresponding functor. This bijection provides a notion of language recognised by an automaton-as-a-functor which is coherent with the associated automaton as shown in the next proposition.

Proposition 3. *Under the bijection of Proposition 2, the restriction of an automaton $\underline{\mathcal{A}} : \mathcal{I}_\Sigma \to \mathcal{E}$ to the sub-$\mathcal{E}$-category $\mathcal{O}_\Sigma$ corresponds to the language $L(\mathcal{A})$ recognised by the automaton $\mathcal{A}$ corresponding to $\underline{\mathcal{A}}$ i.e. $\underline{L(\mathcal{A})} = \underline{\mathcal{A}}|_{\mathcal{O}_\Sigma}$.*

Proof. Let $\mathcal{A} = (Q, i, F, \delta)$ be an automaton over Σ, then

$$L(\mathcal{A}) := \Sigma^* \xrightarrow{\delta^*} Q^Q \xrightarrow{\chi_F^i} \Omega^{\mathbb{1}}$$

corresponds through the natural bijection $\mathcal{E}(\Sigma^*, \Omega^{\mathbb{1}}) \cong \mathcal{E}(\mathbb{1} \times \Sigma^*, \Omega)$ to

$$\begin{aligned}
&\mathbb{1} \times \Sigma^* \xrightarrow{i \times \Sigma^*} Q \times \Sigma^* \xrightarrow{\delta^{*\vdash}} Q \xrightarrow{\chi_F} \Omega \text{ (cf. Remark 1)} \\
=&\mathbb{1} \times \Sigma^* \xrightarrow{\varepsilon_\Sigma \underline{\mathcal{A}}_{\mathrm{in,st}} \times \Sigma^*} Q \times \Sigma^* \xrightarrow{\underline{\mathcal{A}}^{\vdash}_{\mathrm{st,st}}} Q \xrightarrow{\chi_F} \Omega \text{ by definition of } \underline{\mathcal{A}} \\
=&\mathbb{1} \times \Sigma^* \xrightarrow{\varepsilon_\Sigma \times \Sigma^*} \Sigma^* \times \Sigma^* \xrightarrow{\underline{\mathcal{A}}_{\mathrm{in,st}} \times \Sigma^*} Q \times \Sigma^* \xrightarrow{\underline{\mathcal{A}}^{\vdash}_{\mathrm{st,st}}} Q \xrightarrow{\chi_F} \Omega \\
=&\mathbb{1} \times \Sigma^* \xrightarrow{\varepsilon_\Sigma \times \Sigma^*} \Sigma^* \times \Sigma^* \xrightarrow{m_\Sigma} \Sigma^* \xrightarrow{\underline{\mathcal{A}}_{\mathrm{in,st}}} Q \xrightarrow{\chi_F} \Omega \\
=&\mathbb{1} \times \Sigma^* \xrightarrow{\sim} \Sigma^* \xrightarrow{\underline{\mathcal{A}}_{\mathrm{in,st}}} Q \xrightarrow{\chi_F} \Omega
\end{aligned}$$

where $(\underline{\mathcal{A}}_{\mathrm{in,st}} \times \Sigma^*)\underline{\mathcal{A}}^{\vdash}_{\mathrm{st,st}} = m_\Sigma \underline{\mathcal{A}}_{\mathrm{in,st}}$ by $\mathcal{E}$-functoriality of $\underline{\mathcal{A}}$, and $(\varepsilon_\Sigma \times \Sigma^*) m_\Sigma = \mathbb{1} \times \Sigma^* \xrightarrow{\sim} \Sigma^*$ by left unitality in the monoid Σ^*. Then the morphism above is

$$\mathbb{1} \times \Sigma^* \xrightarrow{\sim} \Sigma^* \xrightarrow{\underline{\mathcal{A}}_{\mathrm{in,st}}} Q \xrightarrow{(\varepsilon_F \underline{\mathcal{A}}_{\mathrm{st,out}})^{\vdash}} \Omega$$

by definition of $\underline{\mathcal{A}}$ and corresponds through $\mathcal{E}(\mathbb{1} \times \Sigma^*, \Omega) \cong \mathcal{E}(\mathbb{1}, \Omega^{\Sigma^*})$ to

$$\begin{aligned}
&\mathbb{1} \xrightarrow{\varepsilon_\Sigma} \Sigma^* \xrightarrow{\underline{\mathcal{A}}_{\mathrm{st,out}}} \Omega^Q \xrightarrow{\Omega^{\underline{\mathcal{A}}_{\mathrm{in,st}}}} \Omega^{\Sigma^*} \\
=&\mathbb{1} \xrightarrow{\varepsilon_\Sigma} \Sigma^* \xrightarrow{m_\Sigma^{\dashv}} \Sigma^{*\Sigma^*} \xrightarrow{\underline{\mathcal{A}}_{\mathrm{in,out}}} \Omega^{\Sigma^*} \text{ by } \mathcal{E}\text{-functoriality of } \underline{\mathcal{A}} \\
=&\mathbb{1} \xrightarrow{\mathrm{id}^{\mathcal{E}}_{\Sigma^*}} \Sigma^{*\Sigma^*} \xrightarrow{\underline{\mathcal{A}}_{\mathrm{in,out}}} \Omega^{\Sigma^*} \text{ by right unitality in the monoid } \Sigma^*
\end{aligned}$$

and finally going through the inverse natural bijection $\mathcal{E}(\mathbb{1}, \Omega^{\Sigma^*}) \cong \mathcal{E}(\mathbb{1} \times \Sigma^*, \Omega) \cong \mathcal{E}(\Sigma^*, \Omega^{\mathbb{1}})$, the morphism above is $\underline{\mathcal{A}}_{\mathrm{in,out}} : \Sigma^* \to \Omega^{\mathbb{1}}$.

3.2 Categories of Automata

We follow Colcombet and Petrişan [7] definition of automaton morphism, i.e. the morphisms we consider are coalgebra morphisms that respects the initial state.

Definition 10. *Let $\underline{L} : \mathcal{O}_\Sigma \to \mathcal{E}$ be a language over Σ in a W-topos $\mathcal{E}$. The category* Auto(L) *of* L-automata *has as objects the $\mathcal{E}$-functors extending $\underline{L}$ along the inclusion $\iota_\Sigma : \mathcal{O}_\Sigma \hookrightarrow \mathcal{I}_\Sigma$, and as morphisms $\mathcal{E}$-natural transformations $\alpha : \underline{\mathcal{A}} \Rightarrow \underline{\mathcal{B}}$ restricting to $\mathrm{id}_{\underline{L}}$ on $\mathcal{O}_\Sigma$.*

In some cases we will obtain an automaton recognising the language only up to an automorphism of Ω, so that $\underline{\mathcal{A}}\big|_{\mathcal{O}_\Sigma}$ might only be isomorphic to $\underline{L}$.

Lemma 1 (Strictification of an automaton with respect to a language). *Let $\underline{L}$ be a language and $\underline{\mathcal{A}}$ an automaton, both defined over an alphabet Σ. If there exists an $\mathcal{E}$-natural isomorphism $\varphi : \underline{\mathcal{A}}\big|_{\mathcal{O}_\Sigma} \cong \underline{L}$, then there exists an automaton $\underline{\mathcal{B}} \in$* Auto($L$) *isomorphic as an $\mathcal{E}$-functor to $\underline{\mathcal{A}}$ via $\psi : \underline{\mathcal{A}} \cong \underline{\mathcal{B}}$ such that $\iota_\Sigma * \psi = \varphi$. The automaton $\underline{\mathcal{B}}$ is constructed as the $\mathcal{E}$-functor $\underline{\mathcal{A}}$ with this only difference:*

$$\underline{\mathcal{B}}_{\mathrm{st,out}} = \Sigma^* \xrightarrow{\mathcal{A}_{\mathrm{st,out}}} \Omega^Q \xrightarrow{(\varphi_{\mathrm{out}}^{-1})^Q} \Omega^Q.$$

It was one of the main insights of Colcombet and Petrişan [7, (2.2)] that the minimal automaton recognising a given language can be constructed by factorising the canonical map from the initial automaton to the final automaton. This remains true in our enriched context as we will see in the next subsections.

3.3 Initial and Terminal Automata as Enriched Kan Extensions

In our setting, automata are (strict) extensions of languages seen as $\mathcal{E}$-functors along the fully faithful $\mathcal{E}$-functor $\iota_\Sigma : \mathcal{O}_\Sigma \to \mathcal{I}_\Sigma$. But in the case of (pointwise) Kan extensions along a fully faithful functor, the Kan extensions are genuine extensions up to isomorphism, and therefore provide two automata:

Proposition 4. *In a W-topos $\mathcal{E}$, the initial $\emptyset(L)$ and terminal $\mathbb{1}(L)$ automata exist for any language L over any alphabet Σ; they are respectively the left and the right $\mathcal{E}$-enriched Kan extensions of $\underline{L} : \mathcal{O}_\Sigma \to \mathcal{E}$ along the fully faithful $\mathcal{E}$-functor $\iota_\Sigma : \mathcal{O}_\Sigma \hookrightarrow \mathcal{I}_\Sigma$. They can be explicitly computed as follows:*

$$\begin{array}{rl|ll}
\Sigma^* = & \emptyset(L)(\mathrm{st}) & \mathbb{1}(L)(\mathrm{st}) & = \Omega^{\Sigma^*} \\
\mathrm{id}_{\Sigma^*} = & \emptyset(L)_{\mathrm{in,st}} & \mathbb{1}(L)_{\mathrm{in,st}} & = (m_\Sigma \chi_L)^\dashv : \Sigma^* \to \Omega^{\Sigma^*} \\
{\Sigma^*}^{\Sigma^*} \leftarrow \Sigma^* : {m_\Sigma}^\dashv = & \emptyset(L)_{\mathrm{st,st}} & \mathbb{1}(L)_{\mathrm{st,st}} & = ((\Omega^{m_\Sigma})^\vdash)^\vdash : \Sigma^* \to (\Omega^{\Sigma^*})^{\Omega^{\Sigma^*}} \\
\Omega^{\Sigma^*} \leftarrow \Sigma^* : (m_\Sigma \chi_L)^\dashv = & \emptyset(L)_{\mathrm{st,out}} & \mathbb{1}(L)_{\mathrm{st,out}} & = (\chi_{\ni_{\Sigma^*}})^\dashv : \Sigma^* \to \Omega^{\Omega^{\Sigma^*}}
\end{array}$$

where $\mathbb{1}(L)_{\mathrm{st,st}}$ is obtained from Ω^{m_Σ} by the natural bijection $\mathcal{E}(\Omega^{\Sigma^*}, \Omega^{\Sigma^* \times \Sigma^*}) \cong \mathcal{E}(\Omega^{\Sigma^*}, (\Omega^{\Sigma^*})^{\Sigma^*}) \cong \mathcal{E}(\Omega^{\Sigma^*} \times \Sigma^*, \Omega^{\Sigma^*}) \cong \mathcal{E}(\Sigma^*, (\Omega^{\Sigma^*})^{\Sigma^*})$

As for Colcombet and Petrişan [7, Lemma 2.9], the Kan extensions, now enriched, are computed pointwise; but here the formula (cf. Kelly [10, (2.2) with (4.24)]) only uses exponentials (resp. binary products), finite products (resp. finite coproducts, which are both finite because $\mathcal{I}_\Sigma$ has only three objects) and an equaliser (resp. coequaliser), which all exist in $\mathcal{E}$.

Proposition 5. *In a W-topos $\mathcal{E}$, let $\underline{A}$ be an L-automaton over an alphabet Σ, the unique automaton morphism α from $\emptyset(L)$ to $\underline{A}$ is given by*

$$\alpha_{\mathrm{st}} = \underline{A}_{\mathrm{in,st}} : \Sigma^* \to Q$$

and the unique automaton morphism β from $\underline{A}$ to $\mathbb{1}(L)$ by

$$\beta_{\mathrm{st}} = (Q \times \Sigma^* \xrightarrow{\underline{A}^{\vdash}_{\mathrm{st,out}}} \Omega)^{\dashv} : Q \to \Omega^{\Sigma^*} \quad \mathit{wrt.}\ \mathcal{E}(Q \times \Sigma^*, \Omega) \cong \mathcal{E}(Q, \Omega^{\Sigma^*}).$$

In particular, the unique morphism from $\emptyset(L)$ to $\mathbb{1}(L)$ is

$$(m_\Sigma \chi_L)^{\dashv} : \Sigma^* \to \Omega^{(\Sigma^*)}.$$

3.4 Minimal Automaton

We understand minimality with respect to a factorisation system, cf. Colcombet and Petrişan [7, Subsection 2.2]:

Definition 11. *In a category $\mathscr{C}$ endowed with a factorisation system (E, M), we say an object X (E, M)-*divides *an object Y if there exists a span*

$$X \xleftarrow{e \in E} Z \xrightarrow{m \in M} Y$$

in $\mathscr{C}$. An object is minimal *if it divides any object of $\mathscr{C}$.*

We recall the key idea of Colcombet and Petrişan [7, Lemma 2.3] to compute the minimal automaton:

Proposition 6. *Let $\mathscr{C}$ be a category with a factorisation system (E, M). If $\mathscr{C}$ has an initial and a terminal object, then the object through which the unique arrow from the initial to the terminal object (E, M)-factorises is (E, M)-minimal.*

However, here, we need the factorisation system to be on the category of automata which is a category of enriched functors. Therefore, we have to lift the (epi, mono) factorisation system on $\mathcal{E}$ to the $\mathcal{E}$-functors category $[\mathcal{I}_\Sigma, \mathcal{E}]$. Given two $\mathcal{V}$-functors and a $\mathcal{V}$-natural transformation $\alpha : \mathcal{F} \Rightarrow \mathcal{G}$, the pointwise factorisation of α, according to a given factorisation system, might only give an unenriched functor. Thus, we need the factorisation to have more properties, which leads to the definition of an enriched factorisation system.

Definition 12. *Let $\mathcal{V}$ be a symmetric closed monoidal category, and $\mathscr{C}$ a $\mathcal{V}$-category. A factorisation system (E, M) on $\mathscr{C}$ is $\mathcal{V}$-*enriched *if for all $A \xleftarrow{e \in E} B$ and $X \xrightarrow{m \in M} Y$, the following square*

$$\begin{array}{ccc} \mathscr{C}(A,X) & \xrightarrow{\mathscr{C}(A,m)} & \mathscr{C}(A,Y) \\ {\scriptstyle \mathscr{C}(e,X)}\downarrow & \lrcorner & \downarrow{\scriptstyle \mathscr{C}(e,Y)} \\ \mathscr{C}(B,X) & \xrightarrow[\mathscr{C}(B,m)]{} & \mathscr{C}(B,Y) \end{array}$$

is a pullback in $\mathcal{V}$.

One can characterise enriched factorisation systems amongst unenriched ones using powers or copowers, according to Lucyshyn-Wright [12, Theorem 5.7]:

Proposition 7. *If $\mathscr{C}$ has $\mathcal{V}$-copowers (respectively $\mathcal{V}$-powers), then a factorisation system (E, M) on $\mathscr{C}$ is enriched if and only if E is stable under $\mathcal{V}$-copowers (resp. M is stable under $\mathcal{V}$-powers). This is in particular the case if $\mathcal{E} = \mathscr{C}$ is a topos, and (E, M) is the (epi, mono) factorisation system.*

An enriched factorisation system ensures that factorising enriched natural transformations provides an enriched functor. Proposition 7 can be used to show:

Proposition 8. *Let (E, M) be a $\mathcal{V}$-factorisation system on $\mathcal{V}$, and $\mathscr{I}$ a small $\mathcal{V}$-category. Then the classes $E_{\mathscr{I}}$ of $\mathcal{V}$-natural transformations that are pointwise in E, and $M_{\mathscr{I}}$ of $\mathcal{V}$-natural transformations that are pointwise in M form a $\mathcal{V}$-enriched factorisation system on $[\mathscr{I}, \mathcal{V}]$.*

A lifted factorisation system on $[\mathcal{I}_\Sigma, \mathcal{E}]$ can then be restricted to $\mathrm{Auto}(L)$:

Proposition 9. *Any $\mathcal{E}$-factorisation system on $\mathcal{E}$ can be lifted to* $\mathrm{Auto}(L)$, *so that the factorisation of an automaton morphism is obtained as the pointwise factorisation of the underlying $\mathcal{E}$-natural transformation.*

Proof. Consider a morphism $\alpha : \underline{\mathcal{A}} \to \underline{\mathcal{B}}$ and its factorisation $\underline{\mathcal{A}} \overset{e}{\Rightarrow} F \overset{m}{\Rightarrow} \underline{\mathcal{B}}$ in $[\mathcal{I}_\Sigma, \mathcal{E}]$. Then by unicity of the factorisation in $\mathcal{E}$ we have:

$$\begin{array}{ccccc} & & F(\mathrm{in}) & & \\ & \overset{e_{\mathrm{in}}}{\nearrow} & & \overset{m_{\mathrm{in}}}{\searrow} & \\ \mathbb{1} & & \downarrow\wr & & \mathbb{1} \\ & \underset{\mathrm{id}_{\mathbb{1}}}{\searrow} & & \underset{\mathrm{id}_{\mathbb{1}}}{\nearrow} & \\ & & \mathbb{1} & & \end{array} \quad \text{and} \quad \begin{array}{ccccc} & & F(\mathrm{out}) & & \\ & \overset{e_{\mathrm{out}}}{\nearrow} & & \overset{m_{\mathrm{out}}}{\searrow} & \\ \Omega & & \downarrow\wr & & \Omega \\ & \underset{\mathrm{id}_{\Omega}}{\searrow} & & \underset{\mathrm{id}_{\Omega}}{\nearrow} & \\ & & \Omega & & \end{array}$$

because $\alpha_{\mathrm{in}} = \mathrm{id}_{\mathbb{1}} = e_{\mathrm{in}} m_{\mathrm{in}}$ and $\alpha_{\mathrm{out}} = \mathrm{id}_\Omega = e_{\mathrm{out}} m_{\mathrm{out}}$ so that $F\big|_{\mathcal{O}_\Sigma} \cong \underline{L}$ and by Lemma 1 we can find a factorisation of α through an L-automaton.

The category $\mathrm{Auto}(L)$ is now canonically endowed with the pointwise (epi, mono) factorisation system.

Definition 13. *Let L be a language over an alphabet Σ in a W-topos $\mathcal{E}$. The automaton* $\mathrm{Min}(L)$ *through which the unique arrow from the initial automaton to the terminal automaton factorises with respect to pointwise (epi, mono) factorisation system is* the minimal automaton of L.

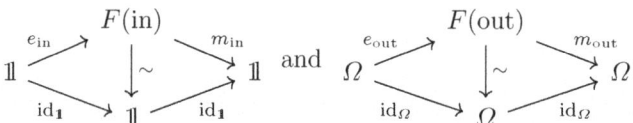

The name "minimal automaton" designates this particular construction, but it is really minimal in the sense of Definition 11:

Corollary 1. *Let L be a language on an alphabet Σ in a W-topos $\mathcal{E}$. The minimal automaton of L is a subquotient of any automaton that recognises L.*

Proof. By Proposition 9 and 4 the category $\mathrm{Auto}(L)$ has pointwise (epi, mono) factorisation system, initial and terminal objects, therefore Proposition 6 ensures $\mathrm{Min}(L)$ is minimal in this context i.e. $\mathrm{Min}(L)$ is a subquotient of every L-automaton.

3.5 Internal Nerode Congruence

In Set, the *Nerode congruence* $\sim_L$ *of a language* L *over* Σ is an equivalence relation on words over Σ defined by $u \sim_L v$ iff for all words x, $ux \in L \Leftrightarrow vx \in L$. It is strictly the same as saying $u \sim_L v$ iff $u^{-1}L = v^{-1}L$. Then the Nerode congruence is merely the kernel pair of left division of L, $u \mapsto u^{-1}L$, which in turn is the adjunct of $(u, w) \mapsto \chi_L(uw)$, namely the composite of monoid multiplication of Σ^* with χ_L.

Definition 14. *Let Σ be an alphabet in a W-topos $\mathcal{E}$. The* Nerode congruence[1] *of a language L* over *Σ is the kernel pair $\equiv_L$ of*

$$(m_\Sigma \chi_L)^{\dashv} : \Sigma^* \to \Omega^{\Sigma^*}$$

where m_Σ is the multiplication of the free monoid Σ^*.

Proposition 10. *In a W-topos $\mathcal{E}$, the states object of* $\mathrm{Min}(L)$ *is the quotient of Σ^* by the internal Nerode congruence.*

Proof. By Corollary 1, $\emptyset(L) \xrightarrow{!} \mathbb{1}(L)$ factorises as

$$\emptyset(L) \hookrightarrow \mathrm{Min}(L) \twoheadrightarrow \mathbb{1}(L)$$

and therefore we have an image factorisation of $\emptyset(L)(\mathrm{st}) \xrightarrow{!_{\mathrm{st}}} \mathbb{1}(L)(\mathrm{st})$

$$\emptyset(L)(\mathrm{st}) \hookrightarrow \mathrm{Min}(L)(\mathrm{st}) \twoheadrightarrow \mathbb{1}(L)(\mathrm{st})$$

and the morphism above is $(m_\Sigma \chi_L)^{\dashv}$ according to Proposition 5. A topos is regular and in a regular category, the coequaliser of the kernel pair of a morphism is canonically isomorphic to its image, so that $\mathrm{Min}(L)(\mathrm{st}) \cong \Sigma^*_{\equiv_L}$.

[1] It is not an internal monoid congruence, it is only a categorical congruence, namely an internal equivalence relation.

3.6 Myhill-Nerode Theorems for Different Finiteness Conditions

The following Myhill-Nerode theorems have two main cases of application: the first is in Set, the classical Myhill-Nerode theorem stating that a language is regular if and only if the Nerode congruence is of finite index (cf. Nerode [15]), and the second in the topos Nom of *nominal sets*, proven for any G-sets topos by Bojańczyk, Klin and Lasota [5] (where G is a discrete group or G is the topological group of permutations of natural numbers acting on discrete spaces), which states that a G-language (resp. nominal language) is regular, in the sense that it is recognised by an orbit-finite deterministic G-automaton (resp. nominal automaton) if and only if the quotient of the nominal set of words on the alphabet by the Nerode congruence is orbit-finite. Our Theorem 1 is a generalisation and another point of view on Bojańczyk, Klin and Lasota [5, Theorems 3.8 and 5.2].

Definition 15. *Each time we consider a finiteness condition (FC), we say an automaton* is (FC) *if its states object is (FC). A language L is* (FC)-regular *if it admits an (FC) automaton that recognises it.*

Theorem 1. *Let L be a language on an alphabet Σ in a W-topos $\mathcal{E}$.*

1. *For any non-empty class of points P of $\mathcal{E}$, L is P-stalkwise-regular iff $\Sigma^*_{/\equiv_L}$ is P-stalkwise finite.*
2. *If $\mathcal{E}$ atomic, L is decomposition-regular iff $\Sigma^*_{/\equiv_L}$ is decomposition-finite.*
3. *If $\mathcal{E}$ is Boolean, L is K-regular iff $\Sigma^*_{/\equiv_L}$ is K-finite.*

Proof. By Corollary 1, $\mathrm{Min}(L)$ exists and by Proposition 10, $\mathrm{Min}(L)(\mathrm{st}) = \Sigma^*_{/\equiv_L}$. If there is some L-automaton $\underline{\mathcal{A}}$ that is (FC) for one of those cases, then because $\mathrm{Min}(L)$ divides $\underline{\mathcal{A}}$, $\mathrm{Min}(L)(\mathrm{st}) = \Sigma^*_{/\equiv_L}$ divides $\underline{\mathcal{A}}(\mathrm{st})$ which is (FC), so by Proposition 1, $\Sigma^*_{/\equiv_L}$ is also (FC).

3.7 The Syntactic Monoid

There exists an algebraic notion of recognition where the recogniser is a monoid morphism (cf Jean-Éric Pin [16, Subsection IV.2.1]). With this point of view, automata are merely presentations of such algebraic recognisers, given by the transition monoid of the automaton. Amongst monoids recognising a language, there is a smallest recogniser with respect to monoid divisibility: the syntactic monoid of a language. It can be defined abstractly as the quotient of the monoid of words by a *syntactic* congruence, or simply by the fact it is the transition monoid of the minimal automaton. We will now describe this (non-functorial) construction in any W-topos and discuss its behaviour with respect to a given finiteness condition.

Definition 16. *Let L be a language on an alphabet Σ in a W-topos $\mathcal{E}$. We say a monoid morphism $\varphi : \Sigma^* \to M$* recognises *L if there exist $\chi : M \to \Omega$ making the following triangle commute:*

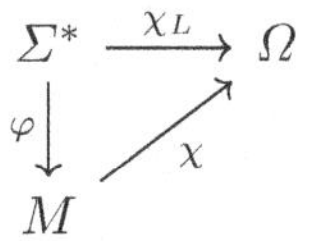

A monoid M recognises L if there exists such a φ with target M.

We call the triple (M, φ, χ) an L-monoid, and an L-monoid morphism from (M, φ, χ) to (M', φ', χ') is a monoid morphism $f : M \to M'$ such that those two triangles commute:

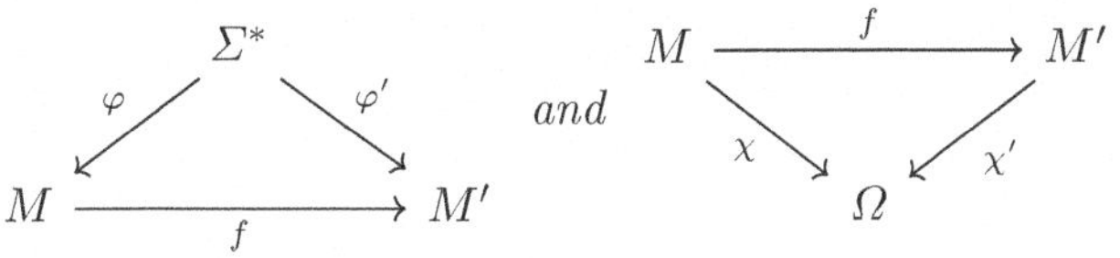

in other words: f is a morphism from φ to φ' in $\Sigma^* / \mathrm{Mon}(\mathcal{E})$ and from χ to χ' in $\mathcal{E}/\Omega$ as well.

This defines a category $\mathrm{Mon}(L)$, and we denote by $\Sigma\,\mathrm{Mon}(L)$ the full subcategory of Σ-generated L-monoids spanned by L-monoids of the form $(M, \varphi, \chi_{\varphi(L)})$ where φ is epic in $\mathcal{E}$, and for those we usually drop the now implicit $\chi_{\varphi(L)}$.

Remark 4. If χ classifies $p : P \hookrightarrow M$ and χ' classifies $q : Q \hookrightarrow M'$, then we have $f\chi' = \chi$ iff $p = f^*(q)$ where f^* is the inverse image of (i.e. pullback along) f.

Lemma 2. *If $\varphi : \Sigma^* \twoheadrightarrow M$ recognises L, then $(\mathrm{Im}\,\varphi, e : \Sigma^* \twoheadrightarrow \mathrm{Im}\,\varphi, \chi_{\varphi(L)})$ is a Σ-generated L-monoid, where e is the image of φ and $\chi_{\varphi(L)}$ is the characteristic morphism of the image inclusion of the composite $L \hookrightarrow \Sigma^* \xrightarrow{e} \mathrm{Im}\,\varphi$ in $\mathrm{Im}\,\varphi$.*

Proof. By pasting law of pullbacks, because the outer rectangle is a pullback ($\varphi\chi = \chi_L$ so by Remark 4, $L = \varphi^*(P)$ where $P \leq M$ is classified by χ i.e. $\chi_P = \chi$) and the right one two,

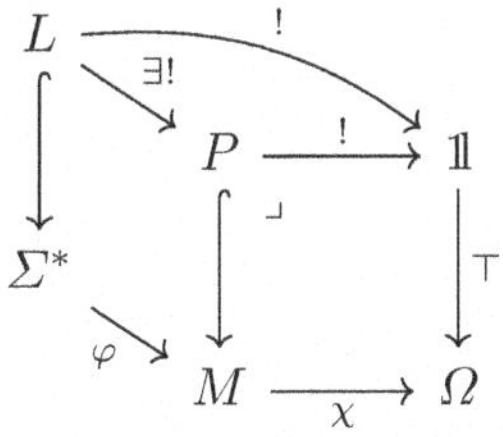

then the left one is also a pullback and by (epi, mono) factorisation of φ and pulling back along the inclusion of P we have a unique filler (by functoriality of factorisation systems)

$$\begin{array}{ccccc} L & \twoheadrightarrow & \varphi(L) & \hookrightarrow & P \\ \downarrow & & \downarrow{\scriptstyle \exists!} & & \downarrow \\ \Sigma^* & \twoheadrightarrow_{e} & \mathrm{Im}\,\varphi & \hookrightarrow & M \end{array}$$

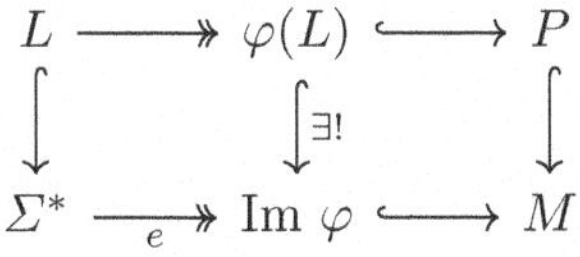

which is also a monomorphism by right cancellation. Then this diagram provides the (epi, mono) factorisation of $L \hookrightarrow \Sigma^* \xrightarrow{\varphi} M$, and because the left square above and the right square below are pullbacks

$$\begin{array}{ccccc} L & \twoheadrightarrow & \varphi(L) & \xrightarrow{!} & 1\!\!1 \\ \downarrow & & \downarrow & & \downarrow{\scriptstyle\top} \\ \Sigma^* & \underset{e}{\twoheadrightarrow} & \mathrm{Im}\,\varphi & \xrightarrow[\chi_{\varphi(L)}]{} & \Omega \end{array}$$

the outer rectangle above is a pullback too stating $\chi_L = e\chi_{\varphi(L)}$.

Consider an L-monoid morphism $f : (M, \varphi, \chi) \to (N, \psi, \chi')$ with φ and ψ epimorphic in $\mathcal{E}$. We have to show that $f\chi_{\psi(L)} = \chi_{\varphi(L)}$. For this consider the following diagram

$$\begin{array}{ccccccc} & & L & = & L & & \\ & \diagup\!\diagup \ \circlearrowleft & \downarrow & \circlearrowleft & \downarrow & & \\ L & \twoheadrightarrow & \varphi(L) & \twoheadrightarrow & \psi(L) & \xrightarrow{!} & 1\!\!1 \\ \downarrow & (1) & \downarrow & (2) & \downarrow & (3) & \downarrow{\scriptstyle\top} \\ \Sigma^* & \underset{\varphi}{\twoheadrightarrow} & M & \underset{f}{\twoheadrightarrow} & N & \xrightarrow[\chi_{\psi(L)}]{} & \Omega \end{array}$$

where the vertical composite morphisms $L \to M$ and $L \to N$ are the (epi, mono) factorisations of, respectively, $i\varphi$ and $i\psi$ (where $i : L \hookrightarrow \Sigma^*$ is the inclusion of L in Σ^*), and $\varphi(L) \to \psi(L)$ is the unique filler. Showing $f\chi_{\psi(L)} = \chi_{\varphi(L)}$ amounts to showing the square (2+3) is a pullback (by Remark 4). But we already know that (3) is a pullback so we have to show, by pullback pasting, that (2) is a pullback, and to do so we use Carboni, Janelidze, Kelly and Paré [6, Lemma 4.6]. As a topos, $\mathcal{E}$ is regular, φ is an epimorphism (by hypothesis) and (1) is a pullback ($\varphi\chi_{\varphi(L)} = \chi_L$ then L is the pullback of $\varphi(L)$ along φ, cf. Remark 4); thus (2) is a pullback iff (1+2) is. But $\varphi f = \psi$ so (1+2+3) is the pullback stating $\psi\chi_{\psi(L)} = \chi_L$ and (3) is a pullback and therefore so is (1+2) by pasting.

Of course, Σ^* always recognises any language L. This can be seen as the consequence of the fact that the initial automaton always exists.

Definition 17. *Let* $\underline{\mathcal{A}} : \mathcal{I}_\Sigma \to \mathrm{Set}$ *be an automaton on an alphabet* Σ *in a W-topos* $\mathcal{E}$. *The morphism* $\underline{\mathcal{A}}_{\mathrm{st,st}} : \Sigma^* \to \underline{\mathcal{A}}(\mathrm{st})^{\underline{\mathcal{A}}(\mathrm{st})}$ *is a monoid morphism with image factorisation* $\Sigma^* \xrightarrow{\tau_{\mathcal{A}}} T(\mathcal{A}) \hookrightarrow \underline{\mathcal{A}}(\mathrm{st})^{\underline{\mathcal{A}}(\mathrm{st})}$. *The monoid* $T(\mathcal{A})$ *is called the* transition *monoid of* $\mathcal{A}$.

Proposition 11. *If* $\underline{\mathcal{A}}$ *recognises* L, *then* $(T(\underline{\mathcal{A}}), \tau_{\underline{\mathcal{A}}})$ *is a* Σ*-generated* L*-monoid.*

Proof. We apply Lemma 2 to the following triangle

$$\begin{array}{ccc} & \Sigma^* & \\ {\scriptstyle\underline{\mathcal{A}}_{\mathrm{st,st}}}\swarrow & & \searrow{\scriptstyle\underline{\mathcal{A}}_{\mathrm{in,out}} = \chi_L} \\ \underline{\mathcal{A}}(\mathrm{st})^{\underline{\mathcal{A}}(\mathrm{st})} & \xrightarrow[((\varepsilon_\Sigma \underline{\mathcal{A}}_{\mathrm{st,out}})^{\vdash})^{\varepsilon_\Sigma \underline{\mathcal{A}}_{\mathrm{in,st}}}]{} & \Omega \end{array}$$

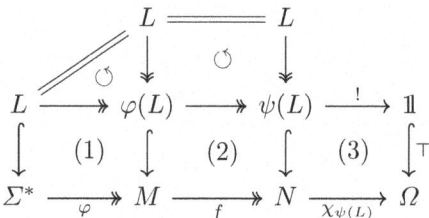

which commutes because of $\mathcal{E}$-functoriality of $\underline{\mathcal{A}}$ and the fact that ε_Σ is the identity of st in the $\mathcal{E}$-category $\mathcal{I}_\Sigma$.

Remark 5. However, the T construction is not functorial; to witness this in Set, consider any finite automaton $\mathcal{A}$ with at least two distinct states q and r on an alphabet with at least two letters a and b, and construct an automaton $\mathcal{B}$ by adding a new state t to $\mathcal{A}$, and such that $\underline{\mathcal{B}}(a)(t) = q$, $\underline{\mathcal{B}}(b)(t) = r$, $\underline{\mathcal{B}}(c)(t) = t$ if $c \in \Sigma \setminus \{a, b\}$ and $\underline{\mathcal{B}}(d)(s) = \underline{\mathcal{A}}(d)(s)$ if $d \in \Sigma, s \in \underline{\mathcal{A}}(\mathrm{st})$. The initial state and final states of $\mathcal{B}$ are those of $\mathcal{A}$ so that the inclusion of states of $\mathcal{A}$ in those of $\mathcal{B}$ defines a monomorphic automaton morphism from $\mathcal{A}$ to $\mathcal{B}$, and the transition monoid of $\mathcal{B}$ contains strictly more endofunctions than those of $\mathcal{A}$. However, an L-monoid morphism between $\tau_\mathcal{A}$ and $\tau_\mathcal{B}$ has to be surjective because $\tau_\mathcal{A}$ and $\tau_\mathcal{B}$ are, which is impossible in that case.

The transition monoid construction might not be functorial but it at least preserves divisibility.

Proposition 12. *The T construction induces two functors : a covariant one $\overrightarrow{T}$ from the wide subcategory* $\mathrm{Auto}_{epi}(L)$ *of* $\mathrm{Auto}(L)$ *of automata and pointwise epic automaton morphisms to* $\Sigma\,\mathrm{Mon}(L)$*, and a contravariant one $\overleftarrow{T}$ from the wide subcategory* $\mathrm{Auto}_{mono}(L)$ *of* $\mathrm{Auto}(L)$ *of automata and pointwise monic automaton morphisms to* $\Sigma\,\mathrm{Mon}(L)$.

Proof. Consider a pointwise epimorphic automaton morphism $e : \underline{\mathcal{A}} \twoheadrightarrow \underline{\mathcal{B}}$. By $\mathcal{E}$-naturality of e, epimorphy of e entailing monomorphy of $\underline{\mathcal{B}}(\mathrm{st})^e$ and $(\mathrm{epi}, \mathrm{mono})$ factorisation we have a unique filler

$$\begin{array}{ccccccc}
\Sigma^* & \xrightarrow{\tau_{\underline{\mathcal{A}}}} & T(\underline{\mathcal{A}}) & \hookrightarrow & & & \underline{\mathcal{A}}(\mathrm{st})^{\underline{\mathcal{A}}(\mathrm{st})} \\
\| & & \downarrow{\scriptstyle \exists! T(e)} & & & & \downarrow{\scriptstyle e^{\underline{\mathcal{A}}(\mathrm{st})}} \\
\Sigma^* & \xrightarrow[\tau_{\underline{\mathcal{B}}}]{} & T(\underline{\mathcal{B}}) & \hookrightarrow & \underline{\mathcal{B}}(\mathrm{st})^{\underline{\mathcal{B}}(\mathrm{st})} & \xhookrightarrow[\underline{\mathcal{B}}(\mathrm{st})^e]{} & \underline{\mathcal{B}}(\mathrm{st})^{\underline{\mathcal{A}}(\mathrm{st})}
\end{array}$$

making the diagram commute, and it also is an epimorphism. By functoriality of orthogonal factorisation systems, this construction is functorial where it makes sense, namely on $\mathrm{Auto}_{\mathrm{epi}}(L)$. A similar argument of functorial factorisation allows constructing the contravariant functor $\overleftarrow{T}$ and uses the fact that monomorphy of m implies monomorphy of $m^{\underline{\mathcal{A}}(\mathrm{st})}$.

Corollary 2. *If* $\underline{\mathcal{A}}$ *divides* $\underline{\mathcal{B}}$*, i.e. we have* $\underline{\mathcal{A}} \overset{e}{\twoheadleftarrow} \underline{\mathcal{C}} \overset{m}{\hookrightarrow} \underline{\mathcal{B}}$*, then* $T(\underline{\mathcal{A}}) \xleftarrow{\overrightarrow{T}(e)} T(\underline{\mathcal{C}}) \xleftarrow{\overleftarrow{T}(m)} T(\underline{\mathcal{B}})$*, so in particular,* $T(\underline{\mathcal{A}})$ *divides* $T(\underline{\mathcal{B}})$.

An automaton recognising L can be seen as a presentation of an L-monoid. But in fact, each Σ-generated L-monoid can be seen as the transition monoid of an automaton.

Lemma 3. *The covariant functor* $\overrightarrow{T} : \mathrm{Auto}_{epi}(L) \to \Sigma\,\mathrm{Mon}(L)$ *has a section (up to natural isomorphism) A defined by*

$$\underline{A(M,\varphi)}(\mathrm{st}) = M,\ \underline{A(M,\varphi)}_{\mathrm{in,st}} = \varphi,\ \underline{A(M,\varphi)}_{\mathrm{st,st}} = \Sigma^* \xrightarrow{\varphi} M \xrightarrow{m^{\dashv}} M^M,$$

$$\underline{A(M,\varphi)}_{\mathrm{st,out}} = \Sigma^* \xrightarrow{\varphi} M \xrightarrow{m^{\dashv}} M^M \xrightarrow{\chi^M_{\varphi(L)}} \Omega^M$$

and sends an L-monoid morphism $f : (M,\varphi) \to (N,\psi)$ *to an automaton morphism* $(\mathrm{id}_{\mathbb{1}}, f, \mathrm{id}_\Omega) : \underline{A(M,\varphi)} \to \underline{A(N,\psi)}$.

Proof. Denote by (M, m, e) the monoid M, $\underline{A(M,\varphi)}_{\mathrm{in,out}} = (\underline{A(M,\varphi)}_{\mathrm{in,st}} \times \varepsilon_\Sigma \underline{A(M,\varphi)}_{\mathrm{st,out}})\,\mathrm{ev}^M_\Omega$ by $\mathcal{E}$-functoriality, but $\varepsilon_\Sigma \underline{A(M,\varphi)}_{\mathrm{st,out}} = \varepsilon_\Sigma \varphi m^{\dashv} \chi^M_L = e m^{\dashv} \chi_L$ because φ is a monoid morphism and $em^{\dashv} = \mathrm{id}^{\mathcal{E}}_M$ by right unitality in the monoid M. Therefore, $\underline{A(M,\varphi)}_{\mathrm{in,out}} = (\varphi \times \chi^{\dashv}_L)\,\mathrm{ev}^M_\Omega : \Sigma^* \times \mathbb{1} \to \Omega$ corresponds to $\varphi\chi_{\varphi(L)} : \Sigma^* \to \Omega$, itself equal to χ_L because (M,φ) is a Σ-generated L-monoid. We have to show that M is isomorphic to a submonoid of M^M in a natural way; it is sort of an internal Cayley theorem. Recall that by definition $\Sigma^* \xrightarrow{\tau_{A(M,\varphi)}} T(\underline{A(M,\varphi)}) \hookrightarrow M^M$ is the (regular epi, mono)-factorisation of the monoid morphism $\underline{A(M,\varphi)}_{\mathrm{st,st}} = \varphi m^{\dashv}$ which is also a factorisation : φ is epic by hypothesis and $m^{\dashv}$ is monic because it has a retract M^e by right unitality.

Theorem 2. *Let L be a language over an alphabet* Σ *in a W-topos* $\mathcal{E}$*. The transition monoid of the minimal automaton* $T(\mathrm{Min}(L))$ *is minimal in the category of L-monoids. We then call this monoid the* syntactic monoid *of L and denote it by* $\mathrm{Syn}(L)$.

Proof. Let (M,φ,χ) be any L-monoid, and by Lemma 2, $(\mathrm{Im}\ \varphi, e)$ its reflected Σ-generated sub-L-monoid.

$$\begin{array}{ccccccc}
\mathrm{Min}(L) & \twoheadleftarrow & \underline{\mathcal{B}} & \hookrightarrow & \underline{A(\mathrm{Im}\ \varphi, e)} & & \text{in Auto}(L) \\
 & \downarrow \overrightarrow{T} & & \downarrow \overleftarrow{T} & & & \\
\mathrm{Syn}(L) & \twoheadleftarrow & T(\underline{\mathcal{B}}) & \twoheadleftarrow & (\mathrm{Im}\ \varphi, e) & \hookrightarrow\ (M,\varphi,\chi) & \text{in Mon}(L)
\end{array}$$

In Set, the syntactic monoid provides another characterisation of regularity: a language is regular if and only if its syntactic monoid is finite. We discuss this fact with different finiteness conditions. In the following results we might use the fact that $\mathrm{Min}(L)$ is a *reachable automaton.*

Definition 18. *An L-automaton* $\underline{A}$ *is* reachable *if* $\emptyset(L) \xrightarrow{!} \underline{A}$ *is an epimorphism.*

According to Proposition 5, $\underline{A}$ is then reachable iff $\underline{A}_{\mathrm{in,st}}$ is an epimorphism.

Lemma 4. *If* $\underline{A}$ *is reachable, the states object of* $\underline{A}$ *is a quotient of its transition monoid.*

Proof. Consider the diagram

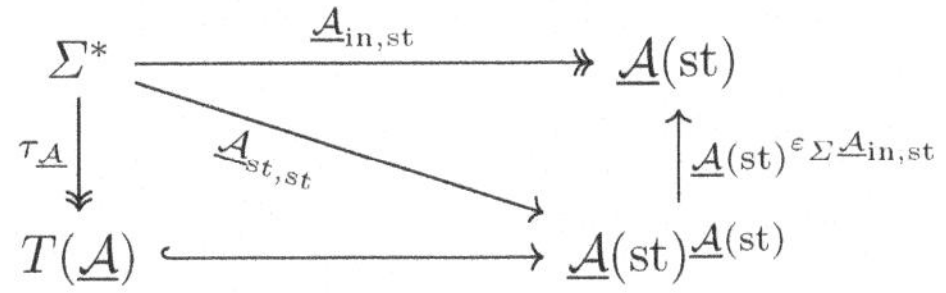

where the top-right triangle commute by $\mathcal{E}$-functoriality of $\underline{\mathcal{A}}$ and the bottom-left one does by definition of $\tau_{\underline{\mathcal{A}}}$. Then, the whole diagram commutes and therefore by left-cancellation of epimorphisms, the composite arrow $T(\underline{\mathcal{A}}) \hookrightarrow \underline{\mathcal{A}}(\mathrm{st})^{\underline{\mathcal{A}}(\mathrm{st})} \to \underline{\mathcal{A}}(\mathrm{st})$ is an epimorphism.

Theorem 3. *Let L be a language in a W-topos $\mathcal{E}$.*

1. *If* $\mathrm{Syn}(L)$ *is K-finite then L is K-regular.*
2. *If $\mathcal{E}$ is Boolean and L is K-regular then* $\mathrm{Syn}(L)$ *is K-finite.*

Proof. 1. By Lemma 4, $\mathrm{Min}(L)(\mathrm{st})$ is a quotient of $\mathrm{Syn}(L)$; the former is K-finite when the latter is according to Johnstone [8, Lemma 5.4.4.ii].
2. As $\mathcal{E}$ is Boolean, $\mathrm{Min}(L)$ is K-finite because is K-regular by Theorem 1, and also by Booleanity, exponentials and subobjects of K-finite objects are K-finite according to Acuña-Ortega and Linton [1, Main Theorem], therefore $\mathrm{Min}(L)^{\mathrm{Min}(L)}$ is finite and so is its subobject $\mathrm{Syn}(L)$.

4 Equivariant Automata

Automata in toposes $\mathbb{B}G$ for a discrete group G were explored by Bojańczyk, Klin and Lasota [5, Section 3] with decomposition-finiteness. Here we also discuss dK-finiteness in this setting.

Definition 19. *An equivariant automaton is an automaton in a topos $\mathbb{B}G$ for a discrete group G.*

Proposition 13. *Let G be a discrete group. An object A of $\mathbb{B}G$ is*

1. *dK-finite iff it is finite as a set, and*
2. *decomposition-finite iff it has a finite number of orbits.*

Proof. 1. See Johnstone [8, Example 5.4.19].
2. Connected G-sets are exactly the transitive ones, therefore orbits.

Example 2. As a first "toy" example we consider an automaton in the topos of sets with an involution, namely $\mathbb{B}\mathbb{Z}_{/2\mathbb{Z}}$, the topos of the actions of the two-element group. For each set with an involution (X, i) we shall denote $i(x) = \overline{x}$. Consider the two-letter alphabet $\Sigma = \{a, \overline{a}\}$ where the involution exchanges the two letters. The free (internal) monoid is simply the free monoid Σ^* where the involution swaps the two letters. We define the language $L = \{lu\overline{l} | l \in A, u \in A^*\}$

of words of length at least two whose first and last letters are different. The Nerode quotient $\Sigma^*_{/\equiv_L}$ is the five elements set

$$\{L, a^{-1}L, \overline{a}^{-1}L, (a\overline{a})^{-1}L, (\overline{a}a)^{-1}L\}$$

so that its only fixed point is $L \in \Sigma^*_{/\equiv_L}$. This allows us to describe the minimal $\mathbb{B}\,\mathbb{Z}_{/2\mathbb{Z}}$-automaton of this $\mathbb{B}\,\mathbb{Z}_{/2\mathbb{Z}}$-language (denote $u^{-1}L$ by $[u]$):

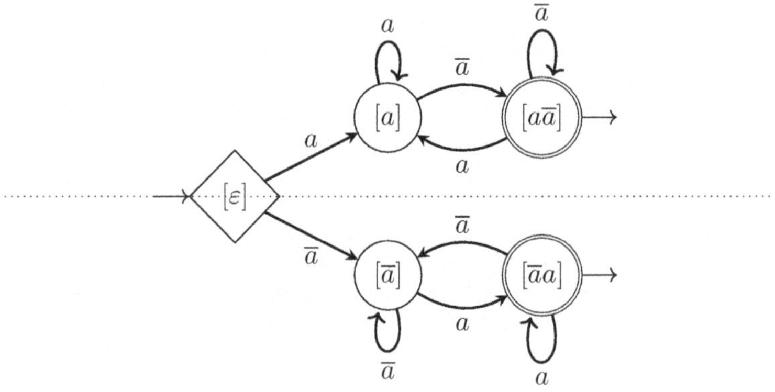

5 Nominal Automata

We define the Myhill-Schanuel topos $\mathbb{B}\,\mathrm{Aut}(\mathbb{N})$ of *nominal sets* and *equivariant functions* to be the category of continuous actions on discrete spaces of the topological group $\mathrm{Aut}(\mathbb{N})$ of permutations of a countable set $\mathbb{N}$ of *names*, where the topology is induced by the inclusion $\mathrm{Aut}(\mathbb{N}) \subset \mathbb{N}^{\mathbb{N}}$ (where $\mathbb{N}^{\mathbb{N}}$ carries the product topology); equivariant functions are those functions that commute with the action. According to Mac Lane and Moerdijk [13, Theorem 3.9.2], this indeed form an atomic W-topos.

Definition 20. *A nominal automaton is an automaton in the topos* $\mathbb{B}\,\mathrm{Aut}(\mathbb{N})$.

In the topos $\mathbb{B}\,\mathrm{Aut}(\mathbb{N})$, point 2 of Theorem 1 becomes:

Theorem 4 (Cf. Bojańczyk, Klin and Lasota [5, Theorem 5.2]). *Let L be a language over Σ in the topos $\mathbb{B}\,\mathrm{Aut}(\mathbb{N})$. The language L is decomposition-regular iff $\Sigma^*_{/\equiv_L}$ is orbit-finite.*

Example 3. Consider on the alphabet $\mathbb{N}$ the classical example of the language L of words where the first letter appears at least once again further in the word

$$L = \{ab_1b_2\cdots b_n \in \mathbb{N}^* | n \in \mathbb{N}, a, b_i \in \mathbb{N}, \exists i \in \mathbb{N}, 1 \leq i \leq n, b_i = a\}$$

which is a nominal set. Indeed, it is stable under permutations of letters, and each word is finitely supported by the finite set of letters that appears in it.

Let us compute the minimal automaton for this language. Recall that for any nominal set A, $\Omega^A = \{P \subset A | P \text{ is finitely supported}\}$. The states object is $\mathbb{N}^*_{/\equiv_L}$ and here it is therefore the nominal set

$$\{u^{-1}L \subset \mathbb{N}^* \big| u \in \mathbb{N}^*\} = \{L\} \cup \{\mathbb{N}^* a \mathbb{N}^* | a \in \mathbb{N}\} \cup \{\mathbb{N}^*\}$$

so that by Theorem 1, L is decomposition-regular.
To finally describe the minimal automaton, recall that the initial state, a fixed point, is simply the equivalence class L of ε, and an equivalence class is a final state if and only if it contains a language that contains the empty string, ε. The only such class is $\mathbb{N}^*$, therefore it is the only final state of the automaton. Then, the action $- \cdot a$ of a letter $a \in \mathbb{N}$ is given by $K \cdot a = a^{-1}K$. The following diagram sums up the construction, and the register automaton counterpart of this nominal automaton can be found in Francez and Kaminski [9, Figure 7]:

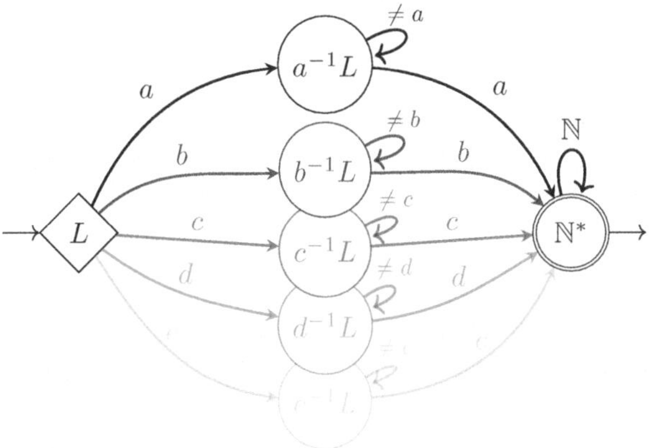

where the diamond state is the initial state and the double circle is a final state (in fact the only one in this case). Observe that states in the same column are in the same orbit. The orbit $\{a^{-1}L \big| a \in \mathbb{N}\}$ can be thought of as a single state such that a transition from the initial state to this state-orbit writes the read letter (which is the first letter of the word) in a register. Reading the rest of the word, we loop on this state-orbit until we read a letter that is other than the one in the register. In that case we reach the final state on which we loop until the word is finished reading.

The syntactic monoid $\mathrm{Syn}(L)$ of L, however, is not decomposition-finite, cf. [4, Solution to Exercise 91].

In nominal automata theory, dK-finiteness is not an interesting notion:

Proposition 14. *The dK-finite objects in* $\mathbb{B}\,\mathrm{Aut}(\mathbb{N})$ *are exactly finite sets with the trivial action.*

Proof. The topos $\mathbb{B}\,\mathrm{Aut}(\mathbb{N})$ is a subtopos of $[\mathrm{Aut}(\mathbb{N})^{\mathrm{op}}, \mathrm{Set}]$, so dK-finite objects of $\mathbb{B}\,\mathrm{Aut}(\mathbb{N})$ have to be dK-finite objects of $[\mathrm{Aut}(\mathbb{N})^{\mathrm{op}}, \mathrm{Set}]$ (cf. Johnstone [8, Corollary 5.4.12]), which entails they have a finite underlying set according to Proposition 13. Because $\mathbb{B}\,\mathrm{Aut}(\mathbb{N})$ is atomic, we restrict to connected objects, i.e.

non-empty transitive nominal sets, and show that the only transitive nominal set with finite underlying set is $\mathbb{1}$. According to Bojańczyk [4, Theorem 6.3], transitive nominal sets are of the form $\mathbb{N}^{[F]}/_G$ where F is a finite set, $\mathbb{N}^{[F]}$ is the set of injections from F to $\mathbb{N}$, and G a subgroup of $\mathfrak{S}_F$. But then the only case for which $\mathbb{N}^{[F]}_{/\mathfrak{S}_F}$ is a finite set is when $F \cong \emptyset$, and in that case $\mathbb{N}^{[F]}/_G \cong \mathbb{1}$.

6 Conclusion and Future Perspectives

Because the subobject classifier in Set played a crucial rôle in the Colcombet and Petrişan functorial viewpoint of automata, we adapted it to a wide class of toposes, recovering minimisation results, and adding Myhill-Nerode type theorems to it, as well as some discussions about the syntactic monoid of a language, everything internally to a given topos. The results still make sense for sets, and can be applied to the Myhill-Schanuel topos of nominal sets.

We would like to make more use of the enriched Colcombet and Petrişan functorial point of view of automata. For example, because everything is done using enriched category theory, a generalisation to non-cartesian monoidal closed categories is immediate, and should at least encompass the categories of Adámek, Milius and Urbat [2].

Acknowledgments. The author would like to thank Clemens Berger for the discussions and the help for the redaction of this article.

References

1. Acuña-Ortega, O., Linton, F.E.J.: Finiteness and decidability: I. In: Fourman, M., Mulvey, C., Scott, D. (eds.) Applications of Sheaves. LNM, vol. 753, pp. 80–100. Springer, Heidelberg (1979). https://doi.org/10.1007/BFb0061813
2. Adámek, J., Milius, S., Urbat, H.: Syntactic monoids in a category. In: Proceedings CALCO 2015, p. 1504.02694, June 2015
3. Arbib, M.A., Manes, E.G.: Adjoint machines, state-behavior machines, and duality. J. Pure Appl. Algebra **6**(3), 313–344 (1975). https://doi.org/10.1016/0022-4049(75)90028-6
4. Bojańczyk, M.: Slightly Infinite Sets, September 2019. https://www.mimuw.edu.pl/~bojan/paper/atom-book
5. Bojańczyk, M., Klin, B., Lasota, S.: Automata theory in nominal sets. Log. Methods Comput. Sci. **103**(4), 1402.0897 (2014). https://doi.org/10.2168/LMCS-10(3:4)2014
6. Carboni, A., Janelidze, G., Kelly, G.M., Paré, R.: On localization and stabilization for factorization systems. Appl. Categ. Struct. **5**(1), 1–58 (1997). https://doi.org/10.1023/A:1008620404444
7. Colcombet, T., Petrişan, D.: Automata minimization: a functorial approach. Log. Methods Comput. Sci. **16**, 1712.07121 (2020). https://doi.org/10.23638/LMCS-16(1:32)2020
8. Johnstone, P.T.: Sketches of an Elephant. Oxford University Press, Oxford (2002)
9. Kaminski, M., Francez, N.: Finite-memory automata. Theoret. Comput. Sci. **134**, 329–363 (1994)

10. Kelly, G.M.: Basic Concepts of Enriched Category Theory, London Mathematical Society Lecture Note Series, vol. 64. Cambridge University Press (1982). http://www.tac.mta.ca/tac/reprints/articles/10/tr10abs.html
11. Kleene, S.C.: Representation of events in nerve nets and finite automata. Automata Stud. **34**, 3–42 (1956)
12. Lucyshyn-Wright, R.B.B.: Enriched factorization systems. Theory Appl. Categ. **29**(18), 1401.0315 (2014)
13. Mac Lane, S., Moerdijk, I.: Sheaves in Geometry and Logic. Universitext, Springer, New York (1994). https://doi.org/10.1007/978-1-4612-0927-0
14. Moerdijk, I., Palmgren, E.: Wellfounded trees in categories. Ann. Pure Appl. Logic **104**(1), 189–218 (2000). https://doi.org/10.1016/S0168-0072(00)00012-9
15. Nerode, A.: Linear automaton transformations. Proc. Am. Math. Soc. **9**(4), 541–544 (1958). https://doi.org/10.2307/2033204
16. Pin, J.É.: Mathematical Foundations of Automata Theory, February 2022. https://www.irif.fr/~jep/PDF/MPRI/MPRI.pdf

Graded Semantics and Graded Logics for Eilenberg-Moore Coalgebras

Jonas Forster[1(✉)], Lutz Schröder[1], Paul Wild[1], Harsh Beohar[2], Sebastian Gurke[3], and Karla Messing[3]

[1] Friedrich-Alexander-Universität Erlangen-Nürnberg, Erlangen, Germany
{jonas.forster,lutz.schroeder,paul.wild}@fau.de

[2] University of Sheffield, Sheffield, UK
h.beohar@sheffield.ac.uk

[3] Universität Duisburg-Essen, Duisburg, Germany
{sebastian.gurke,karla.messing}@uni-due.de

Abstract. Coalgebra, as the abstract study of state-based systems, comes naturally equipped with a notion of behavioural equivalence that identifies states exhibiting the same behaviour. In many cases, however, this equivalence is finer than the intended semantics. Particularly in automata theory, behavioural equivalence of nondeterministic automata is essentially bisimilarity, and thus does not coincide with language equivalence. Language equivalence can be captured as behavioural equivalence on the determinization, which is obtained via the standard powerset construction. This construction can be lifted to coalgebraic generality, assuming a so-called Eilenberg-Moore distributive law between the functor determining the type of accepted structure (e.g. word languages) and a monad capturing the branching type (e.g. nondeterministic, weighted, probabilistic). Eilenberg-Moore-style coalgebraic semantics in this sense has been shown to be essentially subsumed by the more general framework of graded semantics, which is centrally based on graded monads. Graded semantics comes with a range of generic results, in particular regarding invariance and, under suitable conditions, expressiveness of dedicated modal logics for a given semantics; notably, these logics are evaluated on the original state space. We show that the instantiation of such graded logics to the case of Eilenberg-Moore-style semantics works extremely smoothly, and yields expressive modal logics in essentially all cases of interest. We additionally parametrize the framework over a quantale of truth values, thus in particular covering both the two-valued notions of equivalence and quantitative ones, i.e. behavioural distances.

J. Forster, L. Schröder, P. Wild, S. Gurke, K. Messing—Supported by the Deutsche Forschungsgemeinschaft (DFG, German Research Foundation) – project number 434050016 (SpeQt).

H. Beohar—Supported by the EPSRC NIA Grant EP/X019373/1.

Published by Springer Nature Switzerland AG 2024
B. König and H. Urbat (Eds.): CMCS 2024, LNCS 14617, pp. 114–134, 2024.
https://doi.org/10.1007/978-3-031-66438-0_6

1 Introduction

When dealing with the logical aspects of state-based systems, one is particularly interested in the property of *expressiveness*, that is, the ability of a logic to differentiate between states that behave in different ways. The prototypical example of this property is captured by the *Hennessy-Milner theorem* [15], with modal logic distinguishing states in finitely branching transition systems precisely up to bisimilarity. There is, however, a wide array of equivalences of interest that are coarser than bisimilarity [13], each necessitating a different type of logic to stay invariant under the semantics while ensuring expressiveness. A similar story unfolds when state-based systems are studied abstractly as coalgebras for a given functor that encapsulates the transition type of systems [32]: The finest and mathematically most convenient type of equivalence is given by coalgebraic behavioural equivalence, with much of the literature on coalgebraic logic focusing on expressiveness with respect to this type of equivalence (e.g. [11,23,27,30,33,39]), though this might not necessarily be the equivalence the application demands. Consider for example nondeterministic automata, i.e. coalgebras for the **Set**-endofunctor $2 \times (\mathcal{P}-)^{\Sigma}$. The equivalence of interest in these systems is language equivalence, and as such is potentially much coarser than the coalgebraic notion of behavioural equivalence, which in this case instantiates to bisimilarity. A possible way to deal with this mismatch is to first transform the nondeterministic automaton into a deterministic one, that is, a coalgebra for the **Set**-endofunctor $F = 2 \times (-)^{\Sigma}$, via the powerset construction, obtaining language equivalence as behavioural equivalence in the determinized automaton. The powerset construction generalizes to coalgebras for functors of the form FT where F is a functor capturing the type of accepted structure (e.g. word languages for $F = 2 \times (-)^{\Sigma}$ as above) and a monad T capturing the branching type of systems ($T = \mathcal{P}$ as above captures nondeterminism; other choices of T capture, e.g., probabilistic or weighted branching). To be applicable, this approach requires a so-called Eilenberg-Moore distributive law of T over F [34]; it then equips FT-coalgebras with a language-type semantics determined by F, to which we refer as *Eilenberg-Moore semantics*.

Our present aim is to obtain modal logics that are expressive and invariant for Eilenberg-Moore semantics, and at the same time can be seen as fragments of the standard expressive branching-time coalgebraic modal logics (in analogy to logics for the linear-time/branching-time spectrum of labelled transition system semantics [13], which are fragments of standard Hennessy-Milner logic). To this end, we exploit the machinery of *graded semantics* [8,28], in which notions of behavioural equivalence are modelled by mapping into a *graded monad* [35]; it has been shown that Eilenberg-Moore semantics can essentially be cast as a graded semantics [25,28]. Graded semantics comes with a general notion of invariant *graded logic* and a criterion for a graded logic to be expressive [8,28].

Contribution. By instantiating the expressivity criterion of the graded semantics framework to Eilenberg-Moore semantics, we show that it is sufficient to provide a set of modal operators that separate the elements of FX, while the treatment of T is automatically provided by the framework. Separation of FX is typically easy to ensure, justifying the slogan that Eilenberg-Moore semantics essentially always admits an expressive invariant logic. We parametrize our results over the choice of a quantale that serves as a domain of both distances and truth values, allowing an instantiation to both the two-valued setting, where states are either equivalent or not and formulae take binary truth values, and to quantitative settings, where similarity of states is a continuum and formulae may take intermediate values, for instance in the real unit interval. We thus in particular cover notions of *behavioural distance* (e.g. [2,3,38]), providing logics that are expressive in the sense that the behavioural distance between states is always witnessed by differences in the evaluation of suitable formulae. We discuss a range of examples, in some cases obtaining new characteristic modal logics, e.g. for probabilistic trace equivalence of reactive probabilistic automata with black-hole termination.

Omitted proofs may be found in the full version [12].

Related Work. There has been a fair amount of work on the coalgebraic treatment of system semantics beyond branching time. Approaches using Kleisli-type distributive laws [14] and Eilenberg-Moore distributive laws (e.g. [6,18,20,34]) are subsumed by graded semantics [25,28]. The Kleisli approach has also been applied to infinite-trace semantics (e.g. [7,16,19,37]) and to trace semantics via steps [31]. Klin and Rot [21] present a notion of semantics based on selecting a modal logic, which is then expressive by definition of the semantics. For our present purposes, the most closely related piece of previous work uses corecursive algebras as a unifying concept subsuming the Kleisli-based, Eilenberg-Moore-based, and logic-based approaches to coalgebraic trace semantics [31]. In particular, the comparison between the Eilenberg-Moore-based and the logic-based semantics in this framework [31, Section 7.1] can be read as an expressiveness criterion for logics over Eilenberg-Moore semantics. In relation to this criterion, the distinguishing feature of our present main result lies in the concreteness of the construction of the logic in terms of modal and propositional operators, as well as the ease of checking our expressiveness criterion, which comes essentially for free in all cases of interest. We note also that our criterion works in quantalic generality, and thus applies also to notions of behavioural distance, which so far are not covered in the approach via corecursive algebras.

Through its applicability to behavioural distances, our work relates additionally to a spate of recent work on the coalgebraic treatment of characteristic logics for behavioural distances. For the most part, such results have been concerned mainly with branching-time distances (e.g. [11,22,23,39]).

Kupke and Rot [24] study logics for *coinductive predicates*, generalizing branching-time behavioural distances. Our overall setup differs from the one used in [24] by working with coalgebras for functors that live natively on metric spaces, including such functors that are not liftings of a set functor.

In recent work by König and (some of) the present authors [4,5], expressive logics for coalgebraic trace-type behavioural distances have been approached by setting up Galois connections between logics and distances. This concept is highly general (and in fact not even tied to models being coalgebras) but requires a comparatively high amount of effort for concrete instantiations. Moreover, its focus is on fixpoint characterizations of logical distance rather than on expressiveness w.r.t. a given notion of behavioural distance, and in fact the behaviour function inducing behavioural distance is defined directly via the logic.

2 Preliminaries

We assume basic familiarity with category theory (e.g. [1]). In the following, we recall requisite definitions and facts on universal coalgebra, quantales, and lifting functors to categories of monad algebras.

2.1 Universal Coalgebra

State-based systems of various types, such as non-deterministic, probabilistic, weighted, or game-based transition systems, are treated uniformly in the framework of *universal coalgebra* [32]. The branching type of a system is encapsulated as a functor $G\colon \mathbf{C} \to \mathbf{C}$ on a suitable base category $\mathbf{C}$, for instance on the category **Set** of sets and maps. A *G-coalgebra* (C, c) then consists of a $\mathbf{C}$-object C, thought of as an object of *states*, and a morphism $c\colon C \to GC$, thought of as a *transition map* that assigns to each state a structured collection of successor states, with structure determined by G. For instance, on $\mathbf{C} = \mathbf{Set}$, a $\mathcal{P}$-coalgebra for the covariant powerset functor is just a nondeterministic transition system, while a G-coalgebra for the functor G given by $GX = 2 \times X^{\Sigma}$, with Σ a fixed *alphabet*, is a deterministic automaton (without initial state), assigning to each state a finality status and a tuple of successors, one for every letter in Σ.

A *morphism* $h\colon (C, c) \to (D, d)$ of G-coalgebras is a $\mathbf{C}$-morphism $h\colon C \to D$ that is compatible with the transition maps in the sense that $d \cdot h = Gh \cdot c$. States $x, y \in C$ in a coalgebra (C, c) are *behaviourally equivalent* if there exist a G-coalgebra (D, d) and a morphism $h\colon (C, c) \to (D, d)$ such that $h(x) = h(y)$. For instance, two states in a labelled transition system (i.e. a coalgebra for $G = \mathcal{P}(\Sigma \times (-))$ where Σ is the set of labels) are behaviourally equivalent iff they are bisimilar in the usual sense.

The (initial ω-segment of) the *final chain* of G is the sequence $(G^n 1)_{n<\omega}$ of $\mathbf{C}$-objects. Given a G-coalgebra (C, c), we have the *canonical cone* of maps $c_n\colon C \to G^n 1$, defined by c_0 being the unique map $C \to 1$ and by $c_{n+1} = C \xrightarrow{c} GC \xrightarrow{Gc_n} G^{n+1}1$. When $\mathbf{C}$ is a concrete category over **Set**, states $x, y \in C$ are termed *finite-depth behaviourally equivalent* if $c_n(x) = c_n(y)$ for all $n \in \mathbb{N}$. For finitary set functors, finite-depth behavioural equivalence and behavioural equivalence coincide [40].

2.2 Quantales

We use (symmetrized) *quantale-enriched categories* as a joint generalization of equivalence relations and pseudometric spaces; this enables us to cover both two-valued and quantitative semantics and logics uniformly in one framework. In a nutshell, a quantale is a monoid in the category of complete join semilattices. Explicitly, this notion expands as follows:

Definition 1. A (commutative unital) *quantale* $\mathcal{V} = (V, \otimes, k, \leq)$ consists of a set V that carries both the structure of a complete lattice $(V, \leq)$ and the structure of a commutative monoid $(V, \otimes, k)$ such that for all $v \in V$, the operation $- \otimes v$ is join-continuous; that is,

$$\left(\bigvee_{i\in I} u_i\right) \otimes v = \bigvee_{i\in I} (u_i \otimes v)$$

where we use $\bigvee$ to denote joins.

By the standard equivalence between join preservation and adjointness for functions on complete lattices, it follows that for every $b \in V$, the map $- \otimes b$ has a right adjoint $[b, -]$, with defining property

$$a \otimes b \leq c \Leftrightarrow a \leq [b, c]$$

As first observed by Lawvere [26], metric spaces can be seen as enriched categories, which leads to the notion of categories enriched in a quantale $\mathcal{V}$, or briefly $\mathcal{V}$-categories, as a generalized notion of (pseudo-)metric space:

Definition 2. A $\mathcal{V}$*-category* is a pair (X, d_X) consisting of a set X and a function $d_X : X \times X \to V$ such that for all $x, y, z \in X$ we have $d_X(x, y) \otimes d_X(y, z) \leq d_X(x, z)$, as well as $k \leq d_X(x, x)$. A $\mathcal{V}$-category (X, d_X) is *symmetric* if $d_X(x, y) = d_X(y, x)$ for all $x, y \in X$, and *separated* if $k \leq d_X(x, y)$ implies $x = y$. A function $f : X \to Y$ is a $\mathcal{V}$*-functor* between $\mathcal{V}$-categories (X, d_X) and (Y, d_Y) if $d_X(a, b) \leq d_Y(f(a), f(b))$ for all $a, b \in X$.

We fix a quantale $\mathcal{V}$ for the rest of the technical development. We write $\mathbf{DPMet}_{\mathcal{V}}$ for the category of $\mathcal{V}$-categories and $\mathcal{V}$-functors, which we view as generalized directed pseudometric spaces, with distance values in $\mathcal{V}$. Further, we write $\mathbf{PMet}_{\mathcal{V}}$ for the full subcategory of symmetric $\mathcal{V}$-categories, viewed as generalized pseudometric spaces, and $\mathbf{Met}_{\mathcal{V}}$ for the full subcategory of symmetric and separated $\mathcal{V}$-categories, viewed as generalized metric spaces. The quantale $\mathcal{V}$ itself has the structure of an object in $\mathbf{DPMet}_{\mathcal{V}}$, where $d(x, y) = [x, y]$ for all $x, y \in \mathcal{V}$. It may also be viewed as an object in $\mathbf{Met}_{\mathcal{V}}$ through symmetrization: $d_{\mathrm{sym}}(x, y) = [x, y] \wedge [y, x]$. In this way, we will often use $\mathcal{V}$ as the codomain of evaluation morphisms of our logics. We will focus on the following two examples:

Example 3. 1. The lattice $2 = \{\bot, \top\}$ carries a quantale $\mathbf{2} = (2, \wedge, \top, \leq)$. In this case, $[b, c]$ is just the Boolean implication $b \to c$. The category $\mathbf{PMet_2}$ is isomorphic to the category of setoids, i.e. of equivalence relations and equivalence-preserving maps, while the category $\mathbf{Met_2}$ is isomorphic to the category of sets and functions. We use this quantale to cover two-valued equivalences, used in situations where one is only interested in determining whether states behave in precisely the same way or not.

2. We use the quantale $[0,1]_\oplus = ([0,1], \oplus, 0, \geq)$, where $\oplus$ is truncated addition ($a \oplus b = \min(a+b, 1)$), to cover cases where one wishes to measure differences in the behaviour of states in a continuous manner. In this case, $[-, -]$ is truncated subtraction ($[b, c] = \max(c-b, 0)$). Indeed, taking $[-, 1]$ as negation makes $[0,1]_\oplus$ into an MV-algebra, providing a domain of truth values for multi-valued Łukasiewicz logic. The category $\mathbf{Met}_{[0,1]_\oplus}$ is isomorphic to the usual category of 1-bounded metric spaces and non-expansive maps [26], while $\mathbf{PMet}_{[0,1]_\oplus}$ is isomorphic to the category of pseudometric spaces (that is, distinct elements may take distance 0). Note that the ordering on the set $[0, 1]$ is reversed compared to its natural ordering. This is necessary, since otherwise $\oplus$ does not distribute over the empty join.

We will use the concept of initiality (in the concrete case of $\mathcal{V}$-categories) to describe the fact that a set of morphisms is large enough to witness the distances in its domain. Later, expressivity demands that the set of evaluation morphisms of formulae form an initial source.

Definition 4. A source $\mathfrak{A}$ of $\mathcal{V}$-functors $f_i \colon (X, d_X) \to (Y_i, d_{Y_i})$ is *initial* if $d_X(x, y) = \bigwedge_{i \in I} d_{Y_i}(f_i(x), f_i(y))$ for all $x, y \in X$.

2.3 Lifting Functors to Eilenberg-Moore Categories

Recall that a *monad* (T, μ, η), denoted just T by abuse of notation, on a base category $\mathbf{C}$ consists of a functor $T \colon \mathbf{C} \to \mathbf{C}$ and natural transformations $\mu \colon TT \Rightarrow T$, as well as $\eta \colon \mathrm{Id} \Rightarrow T$ (the *multiplication* and *unit* of T) satisfying natural laws. Monads on $\mathbf{Set}$ may be thought of as encapsulating algebraic theories, with TX being terms over X modulo provable equality, μ collapsing layered terms into terms, thus abstracting substitution, and η converting variables into terms. We call a monad T *affine* [17] when T preserves the terminal object, that is $T1 \cong 1$. For instance, the *distribution monad* $\mathcal{D}$, given by $\mathcal{D}X$ being the set

$$\{f \colon X \to [0,1] \mid f(x) = 0 \text{ for almost all } x \in X,\ \textstyle\sum_{x \in X} f(x) = 1\}$$

of finitely supported probability distributions on X, is affine. Monads induce a natural notion of algebra: A *monad algebra* or *Eilenberg-Moore algebra* (A, a) for T consists of a $\mathbf{C}$-object A and a morphism $a \colon TA \to A$ making the left and middle diagrams below commute.

$$\begin{array}{ccc}
A \xrightarrow{\eta_A} TA & TTA \xrightarrow{Ta} TA & TA \xrightarrow{Tf} TB \\
{}_{\mathrm{id}_A}\searrow \downarrow a & \mu_A \downarrow \quad \downarrow a & a \downarrow \quad \downarrow b \\
A & TA \xrightarrow{a} A & A \xrightarrow{f} B
\end{array}$$

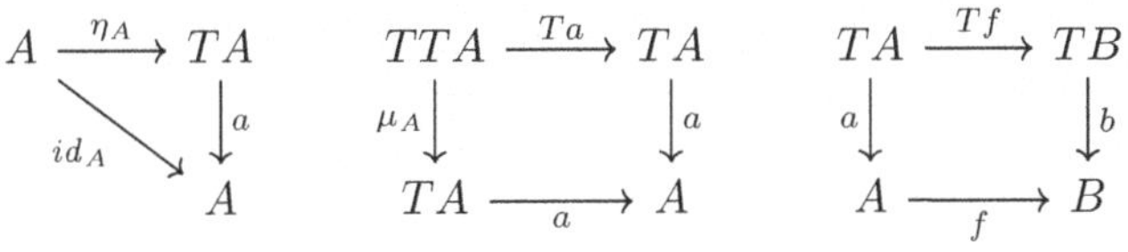

A $\mathbf{C}$-morphism $f \colon A \to B$ is a morphism between algebras $f \colon (A, a) \to (B, b)$ if the right diagram commutes. We write $\mathbf{EM}(T)$ for the category of Eilenberg-Moore algebras for T and their morphisms. We denote the functor that takes a $\mathbf{C}$-object A to the free T-algebra (TA, μ) over A by $L \colon \mathbf{C} \to \mathbf{EM}(T)$. This functor is left adjoint to the forgetful functor $R\colon \mathbf{EM}(T) \to \mathbf{C}$ that takes algebras (A, a) to their carrier A. The category $\mathbf{EM}(T)$ has all limits that $\mathbf{C}$ has [1, Proposition 20.12]. We occasionally need the n-fold power $(A, a)^n$ of an algebra (A, a), whose carrier is the $\mathbf{C}$-object A^n. We denote its algebra structure by $a^{(n)}\colon T(A^n) \to A^n$.

Coalgebraic determinization [34] is concerned with coalgebras for functors of the form $G = FT$ where T is a monad, thought of as capturing the branching type of systems, and F is a functor determining the system semantics. As indicated in the introduction, the basic example is given by nondeterministic automata over an alphabet Σ, which are coalgebras for the set functor $G = FT$ with $FX = 2 \times X^\Sigma$ and $T = \mathcal{P}$, while F-coalgebras are deterministic automata. The coalgebraic generalization of the powerset construction that determinizes nondeterministic automata relies on having a suitable type of distributive law between F and T:

Definition 5. An *Eilenberg-Moore distributive law*, or just *EM law*, of a monad (T, μ, η) over a functor F is a natural transformation $\zeta\colon TF \Rightarrow FT$ such that the following diagrams commute:

$$\begin{array}{ccc} F & \xrightarrow{\eta F} & TF \\ & \underset{F\eta}{\searrow} & \downarrow{\scriptstyle\zeta} \\ & & FT \end{array} \qquad \begin{array}{ccccc} TTF & \xrightarrow{T\zeta} & TFT & \xrightarrow{\zeta T} & FTT \\ \downarrow{\scriptstyle\mu F} & & & & \downarrow{\scriptstyle F\mu} \\ TF & & \xrightarrow{\qquad\zeta\qquad} & & FT \end{array}$$

It is well-known (cf. [29]) that EM laws $\zeta\colon TF \Rightarrow FT$ are in 1–1 correspondence with liftings $\tilde{F}$ of the functor F to the Eilenberg-Moore category $\mathbf{EM}(T)$. Given ζ, the functor $\tilde{F}$ maps the T-algebra (A, a) to $(FA, Fa \cdot \zeta_A)$. As a result, every FT-coalgebra $c\colon X \to FTX$ can be determinized in the presence of an EM law [34], yielding an $\tilde{F}$-coalgebra $c^\#\colon TX \to FTX$ in $\mathbf{EM}(T)$ as follows:

$$TX \xrightarrow{Tc} TFTX \xrightarrow{\zeta_{TX}} FTTX \xrightarrow{F\mu_X} FTX$$

Taking a more abstract perspective, $c^\#$ is the adjoint transpose of c under $L \dashv R$. We say that states $c, d \in C$ are *EM-equivalent* if $\eta_C(c), \eta_C(d) \in TC$ are behaviourally equivalent in the $\tilde{F}$-coalgebra $(TC, c^\#)$. We refer to this equivalence as *EM semantics*; when this equivalence can be captured as the kernel of a suitable map (in this case, the map assigning to each state its accepted language), we also refer to this map as the EM semantics. We will later encounter situations where the codomain of the semantics carries a generalized metric structure, in which case we will also subsume the induced generalized pseudometric on C under the moniker 'EM semantics'.

The standard powerset construction for determinizing nondeterministic automata is recovered by the following EM law:

Example 6. In **Set** (i.e. $\mathbf{Met_2}$), let $T = \mathcal{P}$ be the powerset monad and $F = 2 \times -^\Sigma$. The determinization $(\mathcal{P}C, c^\#)$ of an FT-coalgebra (C, c) w.r.t. the EM law $\zeta\colon TF \Rightarrow FT$ defined by

$$\zeta(t) = \Big(\bigvee\nolimits_{(v,f)\in t} v, \quad \lambda a.\{f(a) \mid (v, f) \in t\}\Big)$$

for $t \in \mathcal{P}(2 \times X^\Sigma)$ is precisely the powerset construction. Thus, the language semantics of nondeterministic automata is an instance of EM semantics.

Example 7. More generally, let $F = A \times -^\Sigma$ where Σ is discrete, let T be a monad on **Set**, and suppose that A carries a T-algebra structure $a\colon TA \to A$. Define a natural transformation $\delta\colon T(-^\Sigma) \Rightarrow (T-)^\Sigma$ by $\delta(t)(\sigma) = T(\lambda f.f(\sigma))(t)$. We then have an EM law $\zeta\colon T(A \times -^\Sigma) \Rightarrow A \times (T-)^\Sigma$ given (componentwise) by

$$\pi_1 \cdot \zeta_X = (T(A \times X^\Sigma) \xrightarrow{T\pi_1} TA \xrightarrow{a} A)$$
$$\pi_2 \cdot \zeta_X = (T(A \times X^\Sigma) \xrightarrow{T\pi_2} T(X^\Sigma) \xrightarrow{\delta} (TX)^\Sigma).$$

The arising EM semantics assigns to each state x a map $\Sigma^* \to A$, which may be thought of as assigning to each word $w \in \Sigma^*$ the degree (a value in A) to which x accepts w.

3 Graded Semantics and Graded Logics

Graded semantics [8,28] uniformly captures a wide range of semantics on various system types and of varying degrees of granularity as found, for instance, on the linear-time/branching-time spectrum of labelled transition system semantics [13]. Here, we are interested primarily in applying general results provided by the framework of graded semantics to the setting of EM semantics, which is, in essence, subsumed by graded semantics [25,28]. We recall the basic definition of graded semantics as such, and then give a new perspective on a general notion of characteristic modal logics for graded semantics, so-called graded logics.

3.1 Graded Semantics

The concepts central to graded semantics are those of graded monads and graded algebras. These are very similar to those of monads and monad algebras as recalled in Sect. 2.3 but, in the mentioned analogy with universal algebra, equip operations and terms with a *depth* that, in the application to system semantics, records the depth of look-ahead; that is, the depth corresponds to the (exact) number of transition steps considered. We briefly review the formal definitions.

Definition 8 (Graded Monad). A *graded monad* $\mathbb{M}$ on a category $\mathbf{C}$ consists of a family of functors $M_n\colon \mathbf{C} \to \mathbf{C}$ for $n \in \mathbb{N}$, a natural transformation $\eta\colon Id \Rightarrow M_0$ (the *unit*), and a family of natural transformations $\mu^{n,k}\colon M_n M_k \Rightarrow M_{n+k}$ for all $n, k \in \mathbb{N}$ (the *multiplication*) such that for all $n, k, m \in \mathbb{N}$ the following diagrams commute:

$$\begin{array}{ccc}
& M_n & \\
{\scriptstyle \eta M_n}\swarrow & \downarrow{\scriptstyle id_{M_n}} & \searrow{\scriptstyle M_n \eta} \\
M_0 M_n \xrightarrow{\mu^{0,n}} & M_n & \xleftarrow{\mu^{n,0}} M_n M_0
\end{array}
\qquad
\begin{array}{ccc}
M_n M_k M_m & \xrightarrow{M_n \mu^{k,m}} & M_n M_{k+m} \\
{\scriptstyle \mu^{n,k} M_m}\downarrow & & \downarrow{\scriptstyle \mu^{n,k+m}} \\
M_{n+k} M_m & \xrightarrow{\mu^{n+k,m}} & M_{n+k+m}
\end{array}$$

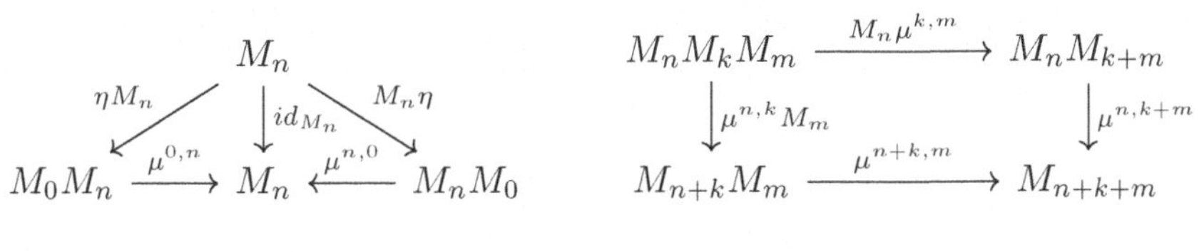

Definition 9 (Graded semantics). A *graded semantics* $(\alpha, \mathbb{M})$ for an endofunctor $G: \mathbf{C} \to \mathbf{C}$ consists of a graded monad $\mathbb{M}$ on $\mathbf{C}$ and a natural transformation $\alpha: G \Rightarrow M_1$. If (C, c) is a G-coalgebra, then we define the n-step behaviour $c^{(n)}: C \to M_n 1$, for $n \in \mathbb{N}$, by

$$c^{(0)} = (X \xrightarrow{M_0! \cdot \eta} M_0 1) \qquad c^{(n+1)}(X \xrightarrow{\alpha \cdot c} M_1 X \xrightarrow{M_1 c^{(n)}} M_1 M_n 1 \xrightarrow{\mu^{1n}} M_{n+1} 1).$$

We think of $c^{(n)}$ as assigning to a state in C its behaviour after n steps. We illustrate this more concretely in Example 12. We are mainly interested in the case where the base category $\mathbf{C}$ is a category of generalized (directed) pseudometric spaces (Sect. 2.2). In this case, a graded semantics induces a notion of behavioural distance:

Definition 10 (Behavioural distance). When $\mathbf{C}$ is $\mathbf{DPMet}_{\mathcal{V}}$ (or $\mathbf{PMet}_{\mathcal{V}}$, $\mathbf{Met}_{\mathcal{V}}$), then we define the *behavioural distance* of two states $x, y \in C$ of a G-coalgebra (C, c) under a graded semantics $(\alpha, \mathbb{M})$ to be

$$d^b(x, y) = \textstyle\bigwedge_{n \in \mathbb{N}} d_{M_n 1}(c^{(n)}(x), c^{(n)}(y)).$$

Remark 11. In case $\mathcal{V} = \mathbf{2}$ (Example 3.1), behavioural distance is two-valued, and thus in fact constitutes either a preorder (if $\mathbf{C}$ is $\mathbf{DPMet}_{\mathcal{V}}$) or an equivalence (if $\mathbf{C}$ is $\mathbf{PMet}_{\mathcal{V}}$).

Example 12. We recall two basic examples of graded monads [28] and associated graded semantics, capturing branching-time semantics and EM semantics, respectively. In both cases, it happens that α is identity; this need not always be the case, however [8].

1. Any functor F induces a graded monad $\mathbb{M}_F$ where the functor parts $M_n = F^n$ are given via repeated application of F and both multiplication and unit are identity. The arising graded semantics of F-coalgebras is branching-time semantics, specifically finite-depth behavioural equivalence (which coincides with behavioural equivalence if F is finitary).
2. Any EM law $\zeta: TF \Rightarrow FT$ induces a graded monad $\mathbb{M}_\zeta$ where $M_n = F^n T$. The unit of $\mathbb{M}_\zeta$ is the unit of T. We define an iterated distributive law $\zeta^{(n)}: TF^n \Rightarrow F^n T$ by putting

$$\zeta^{(0)} = id \quad \text{and} \quad \zeta^{(n+1)} = TF^{n+1} \xrightarrow{\zeta F^n} FTF^n \xrightarrow{F\zeta^{(n)}} F^{(n+1)}T.$$

   The multiplications of the graded monad $\mathbb{M}_\zeta$ are then given by $\mu^{m,n} = F^{n+m}\mu \cdot F^m \zeta^{(n)} T$. The arising graded semantics is essentially EM semantics, in the sense that the latter is obtained by erasing further information by postcomposing the maps $c^{(n)}: C \to F^n T1$ (in the notation of Definition 9) with $F^n!$ where $!$ is the unique map $T1 \to 1$. In particular, the EM semantics and the graded semantics introduced by an EM law for T agree exactly if T is affine (Sect. 2.3). Otherwise, the information erased by $F^n!$ essentially concerns the possibility of executing certain words, without regard to their acceptance [25, Section 5]. For a concrete example where T is affine, consider $T = \mathcal{D}$ (the distribution monad, cf. Sect. 2.3) and $FX = [0,1] \times X^{\Sigma}$, with an EM law ζ as per Example 7. Then FT-coalgebras are reactive probabilistic automata, and for a state x in an FT-coalgebra, $c^{(n)}(x) \in F^n\mathcal{D}1 \cong F^n 1 \cong [0,1]^{\Sigma^{<n}}$ assigns to each word of length $< n$ over Σ its probability of being accepted.

Note that 1. is the special case of 2. where $T = Id$.

We will in fact be interested exclusively in graded monads that are, in the universal-algebraic view [8,28], presented by operations and equations of depth at most 1, which intuitively means that identifications among behaviours do not depend on looking more than one step ahead. Categorically, this notion is captured as follows [28, Proposition 7.3]:

Definition 13. We say that a graded monad is *depth*-1 if for all $n \in \mathbb{N}$, $\mu^{1,n}$ is a coequalizer in the following diagram:

$$M_1 M_0 M_n X \underset{\mu^{1,0} M_n}{\overset{M_1 \mu^{0,n}}{\rightrightarrows}} M_1 M_n X \xrightarrow{\mu^{1,n}} M_{1+n} X.$$

Example 14. All graded monads described in Example 12 are depth-1.

The semantics of modalities in graded logics will rely on a graded variant of the notion of monad algebra:

Definition 15 (Graded algebra). Let $\mathbb{M}$ be a graded monad in $\mathbf{C}$, and $n \in \mathbb{N} \cup \{\omega\}$. A *graded* M_n*-algebra* $A = ((A_k)_{k \le n}, (a^{m,k})_{m+k \le n})$ consists of a family of $\mathbf{C}$-objects A_k and morphisms $a^{m,k}\colon M_m A_k \to A_{m+k}$ satisfying the following conditions: For $m \le n$, we have $a^{0,m} \cdot \eta_{A_m} = id_{A_m}$ and additionally, if $m + r + k \le n$, then the left diagram below commutes:

$$\begin{array}{ccc} M_m M_r A_k & \xrightarrow{M_m a^{r,k}} & M_m A_{r+k} \\ \downarrow{\scriptstyle \mu^{m,r}_{A_k}} & & \downarrow{\scriptstyle a^{m,r+k}} \\ M_{m+r} A_k & \xrightarrow{a^{m+r,k}} & A_{m+r+k} \end{array} \qquad \begin{array}{ccc} M_m A_k & \xrightarrow{M_m f_k} & M_m B_k \\ \downarrow{\scriptstyle a^{m,k}} & & \downarrow{\scriptstyle b^{m,k}} \\ A_{m+k} & \xrightarrow{f_{m+k}} & B_{m+k} \end{array}$$

A *homomorphism* of M_n-algebras A and B is a family of maps $f_k\colon A_k \to B_k$ such that the above right diagram commutes for all $m + k \le n$. For all $n \in \mathbb{N} \cup \{\omega\}$, the collection of M_n-algebras and their morphisms forms a category $\mathrm{Alg}_n(\mathbb{M})$.

The category $\mathrm{Alg}_0(\mathbb{M})$ is the Eilenberg-Moore category $\mathbf{EM}(M_0)$ for the (non-graded) monad $(M_0, \eta, \mu^{0,0})$. The semantics of modalities in graded logics will involve a special type of M_1-algebras [8]:

Definition 16 (Canonical algebras). For $i \in \{0,1\}$, let $(-)_i : \mathrm{Alg}_1(\mathbb{M}) \to \mathrm{Alg}_0(\mathbb{M})$ be the functor taking an M_1-algebra $A = ((A_k)_{k \le 1}, (a^{m,k})_{m+k \le 1})$ to the M_0-algebra $(A_i, a^{0,i})$. We say that an M_1-algebra A is *canonical* if it is free over $(-)_0$, i.e. if for all M_1-algebras B and M_0-homomorphisms $f : (A)_0 \to (B)_0$ there is a unique M_1-homomorphism $g : A \to B$ such that $(g)_0 = f$.

Lemma 17. *([8, Lemma 5.3]) An M_1-algebra A is canonical iff the following diagram is a coequalizer diagram in the category of M_0-algebras:*

$$M_1 M_0 A_0 \underset{\mu^{1,0}}{\overset{M_1 a^{0,0}}{\rightrightarrows}} M_1 A_0 \xrightarrow{a^{1,0}} A_1$$

Combining Definition 13 with Lemma 17 immediately gives us the following fact [8], which is a crucial ingredient for invariance of graded logics:

Proposition 18. *If $\mathbb{M}$ is a depth-1 graded monad, then for every $n \in \mathbb{N}$ and every object X, the M_1-algebra with carriers $M_n X$, $M_{n+1} X$ and multiplications as algebra structure is canonical.*

3.2 Graded Logics as a Fragment of Branching-Time Logic

We proceed to recall the general framework of (branching time) coalgebraic modal logic [30,33] and show that graded logics [8,28] are naturally viewed as a fragment of coalgebraic modal logic.

Syntactically, a logic is a triple $\mathcal{L} = (\Theta, \mathcal{O}, \Lambda)$ where Θ is a set of truth constants, $\mathcal{O}$ is a set of propositional operators, each with associated finite arity, and Λ is a set of modal operators, also each with an associated finite arity. The set of formulae of $\mathcal{L}$ is given by the grammar

$$\phi ::= \theta \mid p(\phi_1, \ldots, \phi_n) \mid \lambda(\phi_1, \ldots, \phi_m)$$

where $p \in \mathcal{O}$ is n-ary, $\lambda \in \Lambda$ is m-ary and $\theta \in \Theta$.

Semantically, formulae are interpreted in coalgebras of some functor $G\colon \mathbf{C} \to \mathbf{C}$, taking values in a truth-value object Ω of $\mathbf{C}$. We assume that $\mathbf{C}$ has finite products and a terminal object. The semantics of a formula ϕ in a G-coalgebra (C, c) is a morphism $[\![\phi]\!]_c\colon C \to \Omega$. The semantics is parametric in the following components:

- For each $\theta \in \Theta$ a $\mathbf{C}$-morphism $\hat{\theta}\colon 1 \to \Omega$.
- For each $p \in \mathcal{O}$ with arity n a $\mathbf{C}$-morphism $[\![p]\!]\colon \Omega^n \to \Omega$
- For each $\lambda \in \Lambda$ a $\mathbf{C}$-morphism $[\![\lambda]\!]\colon G(\Omega^n) \to \Omega$

The semantics of formulae is then defined inductively:

- For $\theta \in \Theta$ we define $[\![\theta]\!]_c = C \xrightarrow{!} 1 \xrightarrow{\hat{\theta}} \Omega$
- For $p \in \mathcal{O}$ we define $[\![p(\phi_1, \ldots, \phi_n)]\!]_c = [\![p]\!] \cdot \langle [\![\phi_1]\!]_c, \ldots, [\![\phi_n]\!]_c \rangle$
- For $\lambda \in \Lambda$ we define $[\![\lambda(\phi_1, \ldots, \phi_m)]\!]_c = [\![\lambda]\!] \cdot G\langle [\![\phi_1]\!]_c, \ldots, [\![\phi_m]\!]_c \rangle \cdot c$

The following definition of logical distance quantifies over all formulae ϕ of *uniform depth*, meaning that all occurrences of truth constants in ϕ are under the same number of nested modal operators. This is a mild restriction; in fact, for the above version of coalgebraic logic, truth constants can always be modelled as 0-ary propositional operators, for which there is no uniformity restriction. Uniform depth does come to play a role once we talk about graded logics, where propositional operators are additionally required to be gM_0-algebra homomorphisms, while truth constants are not. If M_0 is affine, then all **C**-morphisms $1 \to A$ into M_0-algebras A are M_0-algebra homomorphisms.

For the rest of the paper, assume that **C** is one of $\mathbf{Met}_\mathcal{V}$, $\mathbf{PMet}_\mathcal{V}$ or $\mathbf{DPMet}_\mathcal{V}$; in particular, the truth value object Ω carries the structure of a $\mathcal{V}$-category.

Definition 19. The *logical distance* of states $x, y \in C$ in a G-coalgebra (C, c) under the logic $\mathcal{L}$ is

$$d^\mathcal{L}(x, y) = \bigwedge\nolimits_{n\in\mathbb{N}, \phi\in\mathcal{L}_n} d_\Omega([\![\phi]\!]_c(x), [\![\phi]\!]_c(y))$$

where $\mathcal{L}_n$ is the set of all uniform depth-n $\mathcal{L}$ formulae. We say that $\mathcal{L}$ is *invariant* for $(\alpha, \mathbb{M})$ if $d^b \leq d^\mathcal{L}$ and *expressive* if $d^b \geq d^\mathcal{L}$.

It is straightforward to show that the logic defined above is invariant under behavioural equivalence, i.e. the graded equivalence induced by $\mathbb{M}_G$ (Example 12.1). We want to identify logics that are invariant not only under behavioural equivalence, but under an arbitrary graded semantics. To this end, we define graded logics:

Definition 20. Let $(\alpha, \mathbb{M})$ be a graded semantics for G and $o\colon M_0\Omega \to \Omega$ an M_0-algebra structure on Ω. A logic $\mathcal{L}$ is a *graded logic* (for $(\alpha, \mathbb{M})$) if the following hold:

1. For every n-ary $p \in \mathcal{O}$, the morphism $[\![p]\!]$ is an M_0-algebra homomorphism $(\Omega, o)^n \to (\Omega, o)$.
2. For each $\lambda \in \Lambda$ there is a fixed $f\colon M_1\Omega^n \to \Omega$, with the semantics of $\lambda \in \Lambda$ factoring as $[\![\lambda]\!] = f \cdot \alpha_{\Omega^n}$ and the tuple $(\Omega^n, \Omega, o^{(n)}, o, f)$ constituting an M_1-algebra. More concretely, this means that it satisfies $f \cdot \mu^{1,0} = f \cdot M_1 o^{(n)}$ (we refer to this property as *coequalization*), as well as $f \cdot \mu^{0,1} = o \cdot M_0 f$ (*homomorphy*), or written diagrammatically:

$$M_1 M_0 \Omega^n \underset{M_1 o^{(n)}}{\overset{\mu^{1,0}}{\rightrightarrows}} M_1 \Omega^n \xrightarrow{\;f\;} \Omega \qquad\qquad \begin{array}{ccc} M_0 M_1 \Omega^n & \xrightarrow{M_0 f} & M_0 \Omega \\ {\scriptstyle \mu^{0,1}}\downarrow & & \downarrow{\scriptstyle o} \\ M_1 \Omega^n & \xrightarrow[f]{} & \Omega \end{array}$$

In many examples (including those discussed in this work), the factorization in Condition 2 is simplified by the fact that $\alpha = id$, and just requires that $(\Omega^n, \Omega, o^{(n)}, o, [\![\lambda]\!])$ is an M_1-algebra. For readability, we restrict the technical development to unary modalities from now on; treating modalities of arbitrary arity is simply a matter of adding indices. In examples, modalities will have arity either 1 or 0.

Proposition 21. *Let $\mathcal{L}$ be a graded logic for the semantics $(\alpha, \mathbb{M})$ on $G\colon \mathbf{C} \to \mathbf{C}$ and (C, c) a G-coalgebra. For two states $x, y \in C$ we have that $d^b(x, y) \leq d^\mathcal{L}(x, y)$.*

Proof (Sketch). The proof is based on showing, by induction on ϕ, the stronger property that the evaluation functions $[\![\phi]\!]_c$ of depth-n formulae ϕ factor through M_0-homomorphisms

$$[\![\phi]\!]_{\mathbb{M}}\colon M_n 1 \to \Omega, \tag{1}$$

as used in earlier formulations of the semantics [8,9], with canonicity of $M_n 1$ (Lemma 17) being the key property in the step for modalities. □

The proof uses uniformity to enable the factorization of formula evaluation via a single $M_n 1$, which in general is possible only for uniform-depth formulae. In general, non-uniform depth formulae of graded logics fail to be invariant. We provide an example for this fact in the full version [12]. Recall that when M_0 is affine, then uniform depth is not an actual restriction. Having established invariance, we next generalize the expressivity criterion for graded logics [8] to our present quantitative setting:

Definition 22. A graded logic $\mathcal{L}$ consisting of Θ, $\mathcal{O}$, Λ is *depth-0 separating* if the family of maps $\{o \cdot M_0\hat{\theta}\colon M_0 1 \to \Omega \mid c \in \Theta\}$ is initial. Moreover, $\mathcal{L}$ is *depth-1 separating* if for all canonical M_1-algebras A and initial sources $\mathfrak{A}$ of M_0-homomorphisms $(A_0, a^{0,0}) \to (\Omega, o)$, closed under the propositional operators in $\mathcal{O}$, the set

$$\Lambda(\mathfrak{A}) := \{[\![\lambda]\!](g) : A_1 \to \Omega \mid \lambda \in \Lambda, g \in \mathfrak{A}\}$$

is initial, where $[\![\lambda]\!](g)$ is the, by canonicity unique, morphism such that $[\![\lambda]\!](g) \cdot a^{1,0} = f \cdot M_1 g$, with f the fixed morphism associated to λ in Definition 20.

Essentially, the above conditions encapsulate what is needed to push initiality through an induction on the depth of formulae. We thus obtain

Theorem 23. *Suppose that a graded logic $\mathcal{L}$ is both depth-0 separating and depth-1 separating. Then $\mathcal{L}$ is expressive.*

4 Graded Semantics via Coalgebraic Determinization

From now on, fix a **C**-endofunctor F, a monad T on **C**, and an EM law $\zeta\colon TF \Rightarrow FT$. The objective of this section is to show that behavioural equivalences, respectively metrics, on a determinized coalgebra agree with the equivalences/metrics induced by the graded semantics (Lemma 27), and that graded logics for FT may be reduced to coalgebraic logics for F. We recall the notion of predeterminization under a graded semantics [10] and show that this is the same concept as determinization under an EM law, under the condition that the monad T is affine.

Let $\mathbb{M}$ be a graded monad. We have a functor $E\colon \mathrm{Alg}_0(\mathbb{M}) \to \mathrm{Alg}_1(\mathbb{M})$ that takes an M_0-algebra A to the free M_1-algebra over A with respect to $(-)_0$ (which is then canonical, cf. Definition 16). This gives rise to a functor

$$\overline{M}_1 = (\mathrm{Alg}_0(\mathbb{M}) \xrightarrow{E} \mathrm{Alg}_1(\mathbb{M}) \xrightarrow{(-)_1} \mathrm{Alg}_0(\mathbb{M})),$$

which intuitively takes an M_0-algebra of behaviours to the M_0-algebra of behaviours having absorbed one more step. Since $(M_0X, M_1X, \mu_X^{0,0}, \mu_X^{0,1}, \mu_X^{1,0})$ is canonical (Proposition 18), we have $\overline{M}_1(M_0X, \mu^{0,0}) = (M_1X, \mu^{0,1})$, or stated slightly differently, if we denote the free-forgetful adjunction on $\mathrm{Alg}_0(\mathbb{M})$ by $L \dashv R$, then $M_1 = R\overline{M}_1L$. For a graded semantics $(\alpha\colon G \to M_1, \mathbb{M})$ and a coalgebra $c\colon C \to GC$, we have $C \xrightarrow{\alpha \cdot c} M_1C = R\overline{M}_1LC$. The adjunction then yields a unique morphism $c^\dagger : LC \to \overline{M}_1LC$, defining a form of determinization under the graded semantics, similar to the generalized powerset construction. Specifically, if $M_01 = 1$, then for $x, y \in C$, $\eta(x)$ and $\eta(y)$ are behaviourally equivalent in $c^\dagger$ iff x and y are identified under the graded semantics $(\alpha, \mathbb{M})$. We show next that

Lemma 24. *If* $\mathbb{M} = \mathbb{M}_\zeta$ *then* $\overline{M}_1 = \tilde{F}$.

Proof. Let (A, a) be a T-algebra. Then $\tilde{F}(A, a) = (Fa, Fa \cdot \zeta_A)$. On the other hand, by Lemma 17, the 1-part of the canonical algebra of $\overline{M}_1(A, a)$ is given by the following (split) coequalizer:

$$FTTA \underset{FTa}{\overset{F\mu_A}{\rightrightarrows}} FTA \xrightarrow{Fa} FA, \qquad F\eta_{TA}\colon FTA \to FTTA, \qquad F\eta_A\colon FA \to FTA$$

Commutativity of all relevant paths is obvious from the algebra and monad axioms, implying that the diagram is a coequalizer diagram by virtue of being a split coequalizer. Then $(A, FA, a, Fa \cdot \zeta_A, Fa)$ defines a canonical M_1-algebra where coequalization, as well as canonicity (due to Lemma 17), are by the above coequalizer, and homomorphy instantiates to the outer paths of the following diagram:

$$\begin{array}{ccccc} TFTA & \xrightarrow{\zeta_{TA}} & FTTA & \xrightarrow{F\mu} & FTA \\ \downarrow{\scriptstyle TFa} & & \downarrow{\scriptstyle FTa} & & \downarrow{\scriptstyle Fa} \\ TFA & \xrightarrow{\zeta_A} & FTA & \xrightarrow{Fa} & FA \end{array}$$

Commutativity of the outer rectangle follows from the fact that the left square commutes by naturality of ζ and the right square commutes by virtue of (A, a) being a T-algebra. Taking the 1-part of this canonical algebra then leaves us with $(Fa, Fa \cdot \zeta_A)$. On morphisms $h\colon (A, a) \to (B, b)$, the lifting $\tilde{F}$ acts by sending h to Fh. Commutativity of the relevant diagram making Fh a T-algebra morphism between FA and FB is easily checked, as is the fact that (h, Fh) constitutes a morphism between the canonical M_1-algebras. □

Lemma 25. *Let* (C, c) *be an* FT*-coalgebra and* $c^\dagger$ *the predeterminization under the graded semantics* M_ζ*. Then* $c^\# = c^\dagger$.

Proof. This follows from the fact that $c^\#$ can equivalently be defined as the adjoint transpose of c under the free-forgetful adjunction of $\mathbf{EM}(T)$ [34]. Then $c^\#$ and $c^\dagger$ agree by definition. □

Definition 26. Let $T\colon \mathbf{C} \to \mathbf{C}$ be a monad and $H\colon \mathbf{EM}(T) \to \mathbf{EM}(T)$ a functor on the corresponding Eilenberg-Moore category. Further, let $c\colon ((A, d_A), a) \to H((A, d_A), a)$ be an H-coalgebra. The *finite-depth behavioural distance* of two states $x, y \in A$ is given by $d^H(x, y) = \bigwedge_{i \in \mathbb{N}} d_{H^i 1}(f_i(x), f_i(y))$, where the $f_i\colon A \to H^i 1$ are the projections into the final H-chain.

Lemma 27. *Let* $(\alpha\colon G \to M_1, \mathbb{M})$ *be a graded semantics on* $\mathbf{C}$ *with* M_0 *affine, and let* (C, c) *be a* G*-coalgebra. Then* $d^{\overline{M}_1}(\eta(x), \eta(y)) = d^b(x, y)$ *for all* $x, y \in C$*.*

Proof. One shows by induction on n that $f_n \cdot \eta = c^{(n)}$ for all n, where affinity is needed for the base case $n = 0$. □

Remark 28. In the case where the graded monad is $\mathbb{M}_\zeta$, if T is affine, then the final chain of $\overline{M}_1$ lives over the final chain of F. In particular, if F is finitary and $\mathcal{V} = 2$, then finite-depth behavioural equivalence agrees with behavioural equivalence, for both F-coalgebras and $\overline{M}_1$-coalgebras.

Remark 29. As noted in Example 12.2, finite-depth behavioural distance in $\mathbf{EM}(T)$ may be coarser than the graded semantics but may then be canonically recovered from the graded semantics.

From now on, we notationally conflate modalities $\lambda \in \Lambda$ and their interpretations $[\![\lambda]\!]\colon FT\Omega \to \Omega$. The following result completely characterizes the modal operators of graded logics for the semantics $(id, \mathbb{M}_\zeta)$:

Theorem 30. *Let* $\lambda\colon FT\Omega \to \Omega$ *be a modal operator for a graded logic with truth value object* (Ω, o)*. Then* $\lambda = ev_\lambda \cdot Fo$ *for some algebra homomorphism* $ev_\lambda\colon \tilde{F}(\Omega, o) \to (\Omega, o)$*. On the other hand, every algebra homomorphism* $\tilde{F}(\Omega, o) \to (\Omega, o)$ *yields a modal operator in this way.*

As our second main result, we next show that a logic is depth-1 separating for the semantics of $\mathbb{M}_\zeta$ if the F-algebra part of its modal operators is expressive for F. This criterion is typically very easy to establish and can be shown for general classes of functors, which is what we mean by our slogan that expressive graded logics for EM semantics come essentially for free.

Theorem 31. *Let* $\mathcal{L} = (\Theta, \mathcal{O}, \Lambda)$ *be a graded logic for* $\mathbb{M}_\zeta$ *and* $\mathcal{L}' = (\Theta, \mathcal{O}, \Lambda')$ *the (graded) logic for* $\mathbb{M}_F$ *with* $\Lambda' = \{f\colon F\Omega \to \Omega \mid f \cdot Fo \in \Lambda\}$*. Then* $\mathcal{L}$ *is depth-1 separating for* $\mathbb{M}_\zeta$ *if* $\mathcal{L}'$ *is depth-1 separating for* $\mathbb{M}_F$*.*

Proof. Let A be a canonical M_1-algebra. Since $\overline{M}_1 = \tilde{F}$, we know that A has the form $(A_0, FA_0, a^{0,0}, Fa^{0,0} \cdot \zeta, Fa^{0,0})$. For a homomorphism $h\colon (A_0, a^{0,0}) \to (\Omega, o)$ of T-algebras and $\lambda \in \Lambda$ where $\lambda = f \cdot Fo$, $\lambda(h)$ is, by definition, the unique morphism that makes the outer rectangle in the following diagram commute:

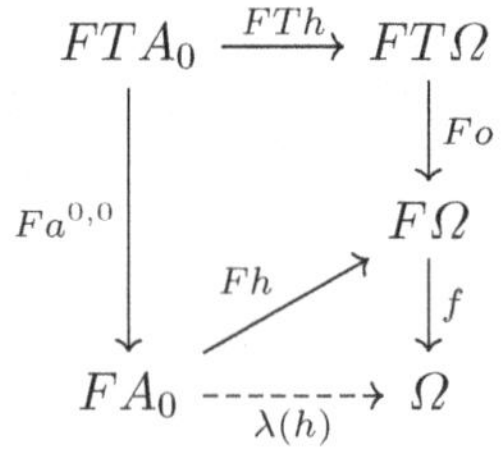

The top square commutes since it is just F applied to the homomorphism square of h. Since $a^{0,0}$ is a split epimorphism (by virtue of being an algebra for a monad), $Fa^{0,0}$ is also a split epimorphism. Therefore, $\lambda(h) = f \cdot Fh$. Let $\mathfrak{A} \subseteq \mathbf{EM}(T)((A_0, a^{0,0}), (\Omega, o))$ be a separating set of algebra homomorphisms; we have to show that $\Lambda(\mathfrak{A})$ is separating. But since $\lambda(h) = f \cdot Fh$ for all $h \in \mathfrak{A}$ and $\lambda = f \cdot Fo \in \Lambda$, we have $\Lambda'(\mathfrak{A}) = \{f \cdot Fh \mid f \in \Lambda', h \in \mathfrak{A}\} = \Lambda(\mathfrak{A})$, and $\Lambda'(\mathfrak{A})$ is spearating by depth-1 separation for $\mathcal{L}'$. □

5 Examples

In our central examples, F takes the form $\mathcal{V} \times (-)^{\Sigma}$ while T varies. In these cases, we always have a set of separating modalities: We have the set $\Lambda' = \{\mathrm{ev}_\sigma \mid \sigma \in \Sigma\} \cup \{\mathrm{ev}_\top\}$ of modalities for F, where $\mathrm{ev}_\sigma \colon \mathcal{V} \times \mathcal{V}^{\Sigma} \to \mathcal{V}$ is a unary operator defined by $(v, f) \mapsto f(\sigma)$, and $\mathrm{ev}_\top \colon \mathcal{V} \times 1^{\Sigma} \to \mathcal{V}$ is the 0-ary operator defined by $(v, f) \mapsto v$. For a monad T and an algebra structure $o \colon T\mathcal{V} \to \mathcal{V}$, the semantics of each $\mathrm{ev}_\lambda \in \Lambda'$ extends to a modal operator $\langle\lambda\rangle$ for FT, given by $\langle\lambda\rangle = \mathrm{ev}_\lambda \cdot Fo$. We thus have coalgebraic logics $\mathcal{L}' = (\emptyset, \emptyset, \Lambda')$ for F and $\mathcal{L} = (\emptyset, \emptyset, \Lambda)$ for FT.

Lemma 32. *Let $F = \mathcal{V} \times -^{\Sigma}$. Let T be a monad and $\tilde{F} \colon \mathbf{EM}(T) \to \mathbf{EM}(T)$ a lifting of F. Moreover, suppose that $\mathcal{V}$ carries a T-algebra structure $o \colon T\mathcal{V} \to \mathcal{V}$. Then, for every $\mathrm{ev}_\lambda \in \Lambda'$, the semantics ev_λ is a homomorphism of algebras $\tilde{F}(\mathcal{V}, o) \to (\mathcal{V}, o)$*

Proof. Since the ev_λ are just product projections, this follows from the fact that the forgetful functor $U \colon \mathbf{EM}(T) \to \mathbf{C}$ creates limits [1, Proposition 20.12]. □

Corollary 33. *Let ζ be defined as in Example 7. The logic $\mathcal{L}$ as defined above is a graded logic for the graded semantics $(id, \mathbb{M}_\zeta)$.*

Lemma 34. *The logic $\mathcal{L}'$ as defined above is depth-1 separating for the graded semantics $(id, \mathbb{M}_F)$.*

Proof. By Proposition 18, canonical M_1-algebras have the form $A = (A_0, FA_0, id, id, id)$. Let A be a canonical M_1-algebra and $\mathfrak{A}$ an initial source $A_0 \to \mathcal{V}$; we then need to show that the lower edges in the following diagram collectively form an initial source, where f ranges over $\mathfrak{A}$:

$$\begin{array}{ccc} FA_0 & \xrightarrow{Ff} & F\mathcal{V} \\ {\scriptstyle id}\downarrow & & \downarrow{\scriptstyle \mathrm{ev}_\lambda} \\ FA_0 & \xrightarrow{\mathrm{ev}_\lambda(f)} & \mathcal{V} \end{array}$$

Since the modal operators ev_λ are precisely the projections of the product $F\mathcal{V}$, they constitute an initial source; moreover, again since F is a product, it preserves initial sources, so the source of all Ff is initial. Thus, $\Lambda(\mathfrak{A})$ is a composite of initial sources, hence itself initial. □

Words in Σ^* can be viewed as formulae of $\mathcal{L}$ in the obvious way. The evaluation $[\![\phi]\!]_c$ then captures the notion of acceptance in the automaton given by the FT-coalgebra (C, c). Logics in general however allow to express far more interesting statements, since on the one hand formulae may specify words only up to a suffix, and on the other hand the logic may include propositional operators. We consider a few concrete examples:

Example 35 (Deadlock-free nondeterministic automata). We take $\mathcal{V} = \mathbf{2}$, and work in the category $\mathbf{Met_2}$ of sets and functions. Concretely, this means that all objects carry the discrete equivalence relation, and initiality of a source is joint injectivity. If T is the nonempty powerset monad $\mathcal{P}^+$, then coalgebras $c\colon C \to 2 \times (\mathcal{P}^+C)^\Sigma$ are deadlock-free nondeterministic automata. With the algebra structure $o\colon \mathcal{P}^+2 \to 2$ defined by $o(X) = \top$ if $\top \in X$ and $o(X) = \bot$ otherwise, we can construct a distributive law ζ as in Example 7. Since, by Lemma 34, $\mathcal{L}$ is depth-1 separating for $2 \times -^\Sigma$, we have that by Theorem 31 the logic $\mathcal{L}$ is expressive for $(id, \mathbb{M}_\zeta)$. We can add disjunction as a propositional operator, preserving invariance of the logic, since disjunction preserves joins (i.e. is a homomorphism of $\mathcal{P}^+$-algebras).

Example 36 (Reactive probabilistic automata). For $\mathcal{V} = [0,1]_\oplus$, we consider reactive probabilistic automata. Let T be the (finitely supported) probability distribution monad $\mathcal{D}$ on $\mathbf{PMet}_{[0,1]_\oplus}$, which equips the set of distributions with the Kantorovich metric (e.g. [2]). We put $\Omega = [0,1]$, equipped with the symmetrized metric $d(x,y) = |x - y|$. We have an algebra $o\colon \mathcal{D}[0,1] \to [0,1]$ taking expected values: $o(\mu) = \sum_{v \in [0,1]} v\mu(v)$. The construction in Example 7 then yields a semantics where, intuitively, the first component of F determines the probability of a state to accept. Upon reading a letter a, the automaton moves to a random successor state according to the probability distribution on states associated with a. The evaluation $[\![\phi]\!]_c(x)$ is then the expected probability of the state $x \in C$ of an automaton (C, c) accepting the word corresponding to ϕ. The distance of two states $x, y \in C$ is the supremum in difference of acceptance across all words in Σ^*. Again we have expressivity of $\mathcal{L}$ by combining Lemma 34 and Theorem 31. The logic remains invariant w.r.t. the semantics when extended with propositional operators that are homomorphisms $[0,1]^n \to [0,1]$ of $\mathcal{D}$-algebras, which in this case means they are affine maps, such as convex combinations or fuzzy negation $x \mapsto 1 - x$.

Example 37 (Reactive probabilistic automata with black hole termination). Going beyond the leading example $F = \mathcal{V} \times (-)^\Sigma$, we add explicit failure in the vein of [36] to reactive probabilistic automata: We now take $\mathcal{V} = 2$, and again view $\mathbf{Met_2}$ as the category of sets and functions (Example 35). Let $\Omega = [0,1]$, equipped with the $\mathcal{D}$-algebra structure $o\colon \mathcal{D}[0,1] \to [0,1]$ that takes expected values. We obtain a distributive law $\mathcal{D}(2 \times - + 1)^\Sigma \Rightarrow 2 \times ((\mathcal{D}-) + 1)^\Sigma$ by composing the distributive law from Example 7 with the law $\lambda\colon \mathcal{D}(- + 1) \Rightarrow (\mathcal{D}-) + 1$ that maps $\mu \in \mathcal{D}(X + 1)$ to $*$ iff $\mu(*) \neq 0$, and to μ otherwise, where $*$ denotes the unique element of 1. The semantics for this type of automaton is like that of probabilistic automata, with the exception that if a run leads to the "state" $*$ with non-zero probability, then the automaton immediately gets stuck and rejects the word. For the logic, we consider the same operators as in the previous examples, with the modification that $\mathrm{ev}_\sigma(v, f) = \bot$ if $f(\sigma) = *$. Additionally we introduce the modal operator $\mathrm{ev}_{\bar{\sigma}}$, which carries the semantics $\mathrm{ev}_{\bar{\sigma}}(v, f) = \top$ if $f(\sigma) = *$ and $\mathrm{ev}_{\bar{\sigma}}(v, f) = \bot$ otherwise. It is straightforward to check that these operations define $\mathcal{D}$-algebra homomorphisms, making them valid modalities according to Theorem 30.

To verify expressivity, it is sufficient by Theorem 31 to prove separation of elements of $2 \times (X + 1)^\Sigma$, so let $\mathfrak{A}$ be an initial set of morphisms of type $X \to 2$. Given $s, t \in 2 \times (X + 1)^\Sigma$ such that $d(s, t) = \bot$, we need to find ev_λ and $h \in \mathfrak{A}$ (or just ev_λ if ev_λ is 0-ary) such that $\mathrm{ev}_\lambda(h)(s) \neq \mathrm{ev}_\lambda(h)(t)$. If $s = (v, f)$ and $t = (w, g)$ differ in their first component $v \neq w$, we can choose $\mathrm{ev}_\top$. If the elements differ in one of the other components σ, we distinguish cases: If $x = f(\sigma) \neq * \neq g(\sigma) = y$, then there is $h \in \mathfrak{A}$ separating x from y, thus $\mathrm{ev}_\sigma(h)(x) \neq \mathrm{ev}_\sigma(h)(y)$. Otherwise, if $f(\sigma) = * \neq g(\sigma)$, we can choose $\mathrm{ev}_{\bar{\sigma}}$ to separate s and t, and similarly for the symmetric case. We thus obtain expressiveness in the two-valued sense, i.e. the logic distinguishes non-equivalent states. Like in the previous example, the logic remains invariant when extended with propositional operators that are affine maps $[0, 1]^n \to [0, 1]$.

6 Conclusion

We have discussed characteristic logics for system semantics arising via determinization in the coalgebraic powerset construction, so-called Eilenberg-Moore semantics, which relies on a distributive law of a functor representing the language type of a system over a monad representing the branching type [34]. Leading examples are languages semantics for various forms of automata. As our main technical tool, we have exploited that Eilenberg-Moore semantics may be cast as an instance of graded semantics, which provides generic mechanisms for designing invariant modal logics and establishing their expressiveness. Our first main result establishes an overview of all graded modalities available for Eilenberg-Moore semantics, showing that these are canoincally obtained from modalities for the language type and a single modality for the branching type. Our second main result shows that expressivity of such a logic follows from branching-time expressivity of the same collection of operators with respect to the language type. Our results are stated in quantalic generality, allowing for instantiation to both two-valued and quantitative types of semantics and logics.

An important next step in the programme of developing graded logics into a verification framework is the question of how graded semantics relates to fixpoint logics. While we have focused on Eilenberg-Moore semantics in the present work, graded semantics does also subsume Kleisli-style trace semantics [14], which poses additional challenges for the design of characteristic modal logics, in particular in the quantitative setting.

References

1. Adámek, J., Herrlich, H., Strecker, G.: Abstract and Concrete Categories. Wiley Interscience: available as Reprints Theory Appl. Cat. **17**(2006), 1–507 (1990)
2. Baldan, P., Bonchi, F., Kerstan, H., König, B.: Behavioral metrics via functor lifting. In: Raman, V., Suresh, S.P. (eds.) Foundation of Software Technology and Theoretical Computer Science, FSTTCS 2014. LIPIcs, vol. 29, pp. 403–415. Schloss Dagstuhl – Leibniz-Zentrum für Informatik (2014). https://doi.org/10.4230/LIPIcs.FSTTCS.2014.403
3. Baldan, P., Bonchi, F., Kerstan, H., König, B.: Coalgebraic behavioral metrics. Logical Methods Comput. Sci. **14**(3) (2018), selected papers of the 6th Conference on Algebra and Coalgebra in Computer Science, CALCO 2015
4. Beohar, H., Gurke, S., König, B., Messing, K.: Hennessy-Milner theorems via Galois connections. In: Klin, B., Pimentel, E. (eds.) Computer Science Logic, CSL 2023. LIPIcs, vol. 252, pp. 12:1–12:18. Schloss Dagstuhl – Leibniz-Zentrum für Informatik (2023). https://doi.org/10.4230/LIPIcs.CSL.2023.12
5. Beohar, H., Gurke, S., König, B., Messing, K., Forster, J., Schröder, L., Wild, P.: Expressive quantale-valued logics for coalgebras: an adjunction-based approach. In: Kupferman, O., Beyersdorff, O., Lokshtanov, D., Kanté, M.M. (eds.) Theoretical Aspects of Computer Science, STACS 2024. LIPIcs, Schloss Dagstuhl – Leibniz-Zentrum für Informatik (2024), to appear
6. Bonchi, F., Bonsangue, M., Caltais, G., Rutten, J., Silva, A.: Final semantics for decorated traces. In: Mathematical Foundations of Programming Semantics, MFPS 2012. ENTCS, vol. 286, pp. 73–86. Elsevier (2012). https://doi.org/10.1016/j.entcs.2012.08.006
7. Cîrstea, C.: From branching to linear time, coalgebraically. Fundam. Informaticae **150**(3–4), 379–406 (2017). https://doi.org/10.3233/FI-2017-1474
8. Dorsch, U., Milius, S., Schröder, L.: Graded monads and graded logics for the linear time - branching time spectrum. In: Fokkink, W.J., van Glabbeek, R. (eds.) Concurrency Theory, CONCUR 2019. LIPIcs, vol. 140, pp. 36:1–36:16. Schloss Dagstuhl – Leibniz-Zentrum für Informatik (2019). https://doi.org/10.4230/LIPIcs.CONCUR.2019.36
9. Ford, C., Milius, S., Schröder, L.: Behavioural preorders via graded monads. In: Logic in Computer Science, LICS 2021, pp. 1–13. IEEE (2021). https://doi.org/10.1109/LICS52264.2021.9470517
10. Ford, C., Milius, S., Schröder, L., Beohar, H., König, B.: Graded monads and behavioural equivalence games. In: Baier, C., Fisman, D. (eds.) Logic in Computer Science, LICS 2022, pp. 61:1–61:13. ACM (2022). https://doi.org/10.1145/3531130.3533374
11. Forster, J., Goncharov, S., Hofmann, D., Nora, P., Schröder, L., Wild, P.: Quantitative Hennessy-Milner theorems via notions of density. In: Klin, B., Pimentel, E. (eds.) Computer Science Logic, CSL 2023. LIPIcs, vol. 252, pp. 22:1–22:20. Schloss Dagstuhl – Leibniz-Zentrum für Informatik (2023). https://doi.org/10.4230/LIPIcs.CSL.2023.22
12. Forster, J., Schröder, L., Wild, P., Beohar, H., Gurke, S., Messing, K.: Graded semantics and graded logics for Eilenberg-Moore coalgebras. CoRR **abs/2307.14826** (2023). https://doi.org/10.48550/ARXIV.2307.14826
13. van Glabbeek, R.: The linear time – branching time spectrum I. In: Bergstra, J., Ponse, A., Smolka, S. (eds.) Handbook of Process Algebra, chap. 1, pp. 3–99. Elsevier (2001)

14. Hasuo, I., Jacobs, B., Sokolova, A.: Generic trace semantics via coinduction. Log. Meth. Comput. Sci. **3** (2007)
15. Hennessy, M., Milner, R.: On observing nondeterminism and concurrency. In: de Bakker, J.W., van Leeuwen, J. (eds.) Automata, Languages and Programming, ICALP 1980. LNCS, vol. 85, pp. 299–309. Springer (1980). https://doi.org/10.1007/3-540-10003-2_79
16. Jacobs, B.: Trace semantics for coalgebras. In: Adámek, J., Milius, S. (eds.) Coalgebraic Methods in Computer Science, CMCS 2004. ENTCS, vol. 106, pp. 167–184. Elsevier (2004). https://doi.org/10.1016/j.entcs.2004.02.031
17. Jacobs, B.: Affine monads and side-effect-freeness. In: Hasuo, I. (ed.) Coalgebraic Methods in Computer Science, CMCS 2016. LNCS, vol. 9608, pp. 53–72. Springer (2016). https://doi.org/10.1007/978-3-319-40370-0_5
18. Jacobs, B., Silva, A., Sokolova, A.: Trace semantics via determinization. J. Comput. Syst. Sci. **81**(5), 859–879 (2015). https://doi.org/10.1016/j.jcss.2014.12.005
19. Kerstan, H., König, B.: Coalgebraic trace semantics for continuous probabilistic transition systems. Log. Methods Comput. Sci. **9**(4) (2013). https://doi.org/10.2168/LMCS-9(4:16)2013
20. Kissig, C., Kurz, A.: Generic trace logics (2011), arXiv preprint 1103.3239
21. Klin, B., Rot, J.: Coalgebraic trace semantics via forgetful logics. In: Pitts, A. (ed.) FoSSaCS 2015. LNCS, vol. 9034, pp. 151–166. Springer, Heidelberg (2015). https://doi.org/10.1007/978-3-662-46678-0_10
22. Komorida, Y., Katsumata, S., Kupke, C., Rot, J., Hasuo, I.: Expressivity of quantitative modal logics : Categorical foundations via codensity and approximation. In: Logic in Computer Science, LICS 2021, pp. 1–14. IEEE (2021). https://doi.org/10.1109/LICS52264.2021.9470656
23. König, B., Mika-Michalski, C.: (Metric) bisimulation games and real-valued modal logics for coalgebras. In: Schewe, S., Zhang, L. (eds.) Concurrency Theory, CONCUR 2018. LIPIcs, vol. 118, pp. 37:1–37:17. Schloss Dagstuhl – Leibniz-Zentrum für Informatik (2018). https://doi.org/10.4230/LIPIcs.CONCUR.2018.37
24. Kupke, C., Rot, J.: Expressive logics for coinductive predicates. Log. Methods Comput. Sci. **17**(4) (2021). https://doi.org/10.46298/lmcs-17(4:19)2021
25. Kurz, A., Milius, S., Pattinson, D., Schröder, L.: Simplified coalgebraic trace equivalence. In: De Nicola, R., Hennicker, R. (eds.) Software, Services, and Systems. LNCS, vol. 8950, pp. 75–90. Springer, Cham (2015). https://doi.org/10.1007/978-3-319-15545-6_8
26. Lawvere, F.W.: Metric spaces, generalized logic, and closed categories. Rendiconti del seminario matématico e fisico di Milano **43**, 135–166 (1973)
27. Marti, J., Venema, Y.: Lax extensions of coalgebra functors and their logic. J. Comput. Syst. Sci. **81**(5), 880–900 (2015). https://doi.org/10.1016/j.jcss.2014.12.006
28. Milius, S., Pattinson, D., Schröder, L.: Generic trace semantics and graded monads. In: Moss, L.S., Sobocinski, P. (eds.) Algebra and Coalgebra in Computer Science, CALCO 2015. LIPIcs, vol. 35, pp. 253–269. Schloss Dagstuhl – Leibniz-Zentrum für Informatik (2015). https://doi.org/10.4230/LIPIcs.CALCO.2015.253
29. Mulry, P.S.: Lifting results for categories of algebras. Theor. Comput. Sci. **278**(1–2), 257–269 (2002). https://doi.org/10.1016/S0304-3975(00)00338-8
30. Pattinson, D.: Expressive logics for coalgebras via terminal sequence induction. Notre Dame J. Formal Log. **45**(1), 19–33 (2004). https://doi.org/10.1305/ndjfl/1094155277
31. Rot, J., Jacobs, B., Levy, P.B.: Steps and traces. J. Log. Comput. **31**(6), 1482–1525 (2021). https://doi.org/10.1093/logcom/exab050

32. Rutten, J.J.M.M.: Universal coalgebra: a theory of systems. Theor. Comput. Sci. **249**(1), 3–80 (2000). https://doi.org/10.1016/S0304-3975(00)00056-6
33. Schröder, L.: Expressivity of coalgebraic modal logic: the limits and beyond. Theor. Comput. Sci. **390**(2–3), 230–247 (2008). https://doi.org/10.1016/j.tcs.2007.09.023
34. Silva, A., Bonchi, F., Bonsangue, M.M., Rutten, J.J.M.M.: Generalizing the powerset construction, coalgebraically. In: Lodaya, K., Mahajan, M. (eds.) Foundations of Software Technology and Theoretical Computer Science, FSTTCS 2010. LIPIcs, vol. 8, pp. 272–283. Schloss Dagstuhl – Leibniz-Zentrum für Informatik (2010). https://doi.org/10.4230/LIPIcs.FSTTCS.2010.272
35. Smirnov, A.: Graded monads and rings of polynomials. J. Math. Sci. **151**, 3032–3051 (2008)
36. Sokolova, A., Woracek, H.: Termination in convex sets of distributions. Log. Methods Comput. Sci. **14**(4) (2018). https://doi.org/10.23638/LMCS-14(4:17)2018
37. Urabe, N., Hasuo, I.: Coalgebraic infinite traces and Kleisli simulations. In: Moss, L.S., Sobocinski, P. (eds.) Algebra and Coalgebra in Computer Science, CALCO 2015. LIPIcs, vol. 35, pp. 320–335. Schloss Dagstuhl – Leibniz-Zentrum für Informatik (2015). https://doi.org/10.4230/LIPIcs.CALCO.2015.320
38. van Breugel, F., Worrell, J.: A behavioural pseudometric for probabilistic transition systems. Theoret. Comput. Sci. **331**, 115–142 (2005)
39. Wild, P., Schröder, L.: Characteristic logics for behavioural hemimetrics via fuzzy lax extensions. Log. Methods Comput. Sci. **18**(2) (2022). https://doi.org/10.46298/lmcs-18(2:19)2022
40. Worrell, J.: Coinduction for recursive data types: partial orders, metric spaces and omega-categories. In: Reichel, H. (ed.) Coalgebraic Methods in Computer Science, CMCS 2000. ENTCS, vol. 33, pp. 337–356. Elsevier (2000). https://doi.org/10.1016/S1571-0661(05)80356-1

Explicit Hopcroft's Trick in Categorical Partition Refinement

Takahiro Sanada[1(✉)], Ryota Kojima[2], Yuichi Komorida[3,4], Koko Muroya[2], and Ichiro Hasuo[3,4]

[1] Fukui Prefectural University, Fukui, Japan
tsanada@fpu.ac.jp
[2] RIMS, Kyoto University, Kyoto, Japan
{kojima,kmuroya}@kurims.kyoto-u.ac.jp
[3] National Institute of Informatics, Tokyo, Japan
komorin@nii.ac.jp, i.hasuo@acm.org
[4] SOKENDAI, Tokyo, Japan

Abstract. Algorithms for *partition refinement* are actively studied for a variety of systems, often with the optimisation called *Hopcroft's trick*. However, the low-level description of those algorithms in the literature often obscures the essence of Hopcroft's trick. Our contribution is twofold. Firstly, we present a novel formulation of Hopcroft's trick in terms of general trees with weights. This clean and explicit formulation—we call it *Hopcroft's inequality*—is crucially used in our second contribution, namely a general partition refinement algorithm that is *functor-generic* (i.e. it works for a variety of systems such as (non-)deterministic automata and Markov chains). Here we build on recent works on coalgebraic partition refinement but depart from them with the use of *fibrations*. In particular, our fibrational notion of *R-partitioning* exposes a concrete tree structure to which Hopcroft's inequality readily applies. It is notable that our fibrational framework accommodates such algorithmic analysis on the categorical level of abstraction.

Keywords: Partition refinement · Category theory · Coalgebra · Fibration · Tree algorithm

1 Introduction

Partition refinement refers to a class of algorithms that computes behavioural equivalence of various types of systems—such as the language equivalence for deterministic finite automata (DFAs), bisimilarity for labelled transition systems (LTSs) and Markov chains, etc.—by a fixed-point iteration. Such algorithms also yield *quotients* of state spaces, making systems smaller and thus easier to analyse e.g. by model checking.

Since its original introduction by Moore [20] for DFAs, partition refinement has been actively studied for enhanced performance and generality. On the performance side, Hopcroft [11] introduced what is now called *Hopcroft's trick* that

B. König and H. Urbat (Eds.): CMCS 2024, LNCS 14617, pp. 135–155, 2024.
https://doi.org/10.1007/978-3-031-66438-0_7

greatly improves the asymptotic complexity. The original paper [11] is famously hard to crack; works such as [6,16] present its reformulation, again focusing on DFAs. On the generality side, partition refinement for systems other than DFAs has been pursued, such as LTSs [15,24], probabilistic transition systems with non-determinism [7], weighted automata [18], and weighted tree automata [9,10]. Hopcroft's trick is used in many of these works for enhanced performance, too.

Such a variety of target systems is uniformly addressed by a recent body of work on *coalgebraic partition refinement* [4,5,14,26]. Here, a target system is identified with a categorical construct called *coalgebra* $c\colon C \to FC$ (see e.g. [13]), where C represents the state space, the *functor* F specifies the type of the system, and c represents the dynamics. By changing the functor F as a parameter, the theory accommodates many different systems such as DFAs and weighted automata. The coalgebraic partition refinement algorithms in [4,5,14,26] are *functor-generic*: they apply uniformly to such a variety of systems.

The current work is inspired by [14] which successfully exploits Hopcroft's trick for generic coalgebraic partition refinement. In [14], their coalgebraic algorithm is described in parallel with its set-theoretic (or even binary-level) concrete representations, letting the latter accommodate Hopcroft's trick. Their experimental results witnessed its superior performance, beating some existing tools that are specialised in a single type of systems.

However, the use of Hopcroft's trick in [14] is formulated in low-level set-theoretic terms, which seems to obscure the essence of the algorithm as well as the optimisation by Hopcroft's trick, much like in the original paper [11]. Therefore, in this paper, we aim at 1) an explicit formulation of Hopcroft's trick, and 2) a categorical partition refinement algorithm that exposes an explicit data structure to which Hopcroft's trick applies.

We achieve these two goals in this paper: 1) an explicit formulation that we call *Hopcroft's inequality*, and 2) a categorical algorithm that uses a *fibration*. Here is an overview.

Hopcroft's Inequality. We identify *Hopcroft's inequality* (Theorem 2.9) as the essence of Hopcroft's trick. Working on general trees with a general notion of vertex weight, it uses the classification of edges into *heavy* and *light* ones and bounds a sum of weights in terms of (only) the root and leaf weights. This inequality can be used to bound the complexity of many tree generation algorithms, including those for partition refinement.

This general theory can accommodate different weights. We exploit this generality to systematically derive partition refinement algorithms with different complexities (Sect. 6.2).

A Fibrational Partition Refinement Algorithm. Hopcroft's inequality does not directly apply to the existing coalgebraic partition refinement algorithms [4,5, 14,26] since the latter do not explicitly present a suitable tree structure. To address this challenge, we found the categorical language of *fibrations* [12] to be a convenient vehicle: it allows us to speak about the relationship between 1) an equivalence relation (an object in a fibre category) and 2) a partitioning of a state space (a mono-sink in the base category). The outcome is a partition refinement algorithm that is both *abstract* (it is functor-generic and applies to a variety of systems) and *concrete* (it explicitly builds a tree to which Hopcroft's inequality applies.) Our development relies on the fibrational theory of bisimilarity [8,17]; yet ours is the first fibrational partition refinement algorithm.

More specifically, in a fibration $p\colon \mathbb{E} \to \mathbb{C}$, an equivalence relation R on a set X is identified with an object $R \in \mathbb{E}_X$ in the fibre over X (consider the well-known fibration **EqRel** $\to$ **Set** of sets and equivalence relations over them). We introduce a categorical notion of *R-partitioning*; it allows $R \in \mathbb{E}_X$ to induce a mono-sink (i.e. a family of monomorphisms) $\{\kappa_i\colon C_i \rightarrowtail C\}_{i\in I}$. The latter is identified with the set of R-equivalence classes.

Figure 1 illustrates one iteration of our fibrational partition refinement algorithm $\mathsf{fPR}^{\mathrm{H}}$ (Algorithm 2). In the last step (Fig. 1c), the object C_{01} is devided into three parts C_{010}, C_{011} and C_{012} along $(c \circ \kappa)^*\overline{F}R$. We call the mono-sink $\{C_{01i} \rightarrowtail C_{01}\}_{i\in\{0,1,2\}}$ the $(c \circ \kappa)^*\overline{F}R$-partitioning of C_{01}. In this manner, a tree structure explicitly emerges in the base category $\mathbb{C}$. Hopcroft's inequality directly applies to this tree, allowing us to systematically present the Hopcroft-type optimisation on the categorical level of abstraction.

We note that, at this moment, our fibrational framework (with a fibration $p\colon \mathbb{E} \to \mathbb{C}$) has only one example, namely the fibration **EqRel** $\to$ **Set** of equivalence relations over sets. While it is certainly desirable to have other examples, their absence does not harm the value of our fibrational framework: we do not use fibrations for additional generality (beyond functor-genericity);[1] we use them to explicate trees in the base category (cf. Fig. 1).

Contributions. Summarising, our main technical contributions are as follows.

- *Hopcroft's inequality* that explicates the essence of Hopcroft's trick.
- A fibrational notion of *R-partitioning* that turns a fibre object into a mono-sink (Sect. 4).
- A fibrational partition refinement algorithm $\mathsf{fPR}^{\mathrm{H}}$ that combines the above two (Sect. 6.1).
- Functor-generic partition refinement algorithms

  $$\mathsf{fPR}^{\textbf{H-ER}}_{w_{\mathrm{C}}}, \quad \mathsf{fPR}^{\textbf{H-ER}}_{w_{\mathrm{P}}}, \quad \mathsf{fPR}^{\textbf{H-ER}}_{w_{\mathrm{R}}}$$

  obtained as instances of $\mathsf{fPR}^{\mathrm{H}}$ but using different weights in Hopcroft's inequality. The three achieve slightly different, yet comparable to the best known, complexity bounds (Sect. 6.2).

2 Hopcroft's Inequality

We present our first contribution, *Hopcroft's inequality*. It is a novel formalisation of Hopcroft's trick in terms of rooted trees. It also generalises the trick,

[1] In this sense, we can say that our use of fibrations is similar to some recent usages of string diagrams in *specific* monoidal categories, such as in [2,22].

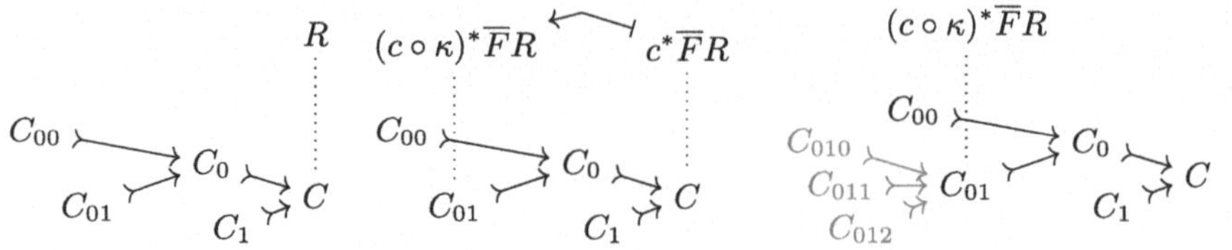

(a) Before the iteration (b) Refine R and restrict to C_{01} (c) Refine C_{01} and expand the tree

Fig. 1. An iteration in our algorithm $\mathsf{fPR}^{\mathrm{H}}$ (Algorithm 2). Figure 1a shows an equivalence relation R over C, and the corresponding partitioning $C_{00}, C_{01}, C_1 \rightarrowtail C$ of the state space C. (The history of refinement is recorded as a tree; this is important for complexity analysis.) In Fig. 1b, the equivalence relation R is refined into $c^*\overline{F}R$ along the one-step transition of the system dynamics c, and is further restricted to the partition C_{01}. In Fig. 1c, the resulting equivalence relation $(c \circ \kappa)^*\overline{F}R$ over C_{01} yields a partitioning of C_{01}, expanding the tree.

accommodating arbitrary *weights* (Definition 2.2) besides the particular one to count the number of items in classes that is typically and widely used (e.g. [11,14,15,24]).

Notation 2.1. Let T be a rooted tree. We denote the set of leaves by $L(T)$, the set of vertices by $V(T)$, the set of edges in the path from v to u by $\mathrm{path}(v, u)$, the set of children of $v \in V(T)$ by $\mathrm{ch}(v)$, and the subtree whose root is $v \in V(T)$ by $\mathrm{tr}(v)$.

Definition 2.2 (weight function). Let T be a rooted finite tree. A *weight function* of T is a map $w\colon V(T) \to \mathbb{N}$ satisfying $\sum_{u\in \mathrm{ch}(v)} w(u) \leq w(v)$ for each $v \in V(T)$. We call a weight function *tight* if $\sum_{u\in \mathrm{ch}(v)} w(u) = w(v)$ for all $v \in V(T) \setminus L(T)$.

Definition 2.3 (heavy child choice). For a weight function w of a tree T, a *heavy child choice* (hcc for short) is a map $h\colon V(T) \setminus L(T) \to V(T)$ satisfying $h(v) \in \mathrm{ch}(v)$ and $w(h(v)) = \max_{u\in \mathrm{ch}(v)} w(u)$ for every $v \in V(T) \setminus L(T)$. We write $h(v)$ as h_v and call the vertex h_v a *heavy child* of v, and a non-heavy child a *light child*. We define $\mathrm{lch}_h(v) = \mathrm{ch}(v) \setminus \{h_v\}$. An edge (v, u) is a *light edge* if $u \in \mathrm{lch}_h(v)$. We define $\mathrm{lpath}(v, u) = \{e \in \mathrm{path}(v, u) \mid e \text{ is a light edge}\}$.

Note that a heavy child choice always exists but is not unique in general.

Examples are in Fig. 2; the weight on the left is not tight while the right one is tight.

In the rest of this section, our technical development is towards Hopcroft's inequality in Theorem 2.9. It gives an upper bound for a sum of weights—we only count those for light children, which is the core of the optimisation in Hopcroft's partition refinement algorithm [11]—in terms of weights of the root and the leaves. This upper bound makes no reference to the tree's height or internal weights, making it useful for complexity analysis of tree generation algorithms.

The following lemma crucially relies on the definition of weight function.

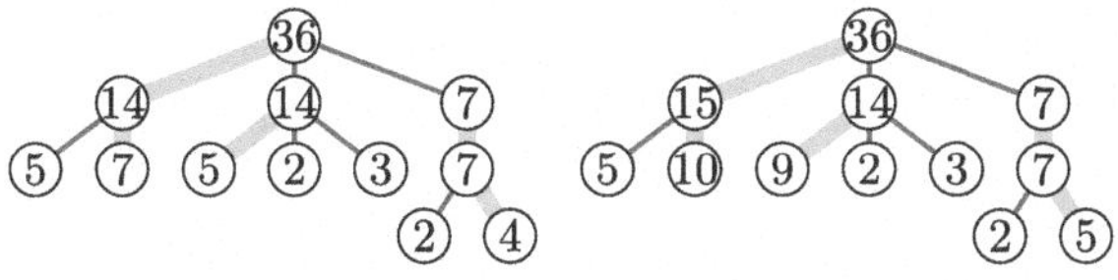

Fig. 2. Examples of rooted trees, each with a weight function and an hcc. The heavy children are indicated by thick edges. Thin edges are light edges.

Lemma 2.4. *Let T be a finite tree with a root r, w be a weight function of T, and S be an arbitrary set of edges of T. Then* $\sum_{v\in V(T)} \sum_{\substack{u\in \mathrm{ch}(v)\\ (v,u)\notin S}} w(u) \geq \sum_{l\in L(T)} \left|\mathrm{path}(r,l) \setminus S\right| \cdot w(l)$ *holds. The equality holds when w is tight.*

Lemma 2.5 is our first key lemma; we use Lemma 2.4 in its proof. It relates the sum of weights of the light children—for which we aim to give an upper bound in Theorem 2.9—with the leaf weights and (roughly) the tree height.

Lemma 2.5. *Let T be a finite tree with a root r, w be a weight function of T, and h be an hcc for w. Then the following inequality holds. The equality holds when w is tight.*

$$\sum_{v\in V(T)} \sum_{u\in \mathrm{lch}_h(v)} w(u) \;\geq\; \sum_{l\in L(T)} \left|\mathrm{lpath}(r,l)\right| \cdot w(l). \tag{1}$$

For the right tree in Fig. 2, the left-hand side of (1) is $(14+7)+5+(2+3)+0+2 = 33$, and the right-hand side is $1\times5+0\times10+1\times9+2\times2+2\times3+2\times2+1\times5 = 33$.

The inequality in (1) is the opposite of what we want (namely an upper bound for the left-hand side). We thus force an equality using *tightening*.

Definition 2.6 (tightening)**.** Let w be a weight function of a rooted finite tree T, and h be its heavy child choice. The *tightening* $w'\colon V(T) \to \mathbb{N}$ of w along h is defined recursively by

$$w'(u) = \begin{cases} w(u) & \text{if } u \text{ is the root of } T \\ w'(v) - \sum_{u'\in \mathrm{lch}_h(v)} w(u') & \text{if } u = h_v \text{ for the parent } v \text{ of } u \\ w(u) & \text{otherwise.} \end{cases}$$

In Fig. 2, the weight function of the right tree is a tightening of that of the left tree. We observe that tightening maintains a heavy child choice:

Lemma 2.7. *Let T be a rooted finite tree, w be a weight function of T, h be an hcc for w, and w' be the tightening of w along h. The following hold.*

1. *The map w' is a tight weight function of T.*
2. *The map h is also a heavy child choice for w'.*
3. *For the root r, $w(r) = w'(r)$ holds, and for each $v \in V(T)$, $w(v) \leq w'(v)$ holds.*

Our second key lemma towards Theorem 2.9 is as follows, bounding $|\operatorname{lpath}(r,v)|$ that occurs on the right in (1). Its proof is by what is commonly called *Hopcroft's trick* [1,11]: it observes that, along a light edge, weights decay at least by 1/2.

Lemma 2.8. *Let T be a finite tree with a root r, w be a weight of T, and h be an hcc for w. For each vertex $v \in V(T)$ with $w(v) \neq 0$, the following inequality holds:* $|\operatorname{lpath}(r,v)| \leq \log_2 w(r) - \log_2 w(v)$.

We combine Lemma 2.8 and Lemma 2.5 (its equality version; we can use it via tightening) to obtain Hopcroft's inequality. It bounds a sum of weights by the root and leaf weights.

Theorem 2.9 (Hopcroft's inequality). *Let T be a finite tree with root r, w be a weight function of T, and h be a heavy child choice for w. The following inequality holds.*

$$\sum_{v \in V(T)} \sum_{u \in \operatorname{lch}_h(v)} w(u) \leq w(r)\log_2 w(r) - \sum_{\substack{l \in L(T) \\ w(l) \neq 0}} w(l)\log_2 w(l) \tag{2}$$

For complexity analysis, we use Hopcroft's inequality in the following form. Assume that a tree generation algorithm takes $t(v)$ time to generate all the children (both heavy and light) of v. If there exists K such that $t(v)$ is bounded by K times the sum of all light children, then the time to generate the whole tree is bounded by $Kw(r)\log_2 w(r)$.

Corollary 2.10. *Let T be a rooted finite tree with root r, w be a weight function of T, and h be a heavy child choice for w. If a map $t\colon V(T) \to \mathbb{N}$ satisfies that there exists a constant $K \in \mathbb{N}$ such that $t(v) \leq K\sum_{u \in \operatorname{lch}_h(v)} w(u)$ for every $v \in V(T)$, then the sum of $t(v)$ is bounded by $Kw(r)\log_2 w(r)$, that is $\sum_{v \in V(T)} t(v) \leq Kw(r)\log_2 w(r)$.*

Remark 2.11. Further adaptations of Hopcroft's trick are pursued in the literature, e.g. in [25], where the notion of heavy child choice is relaxed with an extra parameter $\alpha \in [1/2, 1)$. Our theory can easily be extended to accommodate α, in which case the above description corresponds to the special case with $\alpha = 1/2$. Details are deferred to another venue.

3 Categorical Preliminaries

The rest of the paper is about our second contribution, namely a functor-generic partition refinement (PR) algorithm optimised by an explicit use of Hopcroft's inequality (Theorem 2.9). It is given by our novel formulation of coalgebraic PR algorithms in fibrational terms. Here we shall review some necessary categorical preliminaries.

We use categorical formalisation of intersections and unions.

Definition 3.1. For monomorphisms $m\colon A \rightarrowtail C$ and $n\colon B \rightarrowtail C$ in $\mathbb{C}$, the *intersection* $m \cap n\colon A \cap B \rightarrowtail C$ and the *union* $m \cup n\colon A \cup B \rightarrowtail C$ are defined by the pullback and pushout, respectively:

$$\begin{array}{ccc} A \cap B & \xrightarrowtail{\pi_2} & B \\ \pi_1 \downarrow & \lrcorner & \downarrow n \\ A & \xrightarrowtail{m} & C \end{array} \quad \text{and} \quad \begin{array}{ccc} A \cap B & \xrightarrowtail{\pi_2} & B \\ \pi_1 \downarrow & \ulcorner & \downarrow \\ A & \rightarrowtail & A \cup B \end{array}.$$

We say $m\colon A \rightarrowtail C$ and $m'\colon A' \rightarrowtail C$ are *equivalent* if there is an isomorphism $\phi\colon A \to A'$ such that $m = m' \circ \phi$. The set $\mathbf{Sub}(\mathbb{C})_C$ of equivalence classes of monomorphisms whose codomains are C forms a lattice, assuming enough limits and colimits.

3.1 Fibrations

A fibration $p\colon \mathbb{E} \to \mathbb{C}$ is a functor satisfying some axioms. When $p(R) = C$ for an object $R \in \mathbb{E}$ and an object $C \in \mathbb{C}$, we see that R equips C with some information, e.g. a predicate, a relation, a topology, etc. The main example in this paper is the fibration $\mathbf{EqRel} \to \mathbf{Set}$ where C is a set and R is an equivalence relation over C.

Fibrational constructs that are the most relevant to us are the *inverse image* $f^*(R')$ and the *direct image* $f_*(R)$ along a morphism $f\colon S \to S'$ in $\mathbb{C}$. In the case of $\mathbf{EqRel} \to \mathbf{Set}$, these are computed as follows.

$$\begin{array}{c} f^*(R') = \\ \{(x,y) \mid (fx, fy) \in R'\} \end{array} \xleftarrow{f^*} R' \qquad R \xmapsto{f_*} \begin{array}{c} f_*(R) = \\ \{(fx, fy) \mid (x,y) \in R\} \end{array}$$

$$S \xrightarrow{f} S' \qquad\qquad S \xrightarrow{f} S'$$

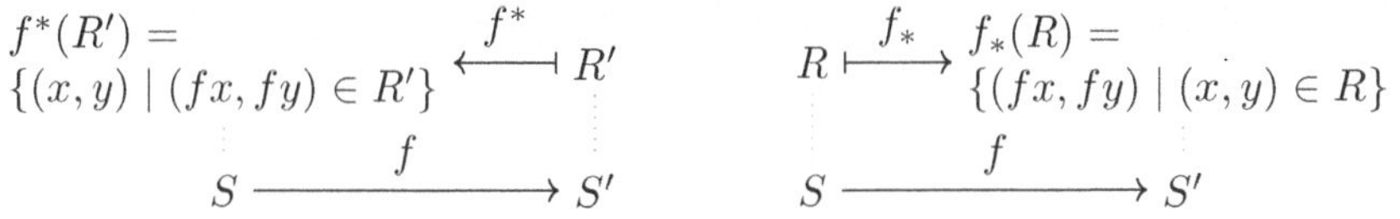

In what follows we introduce some basics of fibrations; they formalise the intuition above. For details, see e.g. [12].

Definition 3.2 (fibration). Let $p\colon \mathbb{E} \to \mathbb{C}$ be a functor. A morphism $f\colon P \to R$ in $\mathbb{E}$ is *Cartesian* if for any $g\colon Q \to R$ in $\mathbb{E}$ with $pg = pf \circ v$ for some $v\colon pQ \to pP$, there exists a unique $h\colon Q \to P$ in $\mathbb{E}$ above v (i.e. $ph = v$) with $f \circ h = g$. The functor p is a *fibration* if for each $R \in \mathbb{E}$ and $u\colon C \to pR$ in $\mathbb{C}$, there are an object u^*R and a Cartesian morphism $\dot{u}(R)\colon u^*R \to R$ in $\mathbb{E}$. See below.

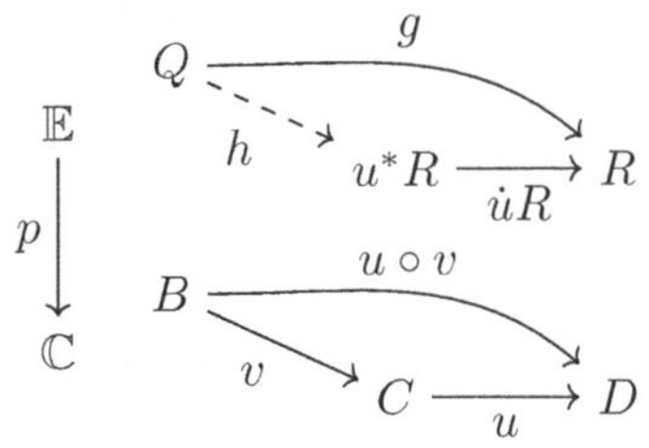

The category $\mathbb{E}$ is called the *total category* and the category $\mathbb{C}$ is called the *base category* of the fibration. For an object $C \in \mathbb{C}$, the objects in $\mathbb{E}$ above C form a category $\mathbb{E}_C$, called the *fibre category* above C. The fibre category $\mathbb{E}_C$ is the category of "equivalence relations" on C.

Definition 3.3 (fibre category). Let $p\colon \mathbb{E} \to \mathbb{C}$ be a fibration and $C \in \mathbb{C}$. The *fibre category* $\mathbb{E}_C$ over C is the subcategory of $\mathbb{E}$ whose objects are defined by $\mathrm{ob}(\mathbb{E}_C) = \{R \in \mathbb{E} \mid pR = C\}$, and morphisms are defined by $\mathbb{E}_C(Q, R) = \{f \in \mathbb{E}(Q, R) \mid pf = \mathrm{id}_C\}$ for $Q, R \in \mathrm{ob}(\mathbb{E}_C)$.

Example 3.4 (**EqRel** $\to$ **Set** is a fibration). Let **EqRel** be the category of equivalence relations. The objects of **EqRel** are pairs (S, R) of a set S and an equivalence relation R on S. A morphism $f\colon (S, R) \to (S', R')$ in **EqRel** is a function $f\colon S \to S'$ satisfying $(f(x), f(y)) \in R'$ for all $(x, y) \in R$. We sometimes write just R for (S, R) when no confusion arises. The functor $p\colon \mathbf{EqRel} \to \mathbf{Set}$ defined by $p(S, R) = S$ is a fibration.

For a morphism $u\colon C \to D$ in the base category of a fibration, the map $u^*\colon \mathrm{ob}(\mathbb{E}_D) \to \mathrm{ob}(\mathbb{E}_C)$ extends to a functor $u^*\colon \mathbb{E}_D \to \mathbb{E}_C$ between fibre categories. We call the functor $u^*\colon \mathbb{E}_D \to \mathbb{E}_C$ an *inverse image* functor.

Given a fibration $\mathbb{E} \to \mathbb{C}$ and an endofunctor $F\colon \mathbb{C} \to \mathbb{C}$ on the base category, if $R \in \mathbb{E}$ is above $C \in \mathbb{C}$, we would like to get an object in $\mathbb{E}$ above FC. A lifting of F specifies the choice of an object above FC.

Definition 3.5 (lifting, fibred lifting). Let $p\colon \mathbb{E} \to \mathbb{C}$ be a fibration and $F\colon \mathbb{C} \to \mathbb{C}$ be a functor. A *lifting* of F is a functor $\overline{F}\colon \mathbb{E} \to \mathbb{E}$ with $p \circ \overline{F} = F \circ p$. A functor $\overline{F}\colon \mathbb{E} \to \mathbb{E}$ is a *fibred lifting* of F if $\overline{F}$ is a lifting of F and preserves Cartesian morphisms.

When $\overline{F}\colon \mathbb{E} \to \mathbb{E}$ is a fibred lifting of $F\colon \mathbb{C} \to \mathbb{C}$, for $f\colon C \to D$ in $\mathbb{C}$ and $R \in \mathbb{E}_D$, we have $\overline{F}(f^*R) = (Ff)^*(\overline{F}R)$ in $\mathbb{E}_{FC}$. An important example of a lifting is a relation lifting.

Definition 3.6 (relation lifting [13]). Let $F\colon \mathbf{Set} \to \mathbf{Set}$ be a weak pullback preserving functor. We define a lifting $\mathrm{Rel}(F)\colon \mathbf{EqRel} \to \mathbf{EqRel}$ of F along the fibration $p\colon \mathbf{EqRel} \to \mathbf{Set}$, called the *relation lifting* of F, as follows. For an object $(C, R) \in \mathbf{EqRel}$, there is the inclusion $\langle r_1, r_2\rangle\colon R \rightarrowtail C \times C$. We define the relation lifting on the object R by $\mathrm{Rel}(F)(R) = \mathrm{Im}\langle Fr_1, Fr_2\rangle$, where $\mathrm{Im}\langle Fr_1, Fr_2\rangle$ is the image factorisation. By the assumption that F preserves weak pullbacks, we can show that $\mathrm{Rel}(F)(R)$ is an equivalence relation. $\mathrm{Rel}(F)$ can be extended to a functor.

$$\begin{array}{ccc} FR & \xrightarrow{\langle Fr_1, Fr_2\rangle} & FC \times FC \\ & \searrow\!\!\!\!\twoheadrightarrow \quad \rightarrowtail\!\!\nearrow & \\ & \mathrm{Im}\langle Fr_1, Fr_2\rangle & \\ & = \mathrm{Rel}(F)(R) & \end{array}$$

In this paper, we deal with a restricted class of fibrations, called $\mathbf{CLat}_\sqcap$-fibrations.

Definition 3.7 ($\mathbf{CLat}_{\sqcap}$-fibration). A fibration $p\colon \mathbb{E} \to \mathbb{C}$ is a $\mathbf{CLat}_{\sqcap}$*-fibration* if each fibre $\mathbb{E}_C$ is a complete lattice and each inverse image functor $u^*\colon \mathbb{E}_D \to \mathbb{E}_C$ preserves meets $\sqcap$.

For a $\mathbf{CLat}_{\sqcap}$-fibration, there always exists the left adjoint $u_*\colon \mathbb{E}_C \to \mathbb{E}_D$ to an inverse image functor u^*, as is well-known (cf. Freyd's adjoint functor theorem). The functor u_* is defined by $u_*(P) = \bigsqcap\{R \in \mathbb{E}_D \mid P \sqsubseteq u^*(R)\}$ on objects. We call u_* a *direct image* functor.

Example 3.8 (EqRel $\to$ Set is a $\mathbf{CLat}_{\sqcap}$-fibration). The functor $p\colon \mathbf{EqRel} \to \mathbf{Set}$ from Example 3.4 is a $\mathbf{CLat}_{\sqcap}$-fibration. We describe the inverse image functor f^* and the direct image functor f_* for a function $f\colon S \to S'$. For an equivalence relation R' on S', the inverse image $f^*(R')$ is the equivalence relation $\{(x,y) \in S \times S \mid (f(x), f(y)) \in R'\}$ on S. For an equivalence relation R on S, the direct image $f_*(R)$ is the *equivalence closure* of the relation $\{(f(x), f(y)) \in S' \times S' \mid (x,y) \in R\}$.

3.2 Coalgebras and Bisimulations

Coalgebras are widely used as a generalisation of state-based systems [13,23].

Definition 3.9 (F-coalgebra)**.** Let $\mathbb{C}$ be a category and $F\colon \mathbb{C} \to \mathbb{C}$ be an endofunctor. An *F-coalgebra* is a pair (C, c) of an object $C \in \mathbb{C}$ and a morphism $c\colon C \to FC$.

For an F-coalgebra $c\colon C \to FC$, F specifies the type of the system, the carrier object C represents the "set of states" of the system, and c represents the transitions in the system. When $\mathbb{C} = \mathbf{Set}$, for an F-coalgebra $c\colon C \to FC$ and a state $x \in C$, the element $c(x) \in FC$ represents properties (e.g. acceptance) and successors of x.

A major benefit of coalgebras is that their theory is *functor-generic*: by changing a functor F, the same theory uniformly applies to a vast variety of systems.

Example 3.10. We describe some F-coalgebras for functors F on **Set**.

1. For the powerset functor $\mathcal{P}$, a $\mathcal{P}$-coalgebra $c\colon C \to \mathcal{P}C$ is a *Kripke frame*. For a state $x \in C$, $c(x) \in \mathcal{P}C$ is the set of successors of x.
2. Let Σ be an alphabet and $N_\Sigma = 2 \times (\mathcal{P}-)^\Sigma$. An N_Σ-coalgebra $c\colon C \to N_\Sigma C$ is a *non-deterministic automaton* (NA). For a state $x \in C$, let $(b,t) = c(x) \in 2 \times (\mathcal{P}C)^\Sigma$. The state x is accepting iff $b = 1$, and there is a transition $x \xrightarrow{a} y$ in the NA iff $y \in t(a)$.
3. The distribution functor $\mathcal{D}$ is defined on a set X to be $\mathcal{D}X = \{d\colon X \to [0,1] \mid \{x \in X \mid f(x) \neq 0\}$ is finite and $\sum_{x \in X} d(x) = 1\}$. A $\mathcal{D}$-coalgebra $c\colon C \to \mathcal{D}C$ is a *Markov chain*. For a state x, $c(x) \in \mathcal{D}C$ is a probability distribution $C \to [0,1]$, which represents the probabilities of transitions to successor states of x.

We are interested in how similar two states of a state-transition system are. We consider two states to be similar if one state can mimic the transitions of the other. *Bisimilarity* by Park [21] and Milner [19] is a notion that captures such behaviour of states. Hermida and Jacobs [8] formulated bisimilarity as a coinductive relation on a coalgebra, using a fibration.

Definition 3.11 (bisimulations and the bisimilarity). Let $p\colon \mathbb{E} \to \mathbb{C}$ be a $\mathbf{CLat}_\sqcap$-fibration, $F\colon \mathbb{C} \to \mathbb{C}$ be a functor, $c\colon C \to FC$ be an F-coalgebra and $\overline{F}$ be a lifting of F. An $(F, \overline{F})$*-bisimulation* is a $c^* \circ \overline{F}$-coalgebra in $\mathbb{E}_C$, that is an object $R \in \mathbb{E}_C$ with $R \sqsubseteq c^*(\overline{F}(R))$ since a morphism in $\mathbb{E}_C$ is a relation ($\sqsubseteq$). By the Knaster–Tarski theorem, there exists the greatest $(F, \overline{F})$-bisimulation $\nu(c^* \circ \overline{F})$ with respect to the order of $\mathbb{E}_C$, and it is called the $(F, \overline{F})$*-bisimilarity.*

In the above definition, the choice of $\overline{F}$ determines a notion of bisimulation. The relation lifting $\mathrm{Rel}(F)$ (Definition 3.6) is often used as a lifting of F. For all the functors we consider, the bisimilarity wrt. $\mathrm{Rel}(F)$ coincides with the *behavioural equivalence*, another well-known notion of bisimilarity [13, §4.5].

Example 3.12. We illustrate $(F, \mathrm{Rel}(F))$-bisimilarities (also called *logical F-bisimilarity* [13]) for F in Example 3.10. Let $C \in \mathbf{Set}$ and $R \in \mathbf{EqRel}_C$.

1. ($F = \mathcal{P}$). The $(\mathcal{P}, \mathrm{Rel}(\mathcal{P}))$-bisimilarity $\nu(c^* \circ \mathrm{Rel}(\mathcal{P}))$ for a $\mathcal{P}$-coalgebra $c\colon C \to \mathcal{P}C$ is the maximum relation R on C such that if $(x, y) \in R$ then
   - for every $x' \in c(x)$, there is $y' \in c(y)$ such that $(x', y') \in R$, and
   - for every $y' \in c(y)$, there is $x' \in c(x)$ such that $(x', y') \in R$.
2. ($F = N_\Sigma$). The $(N_\Sigma, \mathrm{Rel}(N_\Sigma))$-bisimilarity $\nu(c^* \circ \mathrm{Rel}(N_\Sigma))$ for an N_Σ-coalgebra $c\colon C \to N_\Sigma C$ is the ordinary bisimilarity for the NA c, that is the maximum relation R on C such that if $(x, y) \in R$ then $\pi_1(c(x)) = \pi_1(c(y))$ and
   - for each $a \in \Sigma, x' \in \pi_2(c(x))(a)$, there is $y' \in \pi_2(c(y))(a)$ such that $(x', y') \in R$, and
   - for each $a \in \Sigma$ and $y' \in \pi_2(c(y))(a)$, there is $x' \in \pi_2(c(x))(a)$ such that $(x', y') \in R$.
3. ($F = \mathcal{D}$) [3]. The $(\mathcal{D}, \mathrm{Rel}(\mathcal{D}))$-bisimilarity $\nu(c^* \circ \mathrm{Rel}(\mathcal{D}))$ for a $\mathcal{D}$-coalgebra $c\colon C \to \mathcal{D}C$ is the maximum relation R such that if $(x, y) \in R$ then $\sum_{z \in K} c(x)(z) = \sum_{z \in K} c(y)(z)$ for every equivalence class $K \subseteq C$ of R.

4 Fibrational Partitioning

We introduce the notion of fibrational partitioning, one that is central to our algorithm that grows a tree using fibre objects (cf. Fig. 1).

Given an "equivalence relation" R over C—identified with an object $R \in \mathbb{E}_C$ over $C \in \mathbb{C}$ in a suitable fibration $p\colon \mathbb{E} \to \mathbb{C}$—a fibrational R-partitioning is a mono-sink, shown on the right, that is subject to certain axioms. The notion allows us to explicate equivalence classes (namely $\{C_i\}_i$) in the abstract fibrational language.

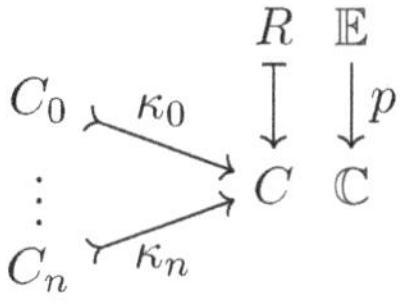

Definition 4.1 (R-partitioning)**.** Let $\mathbb{C}$ be a category with pullbacks and an initial object 0, and $p\colon \mathbb{E} \to \mathbb{C}$ be a $\mathbf{CLat}_\sqcap$-fibration. Let $C \in \mathbb{C}$ and $R \in \mathbb{E}_C$. An *R-partitioning* is a mono-sink (i.e. a family of monomorphisms) $\{\kappa_i\colon C_i \rightarrowtail C\}_{i\in I}$ that satisfies:

1. $\kappa_i^*(R) = \top_{C_i}$ for all $i \in I$,
2. $\bigsqcup_{i\in I}(\kappa_i)_*(\top_{C_i}) = R$, and
3. $C_i \not\cong 0$ and $C_i \cap C_j \cong 0$ for each $i, j \in I$ with $i \neq j$.

We say a $\mathbf{CLat}_\sqcap$-fibration p *admits partitioning* if (1) for each $C \in \mathbb{C}$ and $R \in \mathbb{E}_C$, there is an R-partitioning; and moreover, (2) for each $C \in \mathbb{C}$, $R, R' \in \mathbb{E}_C$ such that $R' \sqsubseteq R$, and each R-partitioning $\{\kappa_i\colon C_i \rightarrowtail C\}_{i\in I}$, we have $\bigsqcup_{i\in I}(\kappa_i)_*(\kappa_i^* R') = R'$.

Cond. 3 asserts that the components C_i are nontrivial and disjoint. Cond. 1 says the partitioning $\{C_i\}_i$ is *not too coarse*—the original equivalence R, when restricted to C_i, should relate all pairs of elements in C_i. Conversely, Cond. 2 means that $\{C_i\}_i$ is *not too fine*—if it were finer than R, then the relation $\bigsqcup_{i\in I}(\kappa_i)_*(\top_{C_i})$ over C would be finer than R. See the concrete description of $(\kappa_i)_*$ in Example 3.8.

Example 4.2 (**EqRel** $\to$ **Set** admits partitioning)**.** **EqRel** $\to$ **Set** admits partitioning. Indeed, given an equivalence relation $R \in \mathbf{EqRel}_C$ over C, the mono-sink $\{\kappa_S\colon S \rightarrowtail C\}_{S\in C/R}$, where $S \in C/R$ is naturally identified with a subset of C, is an R-partitioning. Cond. 1–3 are easily verified following Example 3.8.

An R-partitioning is not necessarily unique. This happens when $R \in \mathbf{EqRel}_C$ has singleton equivalence classes. Let $A \subseteq C$ be an arbitrary subset such that each $x \in A$ composes a singleton R-equivalence class. Then $\{\kappa'_S\colon S \rightarrowtail C\}_{S\in I}$, where $I = (C/R) \setminus \{\,\{x\} \mid x \in A\,\}$, is also an R-partitioning. With this mono-sink (that is "narrower" than the original $\{\kappa_S\}_{S\in C/R}$), Cond. 2 is satisfied since the equivalence closure operation included in the direct images $(\kappa_i)_*(\top_{C_i})$ (see Example 3.8) compensates the absence of $x \in A$.

The fibration **EqRel** $\to$ **Set** is our leading example, and unfortunately, the only example that we know admits partitioning. There are many other examples of $\mathbf{CLat}_\sqcap$-fibrations (see [17]), but they fail to admit partitioning, typically due to the failure of Cond. 2 of Definition 4.1. This absence of examples does not harm the value of our fibrational framework: our goal is to explicate categorical essences of partition refinement; and we do not aim at new instances via categorical abstraction (although such are certainly desirable).

(Beck–Chevalley) (modularity)

$$
\begin{array}{ccc}
A\cap B & \overset{n}{\rightarrowtail} & B \\
{\scriptstyle m}\downarrow & \lrcorner & \downarrow{\scriptstyle \lambda} \\
A & \underset{\kappa}{\rightarrowtail} & C
\end{array}
\quad\Longrightarrow\quad
\begin{array}{ccc}
\mathbb{E}_{A\cap B} & \overset{n^*}{\longleftarrow} & \mathbb{E}_B \\
{\scriptstyle m_*}\downarrow & & \downarrow{\scriptstyle \lambda_*} \\
\mathbb{E}_A & \underset{\kappa^*}{\longleftarrow} & \mathbb{E}_C
\end{array}
\qquad
\begin{array}{ccccc}
\mathbb{E}_A \times \mathbb{E}_B & \overset{\kappa_*\times\lambda_*}{\longrightarrow} & \mathbb{E}_C \times \mathbb{E}_C & \overset{\sqcup}{\longrightarrow} & \mathbb{E}_C \\
 & & {\scriptstyle \lambda^*\times\lambda^*}\downarrow & & \downarrow{\scriptstyle \lambda^*} \\
 & & \mathbb{E}_B \times \mathbb{E}_B & \overset{\sqcup}{\longrightarrow} & \mathbb{E}_B
\end{array}
$$

Fig. 3. Conditions for Assumption 5.1.

5 The Naive Fibrational Algorithm $\mathsf{fPR}^{\mathrm{naive}}$

We introduce a naive fibrational partition refinement algorithm, called $\mathsf{fPR}^{\mathrm{naive}}$, as a preparation step to our main algorithm $\mathsf{fPR}^{\mathrm{H}}$ (Algorithm 2).

In what follows, a prefix-closed set $T \subseteq \mathbb{N}^*$ (where $\mathbb{N}^*$ is the set of strings over $\mathbb{N}$) is identified with a rooted tree. We denote the leaves of T by $L(T)$.

We introduce further conditions that make fibrations "compatible" with partitioning. It is easy to see that $\mathbf{EqRel} \to \mathbf{Set}$ satisfies the conditions on p in Assumption 5.1.

Assumption 5.1. Assume a $\mathbf{CLat}_\sqcap$-fibration $p\colon \mathbb{E} \to \mathbb{C}$ that satisfies the following conditions.

1. For each $C \in \mathbb{C}$, the lattice $\mathbf{Sub}(\mathbb{C})_C$ of subobjects of C in $\mathbb{C}$ is distributive.
2. (Beck–Chevalley) For every pullback diagram along monomorphisms in $\mathbb{C}$, shown in the first diagram in Fig. 3, the induced diagram, the second in Fig. 3, commutes.
3. For any monomorphisms $\kappa\colon A \rightarrowtail C$ and $\lambda\colon B \rightarrowtail C$, the third diagram in Fig. 3 is a fork, where the following diagram is a *fork* if $h_1 \circ g_1 \circ f = h_2 \circ g_2 \circ f$.

$$
\begin{array}{ccccc}
W & \overset{f}{\longrightarrow} & X & \overset{g_1}{\longrightarrow} & Y_1 \\
 & & {\scriptstyle g_2}\downarrow & & \downarrow{\scriptstyle h_1} \\
 & & Y_2 & \underset{h_2}{\longrightarrow} & Z
\end{array}
$$

Definition 5.2. Let $p\colon \mathbb{E} \to \mathbb{C}$ be a $\mathbf{CLat}_\sqcap$-fibration that satisfies Assumption 5.1, $F\colon \mathbb{C} \to \mathbb{C}$ be a functor, and $\overline{F}\colon \mathbb{E} \to \mathbb{E}$ be its lifting along p (Definition 3.5). Algorithm 1 shows our *naive fibrational partition refinement algorithm.* Given a coalgebra $c\colon C \to FC$, it computes a $\nu(c^*\overline{F})$-partitioning of C, i.e. modulo the $(F, \overline{F})$-bisimilarity of c (Definition 3.11).

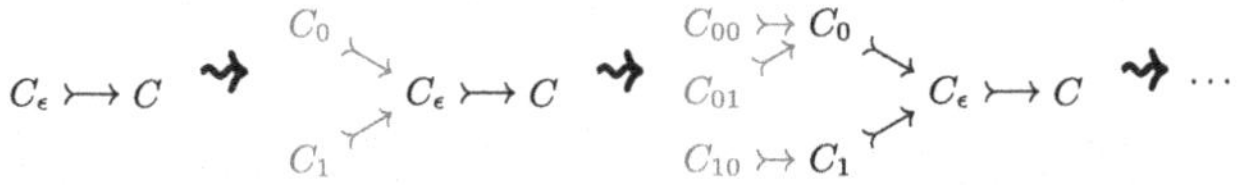

Fig. 4. The R-partitioning gets finer as the algorithm runs.

Algorithm 1 starts with $R = \top_C \in \mathbb{E}_C$ and a singleton family of a monomorphism $\{\kappa_\epsilon\colon C_\epsilon \rightarrowtail C\}$. With each iteration, the object R on C gets smaller

Algorithm 1. The naive fibrational partition refinement algorithm $\mathsf{fPR}^{\mathrm{naive}}$.

Input: A coalgebra $c\colon C \to FC$ in $\mathbb{C}$.
Output: A mono-sink $\{\kappa_i\colon C_i \rightarrowtail C\}_{i\in I}$ for some I.

$T := \{\epsilon\} \subseteq \mathbb{N}^*$; $C_\epsilon := C$; $\kappa_\epsilon := \mathrm{id}_C\colon C_\epsilon \rightarrowtail C$; $R := \top_C$ ▷ initialisation
while $c^*\overline{F}R \neq R$ **do** ▷ the main loop
  $R := c^*\overline{F}R$; $L := L(T)$
  for $\sigma \in L$ **do**
    Take a $\kappa_\sigma^*(R)$-partitioning $\{\lambda_{\sigma,k}\colon C_{\sigma k} \rightarrowtail C_\sigma\}_{k\in\{0,\dots,n_\sigma\}}$ of C_σ
    for $k = 0,\dots,n_\sigma$ **do** $\kappa_{\sigma k} := \kappa_\sigma \circ \lambda_{\sigma,k}\colon C_{\sigma k} \rightarrowtail C$
    $T := T \cup \{\sigma 0,\dots,\sigma n_\sigma\}$
return $\{\kappa_\sigma\colon C_\sigma \rightarrowtail C\}_{\sigma\in L(T)}$

and closer to $\nu(c^*\overline{F})$ and R-partitioning $\{\kappa_\sigma\colon C_\sigma \rightarrowtail C\}_\sigma$ gets finer (see Fig. 4). When the algorithm terminates, R is equal to $\nu(c^*\overline{F})$ and a $\nu(c^*\overline{F})$-partitioning is returned.

Combining the loop invariant (Lemma 5.3) and termination (Lemma 5.4), we can prove the correctness of the naive algorithm.

Lemma 5.3 (loop invariant). *At the beginning of each iteration of the main loop, the following hold.*

1. *The mono-sink $\{\kappa_\sigma\colon C_\sigma \rightarrowtail C\}_{\sigma\in L(T)}$ is an R-partitioning.*
2. *$\nu(c^*\overline{F}) \sqsubseteq R$.*

Lemma 5.4 (termination). *If $\mathbb{E}_C$ is a well-founded lattice, Algorithm 1 terminates.*

Proposition 5.5 (correctness). *If $\mathbb{E}_C$ is well-founded, then Algorithm 1 terminates and returns $\nu(c^*\overline{F})$-partitioning $\{\kappa\colon C_i \rightarrowtail C\}_{i\in I}$.*

6 Optimised Algorithms with Hopcroft's Inequality

Recall that the naive algorithm grows a tree *uniformly* so that every leaf has the same depth (see Fig. 4; note that, even if C_σ is fine enough, we extend the node by a trivial partitioning). By selecting leaves in a smart way and generating a tree selectively, the time cost of each iteration can be reduced, so that Hopcroft's inequality is applicable.

In Sect. 6.1, we present a functor-generic and fibrational algorithm enhanced with the Hopcroft-type optimisation, calling it $\mathsf{fPR}^{\mathrm{H}}$. We use Hopcroft's inequality (Sect. 2) for complexity analysis.

In Sect. 6.2 we instantiate $\mathsf{fPR}^{\mathrm{H}}$ to the fibration **EqRel** → **Set**, obtaining three concrete (yet functor-generic) algorithms $\mathsf{fPR}^{\mathbf{H\text{-}ER}}_{w_\mathrm{C}}, \mathsf{fPR}^{\mathbf{H\text{-}ER}}_{w_\mathrm{P}}, \mathsf{fPR}^{\mathbf{H\text{-}ER}}_{w_\mathrm{R}}$ that use different weight functions. $\mathsf{fPR}^{\mathbf{H\text{-}ER}}_{w_\mathrm{C}}$ is essentially the algorithm in [14]. The other two ($\mathsf{fPR}^{\mathbf{H\text{-}ER}}_{w_\mathrm{P}}, \mathsf{fPR}^{\mathbf{H\text{-}ER}}_{w_\mathrm{R}}$) use the weight functions from the works [6,11,16] on DFA partition refinement. The three algorithms exhibit slightly different asymptotic complexities.

6.1 A Fibrational Algorithm $\mathsf{fPR}^{\mathrm{H}}$ Enhanced by Hopcroft's Inequality

We fix a $\mathbf{CLat}_\sqcap$-fibration $p\colon \mathbb{E} \to \mathbb{C}$, functors $F\colon \mathbb{C} \to \mathbb{C}$ and $\overline{F}\colon \mathbb{E} \to \mathbb{E}$, an F-coalgebra $c\colon C \to FC$, and a map $w\colon \mathrm{ob}(\mathbf{Sub}(\mathbb{C})_C) \to \mathbb{N}$ (which we use for weights). We write $w(C')$ for $w(\lambda\colon C' \rightarrowtail C)$ when no confusion arises.

The following conditions clarify which properties of $\mathbf{EqRel} \to \mathbf{Set}$ are necessary to make our optimised fibrational algorithm $\mathsf{fPR}^{\mathrm{H}}$ work: the last one (Assumption 6.1.10) is for complexity analysis; all the other ones are for correctness.

Assumption 6.1. 1. $\mathbb{C}$ has pullbacks, pushouts along monos, and an initial object 0.
2. The fibre category $\mathbb{E}_0$ above an initial object 0 is trivial, that is $\top_0 = \bot_0$.
3. $\overline{F}$ is a fibred lifting of F along p.
4. $F\colon \mathbb{C} \to \mathbb{C}$ preserves monomorphisms whose codomain is not 0.
5. The fibration p admits partitioning.
6. The fibration $p\colon \mathbb{E} \to \mathbb{C}$ satisfies the three conditions in Assumption 5.1.
7. The fibre category $\mathbb{E}_C$ is a well-founded lattice.
8. If $C' \rightarrowtail C$ and $R \in \mathbb{E}_{C'}$, every R-partitioning $\{\lambda_k\colon D_k \rightarrowtail C'\}_{k\in K}$ is finite ($|K| < \infty$).
9. If $\kappa\colon A \rightarrowtail C$ and $\lambda\colon B \rightarrowtail C$ are monomorphisms and $A \cap B \cong 0$, then the functor $\mathbb{E}_A \times \mathbb{E}_B \xrightarrow{\kappa_* \times \lambda_*} \mathbb{E}_C \times \mathbb{E}_C \xrightarrow{\sqcup} \mathbb{E}_C$ is injective on objects.
10. For a monomorphism $C' \rightarrowtail C$, an object $R \in \mathbb{E}_{C'}$, and an R-partitioning $\{\kappa_i\colon C_i \rightarrowtail C'\}$ of C', $\sum_{i=1}^{n} w(C_i) \le w(C')$ holds.

Assumption 6.1.3 is not overly restrictive. Indeed, the following functors on **Set** have a fibred lifting. The functors described in Example 3.10 are examples of the functor defined by (3).

Proposition 6.2. *Consider the endofunctors on* **Set** *defined by the BNF below.*

$$F ::= \mathrm{Id} \mid A \mid \textstyle\coprod_{b\in B} F_b \mid \prod_{b\in B} F_b \mid \mathcal{P}F \mid \mathcal{D}F \qquad \textit{where } A \textit{ and } B \textit{ are sets.} \quad (3)$$

The relation lifting $\mathrm{Rel}(F)\colon \mathbf{EqRel} \to \mathbf{EqRel}$ *of* F *(Definition 3.6) is fibred.*

Proposition 6.3. *The fibration* $p\colon \mathbf{EqRel} \to \mathbf{Set}$ *with* $\mathrm{Rel}(F)$ *(3) and a coalgebra* $c\colon C \to FC$ *for a finite set* C *satisfies the assumptions 1–9 of Assumption 6.1.*

Definition 6.4. ($\mathsf{fPR}^{\mathrm{H}}$). Let the $\mathbf{CLat}_\sqcap$-fibration $p\colon \mathbb{E} \to \mathbb{C}$, the map w, the functors $F\colon \mathbb{C} \to \mathbb{C}$ and $\overline{F}\colon \mathbb{E} \to \mathbb{E}$, and the object C satisfy Assumption 6.1. Algorithm 2 shows the algorithm $\mathsf{fPR}^{\mathrm{H}}_{(F,\overline{F}),w}$ ($(F,\overline{F})$ and w are omitted when clear from the context). Given a coalgebra $c\colon C \to FC$, it computes a $\nu(c^*\overline{F})$-partitioning of C, like the naive algorithm.

Algorithm 2. An optimised fibrational partition refinement algorithm $\mathsf{fPR}^{\mathrm{H}}_{(F,\overline{F}),w}$.

Input: A coalgebra $c\colon C \to FC$ in $\mathbb{C}$.
Output: A mono-sink $\{\kappa_i\colon C_i \rightarrowtail C\}_{i\in I}$ for some I.

1: $J := \{\epsilon\} \subseteq \mathbb{N}^*$; $C_\epsilon := C$; $C^{\mathsf{cl}}_\epsilon := 0$ ▷ initialisation
2: **while** there is $\rho \in L(J)$ such that $C^{\mathsf{cl}}_\rho \neq C_\rho$ **do** ▷ the main loop
3: $\quad R := \bigsqcup_{\sigma\in L(J)} (\kappa_\sigma)_*(\top_{C_\sigma})$ ▷ **Partitioning** (Line 3–9)
4: $\quad$ Choose a leaf $\rho \in L(J)$ such that $C^{\mathsf{cl}}_\rho \neq C_\rho$
5: $\quad R_\rho := (c \circ \kappa_\rho)^*(\overline{F}(R))$
6: $\quad$ **if** $R_\rho = \top_{C_\rho}$ **then**
7: $\quad\quad C^{\mathsf{cl}}_\rho := C_\rho$
8: $\quad\quad$ **continue**
9: $\quad$ Take an R_ρ-partitioning $\{\kappa_{\rho,k}\colon C_{\rho k} \rightarrowtail C_\rho\}_{k\in\{0,\dots,n_\rho\}}$
10: $\quad$ Choose $k_0 \in \{0,\dots,n_\rho\}$ s.t. $w(C_{\rho k_0}) = \max_{k\in\{0,\dots,n_\rho\}} w(C_{\rho k})$
▷ **Relabelling** (Line 10–12)
11: $\quad$ MarkDirty(ρ, k_0)
12: $\quad J := J \cup \{\rho 0, \dots, \rho n_\rho\}$
13: **return** $\{\kappa_\sigma\colon C_\sigma \rightarrowtail C\}_{\sigma\in L(J)}$

14: **procedure** MarkDirty(ρ, k_0)
15: $\quad$ **for** $k \in \{0,\dots,n_\rho\}$ **do** $C^{\mathsf{cl}}_{\rho k} := C_{\rho k}$
16: $\quad$ Let B be the pullback of the diagram: $\begin{array}{ccc} B & \rightarrowtail & C \\ \downarrow & \lrcorner & \downarrow c \\ F\left(C_{\rho k_0} \cup \left(\bigcup_{\sigma\in L(J)\setminus\{\rho\}} C_\sigma\right)\right) & \rightarrowtail & FC \end{array}$.
17: ▷ the bottom morphism is mono by Assumption 6.1.4
18: $\quad$ **for** $\tau \in L(J \cup \{\rho 0, \dots, \rho n_\rho\})$ **do**
19: $\quad\quad C^{\mathsf{cl}}_\tau := C^{\mathsf{cl}}_\tau \cap B$ ▷ states not in B are marked as dirty

The algorithm $\mathsf{fPR}^{\mathrm{H}}$ exposes a tree structure to which Hopcroft's inequality applies. Table 1 summarises how constructs in $\mathsf{fPR}^{\mathrm{H}}$ fit Sect. 2.

Much like $\mathsf{fPR}^{\mathrm{naive}}$ (Algorithm 1), $\mathsf{fPR}^{\mathrm{H}}$ grows a tree, as shown in Fig. 5. We take the generated tree as $T = (V, E)$ in Sect. 2. Note that, whereas $\mathsf{fPR}^{\mathrm{naive}}$ expands the tree uniformly so that every leaf has the same depth (Fig. 4), $\mathsf{fPR}^{\mathrm{H}}$ expands leaves selectively (Fig. 5).

$\mathsf{fPR}^{\mathrm{H}}$ chooses k_0 so that $w(C_{\rho k_0})$ is maximised (Line 10). These choices constitute a heavy child choice (Definition 2.3), an essential construct in Hopcroft's inequality (Theorem 2.9).

$\mathsf{fPR}^{\mathrm{H}}$ starts with $R = \top_C \in \mathbb{E}_C$, the singleton family of a monomorphism $\{\kappa_\epsilon = \mathrm{id}_C\colon C_\epsilon \rightarrowtail C\}$, and a marking $C^{\mathsf{cl}}_\epsilon = 0$ of states. For each σ, $C^{\mathsf{cl}}_\sigma \rightarrowtail C_\sigma$ is a "subset" of C_σ consisting of *clean* states; the rest of C_σ consists of *dirty* states. Therefore, initially, all states of $C = C_\epsilon$ are marked dirty. The main loop, consisting of **Partitioning** (Line 3–9) and **Relabelling** (Line 10–12), iterates until there is no dirty state (Line 2).

Table 1. Correspondence between constructs in $\mathsf{fPR}^{\mathrm{H}}_{(F,\overline{F}),w}$ and the theory in §2

Constructs in Algorithm 2	Notions in Sect. 2
The tree of subobjects $\{C_\sigma\}_{\sigma \in J}$ of C, cf. Fig. 5	A tree $T = (V, E)$, cf. Fig. 2
A set $\{C_\sigma\}_{\sigma \in J}$ of objects in $\mathbb{C}$	A set V of vertices
A set $\{\kappa_{\sigma,k} \colon C_{\sigma k} \rightarrowtail C_\sigma\}_{\sigma k \in J}$ of monomorphisms	A set E of edges
A map $w \colon \mathrm{ob}(\mathbf{Sub}(\mathbb{C})_C) \to \mathbb{N}$ with Assumption 6.1.10	A weight function $w \colon V \to \mathbb{N}$ (Definition 2.2)
A choice of k_0 made in Line 10 of Algorithm 2	An hcc $h \colon V \setminus L(T) \to V$ (Definition 2.3)
The complexity result of Proposition 6.9	The inequality of Corollary 2.10

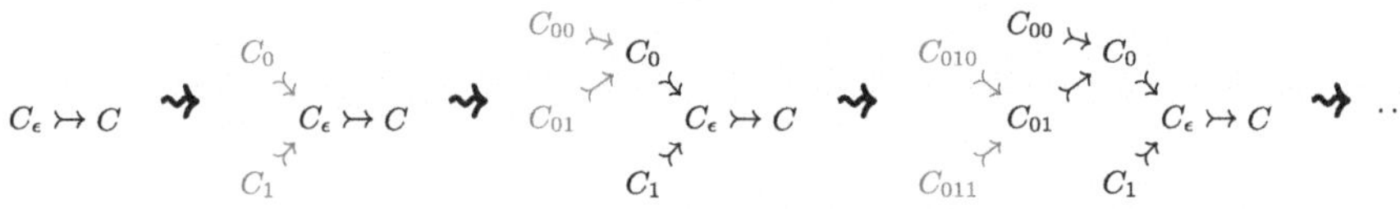

Fig. 5. At each iteration one leaf of the tree is selected and refined.

The **Partitioning** part selects one leaf C_ρ whose states include at least one dirty state (Line 4). The tree is expanded at this selected leaf only. This selection makes Algorithm 2 different from Algorithm 1, which expands the tree at every leaf (cf. Fig. 4 and Fig. 5).

The **Relabelling** part then updates the clean/dirty marking. Firstly, it chooses one "heavy child" $C_{\rho k_0}$ (Line 10) from the leaves generated in **Partitioning**. Then the iteration calls the MarkDirty procedure (Line 14–19). It first collects states (B in Line 16) whose all "successors" with respect to the coalgebra $c \colon C \to FC$ are in the object $C_{\rho k_0} \cup \left(\bigcup_{\sigma \in L(J) \setminus \{\rho\}} C_\sigma\right)$; the latter intuitively consists of states "unaffected" by tree expansion. The procedure marks only states in B as clean (Line 19), which means that the rest of the states are marked dirty.

Towards the correctness theorem of our optimised fibrational algorithm $\mathsf{fPR}^{\mathrm{H}}$ (Theorem 6.8), we first make a series of preliminary observations.

Notation 6.5. We write R_i for R defined at Line 3 of Algorithm 2 at the i-th iteration. We write J_i for J at the beginning of the i-th iteration. We write $C^{\mathsf{cl},i}_\sigma$ and $\kappa^{\mathsf{cl},i}_\sigma$ for C^{cl}_σ and the monomorphism $\kappa^{\mathsf{cl}}_\sigma \colon C^{\mathsf{cl}}_\sigma \rightarrowtail C$, respectively, at Line 16 at the i-th iteration.

We identify loop invariants Proposition 6.6. Termination of $\mathsf{fPR}^{\mathrm{H}}$ follows from Assumption 6.1.7 and 9 (Proposition 6.7). Combining these, we prove the correctness of $\mathsf{fPR}^{\mathrm{H}}$ in Theorem 6.8.

Proposition 6.6 (loop invariants). *At the beginning of the i-th iteration, the following hold.*

1. $(c \circ \kappa_\sigma \circ \kappa_\sigma^{\mathsf{cl},i})^* \overline{F}(R_i) = \top_{C_\sigma^{\mathsf{cl},i}}$ *for each leaf* $\sigma \in L(J_i)$.
2. *The mono-sink* $\{\kappa_\sigma : C_\sigma \rightarrowtail C\}_{\sigma \in L(J_i)}$ *is an* R_i*-partitioning.*
3. $\nu(c^* \overline{F}) \sqsubseteq R_i$.

Therefore, after Algorithm 2 terminates, $(c \circ \kappa_\sigma)^* \overline{F} R = \top_{C_\sigma}$ *holds for each* $\sigma \in L(J)$, $\{\kappa_\sigma \rightarrowtail C\}_{\sigma \in L(J)}$ *is an* R*-partitioning, and* $\nu(c^* \overline{F}) \sqsubseteq R$, *for* $R \in \mathbb{E}_C$ *defined in Line 3.*

Proposition 6.7 (termination). *Algorithm 2 terminates for any input.*

The key observation for the proof of termination is that in each iteration of the main loop either the partition is refined, or it is not but the number of dirty leaves decreases.

Theorem 6.8 (correctness). *Algorithm 2 terminates for any input and returns a* $\nu(c^* \overline{F})$*-partitioning.*

The explicit correspondence between $\mathsf{fPR}^{\mathrm{H}}$ and Sect. 2 (Table 1) allows us to directly use Hopcroft's inequality. The following result, while it does not give a complexity bound for $\mathsf{fPR}^{\mathrm{H}}$ itself, plays a central role in the amortised analysis of its concrete instances in Sect. 6.2.

Proposition 6.9. *If each call of* MarkDirty *in Algorithm 2 takes*

$$\mathcal{O}\left(K \sum_{k=\{0,\ldots,n_\rho\}\setminus\{k_0\}} w(C_{\rho k})\right)$$

time for some K, *the total time taken by the repeated calls of* MarkDirty *is* $\mathcal{O}(K w(C) \log w(C))$.

6.2 Concrete yet Functor-Generic Algorithms $\mathsf{fPR}^{\mathbf{H\text{-}ER}}_{w_{\mathrm{C}}}$, $\mathsf{fPR}^{\mathbf{H\text{-}ER}}_{w_{\mathrm{P}}}$, $\mathsf{fPR}^{\mathbf{H\text{-}ER}}_{w_{\mathrm{R}}}$

We instantiate the fibrational algorithm $\mathsf{fPR}^{\mathrm{H}}_{(F,\overline{F}),w}$ with $\mathbf{EqRel} \to \mathbf{Set}$ as a base fibration. In this situation, the functor F is an endofunctor on $\mathbf{Set}$ and $\overline{F}$ is an endofunctor on $\mathbf{EqRel}$ which is a fibred lifting of F. This instantiation also enables a semantically equivalent reformulation of MarkDirty—its "implementation" is now "predecessor-centric" rather than "successor-centric"—and this aids more refined complexity analysis.

For a weight function w (a parameter of $\mathsf{fPR}^{\mathrm{H}}$), we introduce three examples $w_{\mathrm{C}}, w_{\mathrm{P}}, w_{\mathrm{R}}$, leading to three functor-generic algorithms $\mathsf{fPR}^{\mathbf{H\text{-}ER}}_{(F,\overline{F}),w_{\mathrm{C}}}$, $\mathsf{fPR}^{\mathbf{H\text{-}ER}}_{(F,\overline{F}),w_{\mathrm{P}}}$ and $\mathsf{fPR}^{\mathbf{H\text{-}ER}}_{(F,\overline{F}),w_{\mathrm{R}}}$.

Definition 6.10 ($\mathsf{fPR}^{\mathbf{H\text{-}ER}}_{w}$)**.** Let $\mathbf{Set} \xrightarrow{F} \mathbf{Set}$ and $\mathbf{EqRel} \xrightarrow{\overline{F}} \mathbf{EqRel}$ be functors, $C \xrightarrow{c} FC$ be a coalgebra, and $w \colon \mathcal{P}(C) \to \mathbb{N}$ be a function (which amounts to $w \colon \mathrm{ob}(\mathbf{Sub}(\mathbb{C})_C) \to \mathbb{N}$ in Sect. 6.1), all satisfying Assumption 6.1 (C must be finite, in particular). The algorithm $\mathsf{fPR}^{\mathbf{H\text{-}ER}}_{w}$ is shown in Algorithm 3; it computes a $\nu(c^* \overline{F})$-partitioning of C.

Algorithm 3. The algorithm $\mathsf{fPR}_w^{\text{H-ER}}$, obtained as an instance of fPR^{H} (Algorithm 2) where p is $\mathbf{EqRel} \to \mathbf{Set}$, with a semantically equivalent formulation of MarkDirty (successor-centric in fPR^{H}; predecessor-centric here in $\mathsf{fPR}_w^{\text{H-ER}}$).

Input: A coalgebra $c\colon C \to FC$ in **Set**.
Output: A mono-sink $\{\kappa_i\colon C_i \rightarrowtail C\}_{i \in I}$ for some I.

(the same as Line 1–13 of Algorithm 2)

14: **procedure** MarkDirty(ρ, k_0)
15: **for** $k \in \{0, \dots, n_\rho\}$ **do** $C^{\mathsf{cl}}_{\rho k} := C_{\rho k}$
16: **for** $k \in \{0, \dots, n_\rho\} \setminus \{k_0\}$ and $y \in C_{\rho k}$ **do**
17: **for** x: predecessor of y **do**
18: Find $\tau \in L(J \cup \{\rho 0, \dots, \rho n_\rho\})$ such that $x \in C_\tau$
19: If such τ exists, then $C^{\mathsf{cl}}_\tau := C^{\mathsf{cl}}_\tau \setminus \{x\}$ ▷ mark x as dirty

Line 14–19 of Algorithm 3 uses this categorical notion of predecessor (Line 17), which is in Definition 6.11. Its equivalence to the original definition (Line 14–19 of Algorithm 2) is easy; so $\mathsf{fPR}_w^{\text{H-ER}}$ is correct by Theorem 6.8. The successor-centric description is more convenient in the correctness proof (Theorem 6.8), while the predecessor-centric one is advantageous for complexity analysis.

Definition 6.11 (predecessor [14]). Let $c\colon C \to FC$ be a coalgebra in **Set**. For $x, y \in C$, we say x is a *predecessor* of y if $x \notin B$, where B is a subset of C defined by the following pullback:

$$\begin{array}{ccc} B & \rightarrowtail & C \\ \downarrow & \lrcorner & \downarrow c \\ F(C \setminus \{y\}) & \xrightarrow[F(i_y)]{} & FC \end{array} .$$

Here i_y is the canonical injection.

For w as a parameter of $\mathsf{fPR}_w^{\text{H-ER}}$, we introduce three functions.

Definition 6.12 ($w_{\text{C}}, w_{\text{P}}, w_{\text{R}}$). We define $w_{\text{C}}, w_{\text{P}}, w_{\text{R}}\colon \mathcal{P}(C) \to \mathbb{N}$ as follows: $w_{\text{C}}(A) = |A|$, $w_{\text{P}}(A) = \sum_{x \in A} |\{y \in C \mid y \text{ is a predecessor of } x\}|$, and $w_{\text{R}}(A) = |A \cap C'|$, where $C' = \{x \in C \mid x \text{ is a successor of some } y \in C\}$. The weight functions w_{C}, w_{P} and w_{R} are called the *cardinality*, *predecessor* and *reachability* weights, respectively.

The cardinality weight is the most commonly used one in various partition refinement algorithms, including [14]. The latter two have been used in [6, 11, 16] for DFA partition refinement; we use them for the first time in categorical partition refinement algorithms.

Our algorithm $\mathsf{fPR}^{\text{H-ER}}_{(F,\overline{w}),w}$ induces concrete partition refinement algorithms for various systems as shown in Table 2.

The predecessor-centric MarkDirty and concrete choices of w allow the following fine-grained complexity analysis.

Table 2. Examples of partition refinement algorithms induced by $\mathsf{fPR}^{\textbf{H-ER}}_{(F,\overline{F}),w}$

Functor $F(X)$	Weight function	System	Algorithm: a variation of
$2 \times X^A$	$w_{\mathrm{C}}(A)$, $w_{\mathrm{R}}(A)$	DFA	Hopcroft's algorithm [6,11,16]
$\mathcal{P}(A \times X)$	$w_{\mathrm{C}}(A)$	LTS	[24]
$\mathcal{D}(X)$	$w_{\mathrm{C}}(A)$	Markov chain	[25]
$\mathcal{P}(\mathcal{D}(X))$	$w_{\mathrm{C}}(A)$	Markov decision process	[7]

Theorem 6.13 (complexity of $\mathsf{fPR}^{\textbf{H-ER}}_{w_{\mathrm{C}}}$, $\mathsf{fPR}^{\textbf{H-ER}}_{w_{\mathrm{P}}}$, $\mathsf{fPR}^{\textbf{H-ER}}_{w_{\mathrm{R}}}$).

1. *Let $M = \max_{y \in C} |\{x \in C \mid x \text{ is a predecessor of } y\}|$, the "in-degree" of $c\colon C \to FC$; suppose it takes $\mathcal{O}(f)$ time to compute $c(x) \in FC$ for each $x \in C$. The time complexity of $\mathsf{fPR}^{\textbf{H-ER}}_{w_{\mathrm{C}}}$ is $\mathcal{O}(fM|C|\log|C|)$.*
2. *Let $m = w_{\mathrm{P}}(C) = \sum_{y \in C} |\{x \in C \mid x \text{ is a predecessor of } y\}|$. The time complexity of $\mathsf{fPR}^{\textbf{H-ER}}_{w_{\mathrm{P}}}$ is $\mathcal{O}(fm\log m)$. Since $m \leq |C|^2$, it is also bounded by $\mathcal{O}(fm\log|C|)$.*
3. *The time complexity of $\mathsf{fPR}^{\textbf{H-ER}}_{w_{\mathrm{R}}}$ is $\mathcal{O}(fM|C'|\log|C'|)$, where f and M are from above and C' is from Definition 6.12.*

We note that the complexity bound given in [14] is $\mathcal{O}(fm\log|C|)$ which is the same as 2: their algorithm is essentially $\mathsf{fPR}^{\textbf{H-ER}}_{w_{\mathrm{C}}}$, but their more fine-grained, element-wise analysis derived the aforementioned bound.

We sketch the proof of 1 for illustration. It uses Hopcroft's inequality in its amortised analysis.

Proof (sketch). (of Theorem 6.13.1) We can check that Algorithm 3 (with $w = w_{\mathrm{C}}$) satisfies the premise of Proposition 6.9. By implementing this algorithm properly (preparing a table for x and τ in Line 18 of Algorithm 3, as done in [14]), it takes $\mathcal{O}(1)$ time to execute each iteration of the loop at Line 17. Thus, the loop at Line 16 takes $\mathcal{O}(M|C_{\rho k}|)$ time for each k. Therefore the time taken for each call of MarkDirty is $\mathcal{O}(M\sum_{k\in\{0,\dots,n_\rho\}\setminus\{k_0\}}|C_{\rho k}|)$. By Proposition 6.9, the time taken for the repeated calls of MarkDirty in total is $\mathcal{O}(M|C|\log|C|)$.

The complexity of the other parts of the algorithm is also bounded. We write C^{di}_σ for $C_\sigma \setminus C^{\mathsf{cl}}_\sigma$. The computation of R_ρ (Line 3–5 of Algorithm 2) takes $\mathcal{O}(f|C^{\mathsf{di}}_\rho|)$, and the computation of R_ρ-partitioning (Line 9 of Algorithm 2) takes $\mathcal{O}(|C^{\mathsf{di}}_\rho|)$, using appropriate data structures (see [16]). Hence, it takes $\mathcal{O}(f|C^{\mathsf{di}}_\rho|)$ for each iteration of the main loop except for MarkDirty.

Therefore, the total time for Algorithm 3 except for MarkDirty (let us write $T_{\setminus\mathrm{MarkDirty}}$) is $\mathcal{O}(\sum_{\rho \text{ for each iteration}} f|C^{\mathsf{di}}_\rho|)$. We use amortised analysis to bound this sum. Specifically, it is easy to see that the sum $\sum_\rho f|C^{\mathsf{di}}_\rho|$ is bounded by the number of times that states are marked as dirty, multiplied by f. Throughout the algorithm, the number of times that states are marked as dirty (at Line 19 of Algorithm 3) is at most the time consumed by MarkDirty, which is $\mathcal{O}(M|C| \log |C|)$. Therefore, $T_{\backslash \text{MarkDirty}}$ is $\mathcal{O}(fM|C| \log |C|)$; so is the total time.

Acknowledgments. The authors are supported by ERATO HASUO Metamathematics for Systems Design Project (No. JPMJER1603), JST. T. Sanada and R. Kojima are supported by JST Grant Number JPMJFS2123. Y. Komorida is supported by JSPS KAKENHI Grant Number JP21J13334.

References

1. Berkholz, C., Bonsma, P.S., Grohe, M.: Tight lower and upper bounds for the complexity of canonical colour refinement. Theory Comput. Syst. **60**(4), 581–614 (2017). https://doi.org/10.1007/s00224-016-9686-0
2. Bonchi, F., Holland, J., Piedeleu, R., Sobocinski, P., Zanasi, F.: Diagrammatic algebra: from linear to concurrent systems. Proc. ACM Program. Lang. **3**(POPL), 25:1–25:28 (2019). https://doi.org/10.1145/3290338
3. de Vink, E., Rutten, J.: Bisimulation for probabilistic transition systems: a coalgebraic approach. Theoret. Comput. Sci. **221**(1), 271–293 (1999). https://doi.org/10.1016/S0304-3975(99)00035-3
4. Deifel, H.-P., Milius, S., Schröder, L., Wißmann, T.: Generic partition refinement and weighted tree automata. In: ter Beek, M.H., McIver, A., Oliveira, J.N. (eds.) FM 2019. LNCS, vol. 11800, pp. 280–297. Springer, Cham (2019). https://doi.org/10.1007/978-3-030-30942-8_18
5. Dorsch, U., Milius, S., Schröder, L., Wißmann, T.: Efficient coalgebraic partition refinement. In: Meyer, R., Nestmann, U. (eds.) 28th International Conference on Concurrency Theory (CONCUR 2017). Leibniz International Proceedings in Informatics (LIPIcs), vol. 85, pp. 32:1–32:16. Schloss Dagstuhl–Leibniz-Zentrum fuer Informatik, Dagstuhl, Germany (2017). https://doi.org/10.4230/LIPIcs.CONCUR.2017.32
6. Gries, D.: Describing an algorithm by Hopcroft. Acta Informatica **2**(2), 97–109 (1973)
7. Groote, J.F., Rivera Verduzco, J., De Vink, E.P.: An efficient algorithm to determine probabilistic bisimulation. Algorithms **11**(9) (2018). https://doi.org/10.3390/a11090131
8. Hermida, C., Jacobs, B.: Structural induction and coinduction in a fibrational setting. Inf. Comput. **145**(2), 107–152 (1998). https://doi.org/10.1006/inco.1998.2725
9. Högberg, J., Maletti, A., May, J.: Backward and forward bisimulation minimisation of tree automata. In: Holub, J., Žďárek, J. (eds.) CIAA 2007. LNCS, vol. 4783, pp. 109–121. Springer, Heidelberg (2007). https://doi.org/10.1007/978-3-540-76336-9_12
10. Högberg, J., Maletti, A., May, J.: Bisimulation minimisation for weighted tree automata. In: Harju, T., Karhumäki, J., Lepistö, A. (eds.) DLT 2007. LNCS, vol. 4588, pp. 229–241. Springer, Heidelberg (2007). https://doi.org/10.1007/978-3-540-73208-2_23
11. Hopcroft, J.E.: An $n \log n$ algorithm for minimizing states in a finite automaton. In: Theory of Machines and Computations, pp. 189–196. Academic Press (1971)
12. Jacobs, B.: Categorical Logic and Type Theory. Studies in logic and the foundations of mathematics, Elsevier Science (1999)

13. Jacobs, B.: Introduction to Coalgebra: Towards Mathematics of States and Observation. Cambridge Tracts in Theoretical Computer Science, Cambridge University Press (2016). https://doi.org/10.1017/CBO9781316823187
14. Jacobs, J., Wißmann, T.: Fast coalgebraic bisimilarity minimization. Proc. ACM Program. Lang. **7**(POPL) (Jan 2023). https://doi.org/10.1145/3571245
15. Kanellakis, P.C., Smolka, S.A.: CCS expressions, finite state processes, and three problems of equivalence. Inf. Comput. **86**(1), 43–68 (1990). https://doi.org/10.1016/0890-5401(90)90025-D
16. Knuutila, T.: Re-describing an algorithm by Hopcroft. Theoret. Comput. Sci. **250**(1), 333–363 (2001). https://doi.org/10.1016/S0304-3975(99)00150-4
17. Komorida, Y., Katsumata, S., Hu, N., Klin, B., Humeau, S., Eberhart, C., Hasuo, I.: Codensity games for bisimilarity. New Gener. Comput. **40**(2), 403–465 (2022). https://doi.org/10.1007/s00354-022-00186-y
18. Lombardy, S., Sakarovitch, J.: Morphisms and minimisation of weighted automata. Fundam. Informaticae **186**(1–4), 195–218 (2022). https://doi.org/10.3233/FI-222126
19. Milner, R.: Communication and Concurrency. Prentice-Hall Inc., USA (1989)
20. Moore, E.F.: Gedanken-Experiments on Sequential Machines, pp. 129–154. Princeton University Press, Princeton (1956). https://doi.org/10.1515/9781400882618-006
21. Park, D.: Concurrency and automata on infinite sequences. In: Deussen, P. (ed.) GI-TCS 1981. LNCS, vol. 104, pp. 167–183. Springer, Heidelberg (1981). https://doi.org/10.1007/BFb0017309
22. Piedeleu, R., Kartsaklis, D., Coecke, B., Sadrzadeh, M.: Open system categorical quantum semantics in natural language processing. In: Moss, L.S., Sobocinski, P. (eds.) 6th Conference on Algebra and Coalgebra in Computer Science, CALCO 2015, 24–26 June, 2015, Nijmegen, The Netherlands. LIPIcs, vol. 35, pp. 270–289. Schloss Dagstuhl - Leibniz-Zentrum für Informatik (2015). https://doi.org/10.4230/LIPIcs.CALCO.2015.270
23. Rutten, J.: Universal coalgebra: a theory of systems. Theoret. Comput. Sci. **249**(1), 3–80 (2000). https://doi.org/10.1016/S0304-3975(00)00056-6, modern Algebra
24. Valmari, A.: Bisimilarity minimization in $O(m \log n)$ time. In: Franceschinis, G., Wolf, K. (eds.) PETRI NETS 2009. LNCS, vol. 5606, pp. 123–142. Springer, Heidelberg (2009). https://doi.org/10.1007/978-3-642-02424-5_9
25. Valmari, A., Franceschinis, G.: Simple $O(m \log n)$ time Markov chain lumping. In: Esparza, J., Majumdar, R. (eds.) TACAS 2010. LNCS, vol. 6015, pp. 38–52. Springer, Heidelberg (2010). https://doi.org/10.1007/978-3-642-12002-2_4
26. Wißmann, T., Deifel, H., Milius, S., Schröder, L.: From generic partition refinement to weighted tree automata minimization. Formal Aspects Comput. **33**(4–5), 695–727 (2021). https://doi.org/10.1007/s00165-020-00526-z

Proving Behavioural Apartness

Ruben Turkenburg[1(✉)], Harsh Beohar[2], Clemens Kupke[3], and Jurriaan Rot[1]

[1] Radboud University, Nijmegen, Netherlands
ruben.turkenburg@ru.nl, jrot@cs.ru.nl
[2] University of Sheffield, Sheffield, UK
h.beohar@sheffield.ac.uk
[3] Strathclyde University, Glasgow, UK
clemens.kupke@strath.ac.uk

Abstract. Bisimilarity is a central notion for coalgebras. In recent work, Geuvers and Jacobs suggest to focus on apartness, which they define by dualising coalgebraic bisimulations. This yields the possibility of finite proofs of distinguishability for a wide variety of state-based systems.

We propose *behavioural apartness*, defined by dualising behavioural equivalence rather than bisimulations. A motivating example is the subdistribution functor, where the proof system based on bisimilarity requires an infinite quantification over couplings, whereas behavioural apartness instantiates to a finite rule. In addition, we provide optimised proof rules for behavioural apartness and show their use in several examples.

1 Introduction

In the study of coalgebra as a general approach to state-based systems, a number of notions of equivalence of states have been developed. Among these are: bisimilarity as defined by Aczel & Mendler; behavioural equivalence; and bisimilarity defined using (canonical) relation lifting due to Hermida & Jacobs [1,18]. Alongside these definitions, proof systems have been developed for some of these notions, to give a syntactic way to deduce equivalence of states in a coalgebra. However, due to the coinductive nature of coalgebraic notions of equivalence, these are challenging to develop as they have to deal with the circularity inherent in coinduction [3,7,26,32].

Another (closely related) line of research is coalgebraic modal logic [27], where coalgebras give the semantics of modal formulas. These logics bring with them another notion of equivalence, namely we can call states logically equivalent if they satisfy exactly the same formulas of a given logic. Comparing these equivalences to the aforementioned coinductive ones, we can ask whether bisimilar

This research is partially supported by the NWO grant No. OCENW.M20.053, the EPSRC NIA Grant EP/X019373/1, and the Royal Society Travel Grant IES\R3\223092 and the Leverhulme Trust Research Project Grant RPG-2020-232.

Published by Springer Nature Switzerland AG 2024
B. König and H. Urbat (Eds.): CMCS 2024, LNCS 14617, pp. 156–173, 2024.
https://doi.org/10.1007/978-3-031-66438-0_8

states are logically equivalent (adequacy) and whether logically equivalent states are bisimilar (expressivity). Already (implicitly) present in the study of these properties are dual notions, defining states instead as inequivalent (see, e.g., [8,9,24,36]). For example, expressivity of a logic may more easily be shown by giving, for any pair of non-bisimilar states, a formula distinguishing them.

In recent work, Geuvers & Jacobs [12] focus on such notions of inequivalence (or *distinguishability*), and investigate what they call apartness. This is done first for concrete examples, such as labelled transition systems, before they show how their definitions can be obtained coalgebraically. For this, the starting point is bisimilarity defined using (canonical) relation lifting, which for a coalgebra (in the category Set) is the largest relation R such that if two states are related, then their successor structures must be related by the relation lifting of the behaviour functor applied to R. The dual notion of apartness, again on a Set coalgebra, can then be described as the smallest relation R such that if successor structures of two states are related by the relation obtained by applying *opposite* relation lifting to R, then R must relate those two states. This immediately gives us that states are bisimilar if and only if they are not apart. Furthermore, in *op. cit.*, it is shown how their notion of apartness quite directly gives rise to a derivation system allowing apartness to be determined inductively.

In this work, we continue the investigation of the possible notions of inequivalence related to existing definitions of bisimilarity and behavioural equivalence, and the development of corresponding proof systems. We start by recalling the apartness notion due to Geuvers & Jacobs, before giving an example of a functor for which this does not easily provide a satisfactory proof system for apartness (Sect. 2). Namely, the subdistribution functor, which gives a rule whose premise universally quantifies over the infinite set of couplings. This does not allow apartness to be proved by giving some witness, which can be done for the systems considered in [12].

An alternative characterisation of bisimilarity is present in work on probabilistic systems based on summing successor distributions over equivalence classes [25], which has further been shown to coincide with coalgebraic notions of bisimilarity and behavioural equivalence [29,30,34]. We use this characterisation from the starting point of behavioural equivalence to give a dual notion, which we call *behavioural apartness*. This gives rise to a proof system where the inductive step does not involve a universal quantification, allowing finite proofs of apartness. We present this system, prove it to be sound and complete with respect to the dual of behavioural equivalence for finitary functors on Set, and provide optimised rules which we compare to the basic rule by application to examples (Sects. 3 and 4).

Notation. For an equivalence relation $R \subseteq X \times X$, we denote the quotient map by $q_R \colon X \to X/R$, which sends an element x to its equivalence class $[x]_R$. We extend this to arbitrary relations $R \subseteq X \times X$, by defining q_R as $q_{e(R)}$, where $e(R)$ is the equivalence closure of R. Further, we write $\overline{R}$ for the complement of a relation $R \subseteq X \times X$ on a set X.

2 Cobisimilarity

We start by recalling the well-known notion of coalgebraic bisimulation, defined via relation lifting [18]. The dual of this notion has been investigated in [12], using the so called fibred opposite construction. In this section, we review this approach and show how it can be utilised to obtain a proof system for apartness. We will further show that there exist functors for which this dualised definition does not give rise to a satisfactory proof rule for cobisimilarity, in contrast to the rules provided in *op. cit.*

For an endofunctor B on the category Set of sets and functions, the (canonical) relation lifting $\mathsf{Rel}(B)$ is defined on a relation $R \subseteq X \times X$ as follows

$$\mathsf{Rel}(B)(R) := \{(s,t) | \exists z \in B(R).B(\pi_1)(z) = s \wedge B(\pi_2)(z) = t\}.$$

This is then used to define the following notion of coalgebraic bisimulation:

Definition 1. *A relation $R \subseteq X \times X$ is a coalgebraic bisimulation for a coalgebra $\gamma\colon X \to B(X)$ if $R \subseteq (\gamma \times \gamma)^{-1} \circ \mathsf{Rel}(B)(R)$, that is, it satisfies:*

$$\frac{x\ R\ y}{\gamma(x)\ \mathsf{Rel}(B)(R)\ \gamma(y)}$$

The largest such relation is called bisimilarity, *denoted by* $\underline{\leftrightarrow}$.

Replacing R above with its complement and taking the contrapositive, we obtain a dual definition.

Definition 2. *A* cobisimulation *on $\gamma\colon X \to B(X)$ is a relation R satisfying*

$$\frac{\gamma(x)\ \overline{\mathsf{Rel}(B)(\overline{R})}\ \gamma(y)}{x\ R\ y} \tag{1}$$

Equivalently, R is a cobisimulation if $R \supseteq (\gamma \times \gamma)^{-1} \circ \overline{\mathsf{Rel}(B)(\overline{R})}$, where the right-hand relation is constructed as the following composition of monotone maps:

$$\mathsf{Rel}_X \xrightarrow{\overline{(-)}} \mathsf{Rel}_X \xrightarrow{\mathsf{Rel}(B)} \mathsf{Rel}_{B(X)} \xrightarrow{\overline{(-)}} \mathsf{Rel}_{B(X)} \xrightarrow{(\gamma\times\gamma)^{-1}} \mathsf{Rel}_X$$

where we use Rel_X to denote the lattice of relations on X, ordered by inclusion.

The relation of *cobisimilarity* is then defined as the smallest cobisimulation. We denote this relation using $\#$, and we see that $\underline{\leftrightarrow} = \overline{(\#)}$. For a more general coalgebraic treatment, see [12]. Note that we differ in terminology from *op. cit.* in using *cobisimilarity* rather than apartness, to distinguish from both the dual of equivalence relations, and the notion of behavioural apartness introduced later.

Example 1 (Labelled Transition Systems). In [12] cobisimilarity dual to weak forms of bisimulation are studied. To simplify the presentation, we instantiate the definition of cobisimulation to LTSs, which we model as coalgebras for the functor $\mathcal{P}(-)^A$ for an input alphabet A.

The relation lifting of $\mathcal{P}(-)^A$ acts as follows

$$\begin{aligned}\mathsf{Rel}(\mathcal{P}(-)^A)(R) = \{(f,g) | \forall a \in A.[\forall x \in f(a).\exists y \in g(a).(x,y) \in R] \\ \wedge[\forall y \in g(a).\exists x \in f(a).(x,y) \in R]\}\end{aligned}$$

Applying the dualisation as in (1), we see that a relation R is a cobisimulation on a coalgebra $\gamma \colon X \to \mathcal{P}(X)^A$ if it satisfies the following two rules

$$\frac{x \xrightarrow{a} x' \qquad \forall y'.\, y \xrightarrow{a} y' \implies x' \mathrel{R} y'}{x \mathrel{R} y} \qquad \frac{y \xrightarrow{a} y' \qquad \forall x'.\, x \xrightarrow{a} x' \implies x' \mathrel{R} y'}{x \mathrel{R} y}$$

where $x \xrightarrow{a} x'$ means $x' \in \gamma(x)(a)$. Then cobisimilarity on LTSs is the smallest relation satisfying these rules. This gives an inductive proof system for cobisimilarity, as further explained in [12].

We finish this example by giving a derivation of cobisimilarity for the states x and y in the following LTS:

This derivation goes as follows:

$$\dfrac{x \xrightarrow{b} x_2 \qquad \dfrac{\dfrac{y_2 \xrightarrow{a} y_2 \qquad \dfrac{}{\forall x'.\, x_2 \xrightarrow{a} x'.\, y_2 \# x'}}{x_2 \# y_2}}{\forall y'.\, y \xrightarrow{b} y'.\, x_2 \# y'}}{x \# y}$$

The states x and y can be distinguished by their b-transitions, and their successors by the presence of an a-transition only on y_2.

Example 2 (Subdistributions). We continue by instantiating the rule (1) for cobisimulations to coalgebras for the subdistribution functor. In contrast to LTSs above, the relation lifting approach does not give a pleasant proof system.

The (finitely supported) subdistribution functor $\mathcal{D}_s \colon \mathsf{Set} \to \mathsf{Set}$ is defined as follows, where $\text{supp}(\mu) = \{x \in X | \mu(x) \neq 0\}$.

$$\mathcal{D}_s(X) = \left\{\mu \colon X \to [0,1] | \sum_{x \in X} \mu(x) \leq 1, \text{supp}(\mu) \text{ finite}\right\}$$

We may equivalently write such distributions as formal sums $\sum_{x \in X} \mu(x) |x\rangle$. This allows us to denote the functor's action on arrows as:

$$\mathcal{D}_s(f\colon X \to Y)\colon \left(\sum_{x \in X} \mu(x)\,|x\rangle\right) \mapsto \left(\sum_{x \in X} \mu(x)\,|f(x)\rangle\right)$$

We can now elaborate the action of the relation lifting of the subdistribution functor $\mathcal{D}_s$ on a relation $R \subseteq X \times X$:

$$\begin{aligned}\mathsf{Rel}(\mathcal{D}_s)(R) &= \{(\mathcal{D}_s(\pi_1)(z), \mathcal{D}_s(\pi_2)(z)) | z \in \mathcal{D}_s(R)\} \\ &= \left\{\left[\sum_{x \in X}\left[\sum_{y \in X} \mu(x,y)\right]|x\rangle, \sum_{y \in X}\left[\sum_{x \in X} \mu(x,y)\right]|y\rangle\right] | \mu \in \mathcal{D}_s(R)\right\}\end{aligned}$$

Here, μ is a subdistribution over the elements of R, and the corresponding pair in $\mathsf{Rel}(\mathcal{D}_s)(R)$ consists of the left and right marginals of μ.

Now, we can instantiate the premise of the rule for cobisimulations for this relation lifting on a coalgebra $\gamma\colon X \to \mathcal{D}_s(X)$ to obtain the following rule:

$$\frac{\forall \mu \in \mathcal{D}_s(\overline{R}).\, \gamma(x) \neq \mathcal{D}_s\pi_1(\mu) \vee \gamma(y) \neq \mathcal{D}_s\pi_2(\mu)}{x\ R\ y} \tag{2}$$

However, this quantifies over the infinite set of subdistributions on the complement of R. This is rather unfortunate since we would like to prove apartness by giving a "witness", as was the case for LTSs. There, apartness of states x, y can be shown by giving a successor of x which is apart from all successors of y.

If we consider the example in (3), we see that x and y are cobisimilar, as they have different transition weights to sets of equivalent states. To see this using the rule (2), we could try to reason that given a distribution $\mu \in \mathcal{D}_s(\overline{R})$, the weight of the pair (x_1, y_1) (two states which are clearly equivalent) will always be such that the left and right marginals do not match the transition probabilities from x to x_1 and from y to y_1, as μ should not assign mass to the pair (x_1, y_2) of inequivalent states.

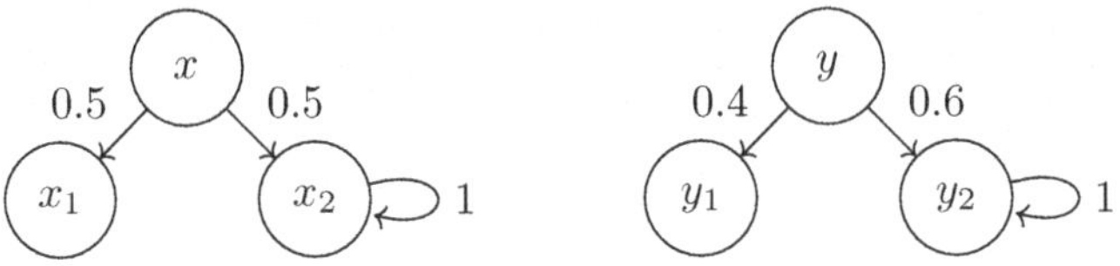

(3)

We see two main issues with rule (2): the reasoning provided (where we choose the pair (x_1, y_1)) is not well reflected in the proof system; and the rule requires reasoning about an infinite set even for a simple *finite* coalgebra. This motivates much of the coming work, in which we move to an apartness notion defined dually to behavioural equivalence rather than coalgebraic bisimulation.

The corresponding proof system will both exhibit the desired existential reasoning, and allow an optimisation giving finite proofs of apartness for both finite and infinite systems.

In the next section, we show the dualisation of behavioural equivalence, before instantiating to examples which will already illustrate the benefit of this approach over cobisimilarity. In Sect. 4, we go on to present the optimised proof rule, and show how this is beneficial by application to the examples of labelled Markov processes and stream systems.

3 Behavioural Apartness

In this section, we use a characterisation of behavioural equivalence based on so called *precongruences* due to Aczel & Mendler [1] to present a basic proof system for the dual: *behavioural apartness*. We will further apply the proof system to coalgebras for the subdistribution functor to show its benefits over the system provided in Sect. 2. Throughout this paper we work with coalgebras living in the category Set for an endofunctor $B\colon \mathsf{Set} \to \mathsf{Set}$.

3.1 Behavioural Equivalence

We start with the general definition of behavioural equivalence on Set coalgebras, based on cospans of coalgebra homomorphisms.

Definition 3. *States x, y of a coalgebra $\gamma\colon X \to B(X)$ are said to be behaviourally equivalent if there is a coalgebra $\delta\colon Y \to B(Y)$ and coalgebra morphisms $f, g\colon (X, \gamma) \to (Y, \delta)$, such that $f(x) = g(y)$.*

Now the following definition (from [1]) will give an alternative characterisation of behavioural equivalence which is amenable to dualisation:

Definition 4. *A precongruence on a coalgebra $\gamma\colon X \to B(X)$ is a relation $R \subseteq X \times X$ satisfying the following rule:*

$$\frac{x\ R\ y}{Bq_R(\gamma(x)) = Bq_R(\gamma(y))} \tag{4}$$

Following Aczel & Mendler we call a precongruence which is an equivalence relation a *congruence*. They further showed that R is a congruence if and only if it is the kernel of a coalgebra morphism, so that the maximal (pre)congruence is equal to behavioural equivalence (also see [14]). The following lemma tells us that this maximal congruence is a greatest fixed point of the operator (implicit in Definition 4) mapping a relation R to the kernel $\ker(Bq_R \circ \gamma)$ [1, Lemma 4.1].

Lemma 1. *For equivalence relations $R, S \subseteq X \times X$, we have $R \subseteq S \implies \forall x, y \in X.[Bq_R(\gamma(x)) = Bq_R(\gamma(y)) \implies Bq_S(\gamma(x)) = Bq_S(\gamma(y))]$.*

We note that Staton has shown similar results for more general categories than Set [33].

3.2 Dualising Behavioural Equivalence

The rule (4) can now be dualised to obtain a definition of behavioural apartness.

Definition 5. Behavioural apartness *is the smallest relation satisfying the following rule:*

$$\frac{Bq_{\overline{R}}(\gamma(x)) \neq Bq_{\overline{R}}(\gamma(y))}{x \, R \, y}$$

where $\overline{R}$ is the complement of the relation R.

Dually to the case of precongruences and behavioural equivalence, this smallest relation will be an *apartness relation.*

Proposition 1. *Behavioural apartness is an apartness relation, i.e., it is irreflexive ($\forall x \in X. \neg(x \; R \; x)$), symmetric, and cotransitive ($\forall x_1, x_2, x_3 \in X.(x_1 \, R \, x_2 \implies x_1 \, R \, x_3 \vee x_2 \, R \, x_3)$).*

In turning the rule of Definition 5 into a proof system, we make a number of changes. First, we make the proof obligations explicit. Second, we require that all pairs of the relation R have been proved apart. Finally, we introduce notation for the inequality $Bq_{\overline{R}}(\gamma(x)) \neq Bq_{\overline{R}}(\gamma(y))$ while also taking the symmetric closure R^s of R in the definition, namely we define

$$t_1 \not\approx^B_R t_2 := Bq_{\overline{R^s}}(t_1) \neq Bq_{\overline{R^s}}(t_2) \tag{5}$$

Note that, in this definition, we are taking the equivalence closure of $\overline{R^s}$, as in Definition 4. This is equivalent to first taking the *apartness interior* $(R^s)^\circ$ of R^s, defined as the largest apartness relation contained in R^s, and then taking $\not\approx^B_{(R^s)^\circ}$ (where the equivalence closure implicit in the definition of q does not change the relation). Also note that symmetric closure does not commute with complement, i.e., taking the symmetric closure first really gives a distinct inequality. Our motivation for taking this symmetric closure will become clear when we come to examples, where it saves us the work of proving symmetric pairs apart.

Our proof rule for apartness is formally stated in the following theorem. To ensure that finite proof trees suffice, we assume that the functor is finitary. It is well-known that this suffices for behavioural equivalence to converge at ω in the final chain. In the proof below, we use the results of Worrell on final sequences of finitary Set functors to detail the case of behavioural apartness [37].

Theorem 1. *Let B be a finitary endofunctor on* Set*, and $\gamma \colon X \to B(X)$ a coalgebra. Then states $x, y \in X$ are behaviourally apart if and only if we have a proof tree of finite height built from the following rule:*

$$\frac{\forall (x', y') \in R. \, x' \# y' \qquad \gamma(x) \not\approx^B_R \gamma(y)}{x \# y} \tag{6}$$

Note that the given "rule" is, strictly speaking, a family of rules indexed by relations $R \subseteq X \times X$. Thus, a proof tree may contain different instances of the rule, involving different choices of R.

Proof. The if direction holds by induction, as then R is contained in behavioural apartness so that $\overline{R^s}$ contains behavioural equivalence. Using Lemma 1, we then see that $\gamma(x) \not\sim^B_R \gamma(y) \implies \gamma(x) \not\sim^B_{\underline{\#}} \gamma(y)$, where we write $\underline{\#}$ for behavioural apartness here. By Definition 5, we thus have $x \,\underline{\#}\, y$.

For the other direction, first recall that a coalgebra $\gamma\colon X \to B(X)$ defines a cone $(f_i\colon X \to B^i 1)_{i \in I}$ over the final sequence of B by (transfinite) induction, with $f_0\colon X \to 1$ the unique map, $f_{i+1} = Bf_i \circ \gamma$, and for a limit ordinal α, $f_\alpha = \lim_{i<\alpha} f_i$. Father recall the notion of n-step behavioural equivalence, defined as the kernel of the n-th map of the above cone, i.e., $\ker(f_n)$. This gives a notion of n-steps behavioural apartness, namely the complement $\overline{\ker(f_n)}$.

We will now first show that for any $n < \omega$ and pair of states $(x, y) \in \overline{\ker(f_n)}$, there is a proof tree of height n with $x \# y$ as its root. The base case is trivial, as $\overline{\ker(f_0)}$ is empty. Now suppose that the property holds for some $N > 0$, and further that we have some $(x, y) \in \overline{\ker(f_{N+1})}$. This means (by definition) that

$$Bf_N(\gamma(x)) \neq Bf_N(\gamma(y))$$

We then need to show that there is some set of pairs $R \subseteq X \times X$ so that

$$Bq_{\overline{R^s}}(\gamma(x)) \neq Bq_{\overline{R^s}}(\gamma(y)) \tag{7}$$

and we claim that $\overline{\ker(f_N)}$ is such an R, and that it gives us a proof tree of height N with $x \# y$ as root.

By the induction hypothesis, for any $(x', y') \in \overline{\ker(f_N)}$, we have proof trees with $x' \# y'$ at the root, and height N. Further, the premise (Eq.(7)) holds by definition of the kernel. Combining these gives us the required proof tree.

For the limit at ω, we have

$$\overline{\ker(f^\omega)} = \overline{\left(\bigcap_{i<\omega} \ker(f^i)\right)} = \bigcup_{i<\omega} \overline{\ker(f^i)}$$

so that for $(x, y) \in \overline{\ker(f^\omega)}$, there is some $i < \omega$ with $(x, y) \in \overline{\ker(f^i)}$. We obtain the required proof tree by (transfinite) induction.

It is further shown by Worrell [37], that the final sequence for finitary functors on Set stabilises at $\omega 2$ and, more importantly here, that the connecting maps $B_\beta \to B_\lambda$ for $\omega \leq \lambda \leq \beta \leq \omega 2$ are injective. Thus, for $\omega \leq \beta \leq \omega 2$, we have $f^\beta(x) \neq f^\beta(y) \implies f^\omega(x) \neq f^\omega(y)$, again giving the required proof tree.

We note that this could be recovered using results from [17], namely that coinductive predicates for finitary functors can be constructed via a final sequence which stabilises after ω steps. Dualising, there should be a correspondence between stages of the initial sequence forming proof trees, and the sequence constructing behavioural apartness as an inductive predicate. However, as we work in Set, we are able to use the earlier results due to Worrell.

Before we show how rule (6) can be improved further, we apply it to the coalgebra for the subdistribution functor of Example 2 to show how it improves on rule (1) of Sect. 2.

3.3 Example: Subdistributions

We saw in Sect. 2 how the definition of cobisimilarity for coalgebras of the subdistribution functor does not allow us to easily produce proofs. Here, we show that our new notion of behavioural apartness is more usable in this regard.

For this, we first elaborate how the functor $\mathcal{D}_s$ acts on the quotient map of an equivalence relation, i.e., a map $q_R \colon X \twoheadrightarrow X/e(R)$ mapping an element to its equivalence class under the equivalence closure $e(R)$ of R. The image of such a map under $\mathcal{D}_s$ is given by

$$\mathcal{D}_s(q_R)(\mu)([z]_R) = \sum_{x \in [z]_R} \mu(x)$$

where, by $[z]_R$ we mean the equivalence class of z under the equivalence closure $e(R)$ of R. We may also write this sum as $\mu[z]_R$, an instance of the general notation $\mu(S \subseteq X) = \sum_{s \in S} \mu(s)$.

Then rule (6), for some coalgebra $\gamma \colon X \to \mathcal{D}_s(X)$, instantiates to

$$\frac{\forall (x', y') \in R.\, x' \# y' \qquad \exists z \in X.\, \gamma(x)[z]_{\overline{R^s}} \neq \gamma(y)[z]_{\overline{R^s}}}{x \# y}$$

Returning to the earlier concrete example of (3), we can now produce a proof which closely matches the reasoning we suggested earlier. We start by noting that for states such as x_1 and y_2 with one state having no outgoing probability weight, we have the following proof of behavioural apartness:

$$\frac{R = \emptyset \qquad \gamma(x_1)[y_2]_{\top_X} = 0 \neq 1 = \gamma(y_2)[y_2]_{\top_X}}{x_1 \# y_2}$$

where $\top_X$ is the total relation on X. In this same way, we can prove $x' \# y'$ for all pairs in the set $S_1 = \{(x_1, x_2), (x_1, y_2), (x_2, y_1), (y_1, y_2)\}$. We go on to prove that both x and y are apart from both of x_1, y_1. For example, we can prove

$$\frac{R = S_1 \qquad \gamma(x)[x]_{\top_X} = 1 \neq 0 = \gamma(y_1)[x]_{\top_X}}{x \# y_1}$$

Although we have already proved a number of apartness pairs, taking the apartness interior still yields the empty set, due to the missing pairs required to make the relation cotransitive. This means we can only distinguish states for which the total outgoing weight is different.

Only once we have proven apartness for all pairs in the set $S := S_1 \cup \{(x, x_1), (x, y_1), (y, x_1), (y, y_1)\}$ can we give the proof of $x \# y$:

$$\frac{R = S \qquad \gamma(x)[x_1]_{\overline{R^s}} = 0.5 \neq 0.4 = \gamma(y)[x_1]_{\overline{R^s}}}{x \# y}$$

Note that here

$$\overline{R^s} = e(\overline{R^s}) = \{(x, y), (x_1, y_1), (x_2, y_2), (x, x_2), (y, y_2), (x, y_2), (y, x_2)\}^s \cup \Delta_X$$

This demonstrates that we are indeed able to provide a proof of behavioural apartness using rule (6) which is finite and built up from inequations exhibiting differences in transition weights. The situation can however still be improved, by reducing the work required in our proofs. As mentioned, we were required to prove for instance the apartness pair $x \# y_1$. As x does not contribute to the transition distributions of the states we are proving apart, this should not be necessary for the proof. Such proof obligations arise because we take the apartness interior/equivalence closure with respect to the entire state space, and (co)transitivity forces that we include many "unnecessary" pairs. In the next section, we show that we can restrict to apartness relations on just those states that are in the supports of the successor distributions, or more generally, states which are reachable in one step.

4 Optimised Rules for Behavioural Apartness

To further improve our proof system, we show that it is sound and complete to prove apartness using an inductive step in which apartness need only be proved on states which are "reachable in one step" from the states which we are proving apart. We make this notion precise in the following definition.

Definition 6. *Given a coalgebra $\gamma\colon X \to B(X)$, we will say that a subobject $m\colon Z \rightarrowtail X$ is a one-step covering of a subobject $s\colon S \rightarrowtail X$ if there exists a $g\colon S \to BZ$ such that $Bm \circ g = \gamma \circ s$.*

Using this definition, we show that inequality of pairs of states under the map $Bq_{\overline{R}} \circ \gamma$ which we have used in all our rules so far, is equivalent to inequality under the map $Bq_{\overline{R}|_Z} \circ g$ where we restrict the relation R to a Z that is a one-step covering of some S via g.

Lemma 2. *Let $\gamma\colon X \to B(X)$ be a Set-coalgebra, $R \subseteq X \times X$ an equivalence relation, and $S \rightarrowtail X$ a subobject. If there is a non-empty $m\colon Z \rightarrowtail X$ which is a one-step covering of S via $g\colon S \to B(Z)$, then*

$$Bq_R(\gamma(x)) \neq Bq_R(\gamma(y)) \iff Bq_{R|_Z}(g(x)) \neq Bq_{R|_Z}(g(y))$$

where $q_R\colon X \to X/R$, $q_{R|_Z}\colon Z \to Z/R|_Z$ are quotient maps on X, Z respectively.

This means, in other words, that we can determine behavioural apartness by looking only at apartness on one-step coverings. The given definition is a generalised version of the notion of the *base of a functor* originally due to Blok [6] which can be stated as the smallest one-step covering. It has been shown that this notion indeed instantiates to one-step reachability [2,35]. As an example, for a coalgebra $\gamma\colon X \to \mathcal{D}_s(X)$ for the subdistribution functor, the base for a singleton $\{x\} \rightarrowtail X$ is exactly the support of the successor distribution of x. More interestingly for our applications are one-step coverings of pairs of states x, y, which will instantiate to the union of the supports of x and y in the subdistribution case. In general, the base may not always exist. Indeed, it requires the base category to be complete and well-powered, and the behaviour functor to preserve wide intersections (see [2, Prop. 12]). The restriction to finitary functors on Set (which we make in Theorem 1) however implies these conditions. Despite this, we choose not to take the base itself in the rule which we give in Theorem 2, as this makes the application of the rule and the proof of completeness simpler.

Proof. Suppose we have a g such that $\gamma \circ s = Bm \circ g$, with $m \colon Z \rightarrowtail X$ non-empty. We form the restriction of a relation to the subobject $m \colon Z \rightarrowtail X$ by the following pullback:

$$\begin{array}{ccc} R|_Z & \rightarrowtail & Z \times Z \\ \downarrow & \lrcorner & \downarrow{\scriptstyle m \times m} \\ R & \rightarrowtail & X \times X \end{array}$$

We then obtain the following diagram:

$$\begin{array}{ccccc} R|_Z & \overset{\pi_1^Z}{\underset{\pi_2^Z}{\rightrightarrows}} & Z & \overset{q_{R|_Z}}{\twoheadrightarrow} & Z/R|_Z \\ \downarrow & & \downarrow{\scriptstyle m} & & \downarrow{\scriptstyle !} \\ R & \overset{\pi_1^X}{\underset{\pi_2^X}{\rightrightarrows}} & X & \overset{q_R}{\twoheadrightarrow} & X/R \end{array}$$

where $! \colon Z/R|_Z \to X/R$ is the unique map which arises from the fact that $q_R \circ m$ coequalizes π_1^Z, π_2^Z by commutativity of the left square. We now consider the commutative diagram:

$$\begin{array}{ccccc} S \overset{s}{\rightarrowtail} X \overset{\gamma}{\to} B(X) & & \xrightarrow{\quad Bq_R \quad} & & B(X/R) \\ {\scriptstyle g}\searrow \quad \nearrow{\scriptstyle Bm} & & & & \nearrow{\scriptstyle B!} \\ BZ & & \xrightarrow[Bq_{(R)|_Z}]{} & B(Z/(R)|_Z) & \end{array}$$

This means that $Bq_R \circ \gamma \circ s = Bq_R \circ Bm \circ g = B! \circ Bq_{(R)|_Z} \circ g$ so that

$$\begin{aligned} & Bq_R(\gamma(s(x))) \neq Bq_R(\gamma(s(y))) \\ \iff & B! \circ Bq_{(R)|_Z} \circ g(x) \neq B! \circ Bq_{(R)|_Z} \circ g(y) \\ \implies & Bq_{(R)|_Z} \circ g(x) \neq Bq_{(R)|_Z} \circ g(y) \end{aligned}$$

Now, if $B!$ is mono, the last implication becomes a bi-implication, which is what we want. In Set this holds as follows: we have $!([z]_{R|_Z}) = [z]_R$ and it is clear that $[z]_R = [z']_R \implies [z]_{R|_Z} = [z']_{R|_Z}$, i.e., ! is mono. Now, as Z is non-empty, $Z/R|_Z$ is non-empty, so that ! is split, and thus its monicity is preserved by B.

We are now in a position to give our final proof rule for behavioural apartness, so that we can prove behavioural apartness $x \# y$ with an inductive step, where only pairs of states in a one-step covering of $\{x, y\}$ need to be proved apart.

Theorem 2. *Let* $\gamma\colon X \to B(X)$*. For all* $x, y \in X$*, and* $m\colon Z \rightarrowtail X$ *a non-empty one-step covering of* $\{x, y\} \rightarrowtail X$ *via* $g\colon S \to B(Z)$*, the following rule is sound and complete for behavioural apartness on* γ*:*

$$\frac{\forall (x', y') \in R.\, x' \# y' \qquad g(x) \not\sim^{B}_{R,Z} g(y)}{x \# y} \tag{8}$$

where we define

$$t_1 \not\sim^{B}_{R,Z} t_2 := Bq_{e(\overline{R^s})|_Z}(t_1) \neq Bq_{e(\overline{R^s})|_Z}(t_2)$$

Note that we now explicitly take the equivalence closure before restricting to Z. This is required for the application of Lemma 2 in the following proof.

Proof. Suppose we have a proof tree with root $x \# y$ built using this rule involving a relation R. Soundness holds by Lemma 2 instantiated to the relation $e(\overline{R^s})$, as for the same R this lemma tells us that $\gamma(x) \not\sim^{B}_{R} \gamma(y)$ will hold, so that the premise of the sound rule (6) holds.

For completeness, we note that the entire state space X is always a non-empty one-step covering of $\{x, y\}$ with $g = \gamma$. The rule (8) then reduces to rule (6), which is complete.

Example 3 (Subdistributions). We return to the example of coalgebras for the subdistribution functor, and show how rule (8) improves on rule (6).

First, note that we can specialise the rule to subdistributions:

$$\frac{\forall (x', y') \in R.\, x' \# y' \qquad \exists z \in Z.\, g(x)[z]_E \neq g(y)[z]_E}{x \# y}$$

where $E = e(\overline{R^s})|_Z$. For the example of (3), we will again prove the apartness $x \# y$, using our new rule. For the last proof step, we will use the one-step covering $Z = \{x_1, x_2, y_1, y_2\}$ (the supports of x and y). Thus, we will have

$$\frac{\begin{array}{c} R = \{(x_1, y_2), (x_1, x_2), (y_1, x_2), (y_1, y_2)\} \\ g(x)[x_1]_E = 0.5 \neq 0.4 = g(y)[x_1]_E \end{array}}{x \# y}$$

where $E = \{(x_1, y_1), (x_2, y_2)\} \cup \Delta_Z$.

We now still need to prove the apartness pairs in this R. Here, we give one pair as an example; the rest are similar.

$$\frac{R = \emptyset \qquad g(x_1)[y_2]_E = 0 \neq 1 = g(y_2)[y_2]_E}{x_1 \# y_2}$$

We have taken $Z = \{y_2\}$ as one-step covering of $\{x_1, y_2\}$. We see that while the proof step for such leaves does not change much when using the rule of Theorem 2, the proof in its totality is easier to provide than what we showed in Sect. 3.3, and fits better with the desired reasoning based on supplying witnesses of apartness and only reasoning about successors.

Example 4 (Streams). Taking $B = A \times (-)$ for some set of symbols A, we can instantiate the rules (6) and (8) to stream systems, by first elaborating the premises. The condition in (6) for a stream system $\langle o, t\rangle : X \to A \times X$ becomes:

$$\begin{aligned} & \mathsf{id}_A \times q_{\overline{R^s}}(\langle o,t\rangle(x)) \neq \mathsf{id}_A \times q_{\overline{R^s}}(\langle o,t\rangle(y)) \\ \Longleftrightarrow\ & (o(x), q_{\overline{R^s}}(t(x))) \neq (o(y), q_{\overline{R^s}}(t(y))) \\ \Longleftrightarrow\ & o(x) \neq o(y) \vee \neg(t(x)\ e(\overline{R^s})\ t(y)) \\ \Longleftrightarrow\ & o(x) \neq o(y) \vee t(x)\ (R^s)^\circ\ t(y) \end{aligned}$$

In words, two states of a stream system behave differently if they have different outputs (the heads of the streams they represent are different) or their successors behave differently (the tails of the streams they represent are different).

Rule (6) can now be instantiated to stream systems:

$$\frac{o(x) \neq o(y)}{x \# y} \qquad \frac{\forall (x', y') \in R.\, x' \# y' \qquad t(x)\ (R^s)^\circ\ t(y)}{x \# y}$$

In this form, the right-hand rule involving successors requires us to take the interior. As we have seen in Sect. 3.3, this will be empty unless we have proven enough apartness pairs. Consider, for instance, the following simple example

$$x_0 \xrightarrow{1} x_1 \xrightarrow{1} x_2 \xrightarrow{2} x_3 \xrightarrow{3} x_4 \xrightarrow{5} \cdots \tag{9}$$

in which $o(x_i)$ is the i-th Fibonacci number, so that x_0 generates the full stream of Fibonacci numbers and x_1 generates the stream of Fibonacci numbers excluding the first one. It should be clear that these states are therefore behaviourally apart. However, in order to prove this with the above rules we must use the rule involving successors, and hence provide a useful relation R. For this R to be an apartness on the whole state space and to be non-empty, we must already prove an infinity of apartness pairs (at least one including each of the states due to cotransitivity). Instantiating instead our optimised rule (8), we obtain

$$\frac{o_g(x) \neq o_g(y)}{x \# y} \qquad \frac{\forall (x', y') \in R.\, x' \# y' \qquad t_g(x)\ \overline{E}\ t_g(y)}{x \# y}$$

where we again write E for $e(\overline{R^s})|_Z$ and where Z is some one-step covering of $\{x, y\}$ via g and we write o_g and t_g for the components of g.

For the example (9), we can take the set $Z = \{x_1, x_2\}$ which is a one-step covering of $\{x_0, x_1\}$ via the map $g\colon \{x_0, x_1\} \to A \times \{x_1, x_2\}$ defined by $g(x_0) = (1, x_1)$ and $g(x_1) = (1, x_2)$. The states x_1, x_2 can be distinguished by their outputs, and so we have the following finite proof of apartness for x_0, x_1, with $\overline{E} = R^s$ so that $(t(x_0), t(x_1)) = (x_1, x_2) \notin E$:

$$\frac{R = \{(x_1, x_2)\} \qquad \dfrac{o(x_1) \neq o(x_2)}{x_1 \# x_2} \qquad t(x_0)\ \overline{E}\ t(x_1)}{x_0 \# x_1}$$

4.1 Inductive Characterisation of $\asymp$

In [12, Appendix A], an inductive definition of cobisimilarity for Kripke polynomial functors is given. Here, we show the instantiation of the relation $\not\asymp_R^B$ to a class of functors extending the Kripke polynomial functors. This will allow us to easily obtain proof rules for coalgebras of functors in this class.

We consider a slight restriction of Kripke polynomial functors on Set as defined by Jacobs [19] extended with the subdistribution functor, the syntax of which we give using the following grammar:

$$B ::= \mathsf{Id} \mid A \mid B \times B \mid B + B \mid B^A \mid \mathcal{P}B \mid \mathcal{D}_s B$$

where A is any set. The restriction is to binary coproducts, which matches the presentation in [12, Appendix A]. Note, also, that we do not restrict to *finite* Kripke polynomial functors as we wish to cover the examples of LTSs and MDPs.

We can now instantiate the definition of the relation $\not\asymp_R^B$ given in (5), with the functor B built from the above grammar. We do this for $\not\asymp_R^B$ here to ease notation. To replace $\not\asymp_R^B$ with $\not\asymp_{R,Z}^B$ we must take the apartness interior in the first statement with respect to Z, and the complement in the case $B = \mathcal{D}_s B_1$ must also be taken with respect to Z. More precisely, we have that $\overline{\not\asymp_{R,Z}^{B_1}} = (Z \times Z) \setminus \not\asymp_{R,Z}^{B_1}$. The proof in each case is a routine calculation.

Lemma 3. *For a relation $R \subseteq X \times X$ and functors $B, B_1, B_2 \colon \mathsf{Set} \to \mathsf{Set}$ we have the following inductive characterisation of $t_1 \not\asymp_R^B t_2$.*

- *If $B = \mathsf{Id}$, then $t_1 \not\asymp_R^B t_2 \iff t_1\ (R^s)^\circ\ t_2$.*
- *If $B = A$, then $t_1 \not\asymp_R^B t_2 \iff t_1 \neq t_2$.*
- *If $B = B_1 \times B_2$, then $(u_1, v_1) \not\asymp_R^B (u_2, v_2) \iff u_1 \not\asymp_R^{B_1} u_2 \vee v_1 \not\asymp_R^{B_2} v_2$.*
- *If $B = B_1 + B_2$, then $t_1 \not\asymp_R^B t_2 \iff [t_1, t_2 \in B_1 X \implies t_1 \not\asymp_R^{B_1} t_2] \wedge [t_1, t_2 \in B_2 X \implies t_1 \not\asymp_R^{B_2} t_2]$*
- *If $B = B_1^A$, then $t_1 \not\asymp_R^B t_2 \iff \exists a \in A. t_1(a) \not\asymp_R^{B_1} t_2(a)$*
- *If $B = \mathcal{P}B_1$, then $t_1 \not\asymp_R^B t_2 \iff [\exists u \in t_1. \forall v \in t_2. u \not\asymp_R^{B_1} v] \vee [\exists v \in t_2. \forall u \in t_1. u \not\asymp_R^{B_1} v]$*
- *If $B = \mathcal{D}_s B_1$, then $t_1 \not\asymp_R^B t_2 \iff \exists z. t_1[z]_{\overline{\not\asymp_R^{B_1}}} \neq t_1[z]_{\overline{\not\asymp_R^{B_1}}}$*

As discussed by Sokolova in [30], the (sub)distribution functor can be seen as an instance of the finitely supported monoid valuations functor $\mathcal{M}_{S,f}^{-}$, a generalisation of the finitely supported multiset functor (see also [15,16]). Due to the construction of our proof system, rules for such a functor can be straightforwardly obtained by calculation of $\not\asymp_R^{\mathcal{M}_{S,f}^{-}}$.

Example: LMPs. There are of course now many examples for which we obtain a proof system. Here, we take coalgebras for the functor $\mathcal{D}_s(-)^A$, with A some finite set of actions, sometimes called Markov Decision Processes (without rewards) or Labelled Markov Processes.

We consider the following LMP:

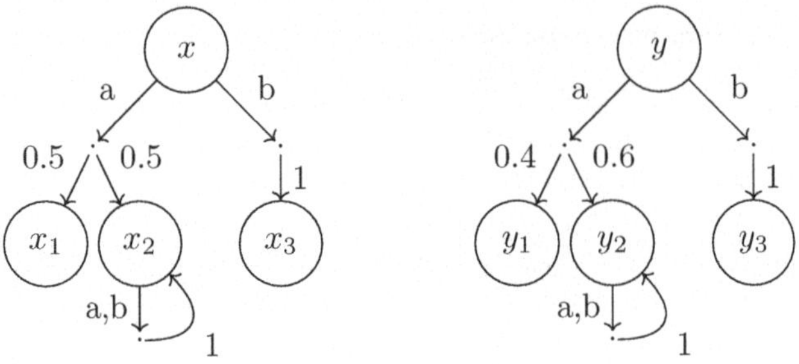

States with no outgoing edge are those s for which $\gamma(s)(\sigma) = 0$ (the zero distribution) for all $\sigma \in A$. Instantiating the generic rule (8) to this setting, gives

$$\frac{\forall (x', y') \in R.\, x' \# y' \qquad \exists a \in A.\, \exists z.\, \gamma(x)(a)[z]_{e(\overline{R^s})|Z} \neq \gamma(y)(a)[z]_{e(\overline{R^s})|Z}}{x \# y}$$

We now take $\sigma = \mathrm{a}$, the one-step covering $Z = \{x_1, x_2, x_3, y_1, y_2, y_3\}$, and

$$R = \{(x_1, x_2), (x_2, x_3), (y_1, y_2), (y_2, y_3), (x_1, y_2), (x_2, y_3), (x_3, y_2), (x_2, y_3)\}$$

Then, for example, $z = x_2$ gives us

$$\frac{\dfrac{\vdots}{\forall (x', y') \in R.\, x' \# y'} \qquad \gamma(x)(a)[x_2]_{e(\overline{R^s})|Z} = 0.5 \neq 0.6 = \gamma(y)(a)[x_2]_{e(\overline{R^s})|Z}}{x \# y}$$

5 Future Work

There are a number of avenues for future work. The first is based on the established link between notions of bisimilarity/behavioural equivalence (and their duals) and (coalgebraic) modal logic [9,11,24,28,36]. We would like to investigate how proofs of behavioural apartness in our system relate to formulas in a corresponding logic. Namely, can we extract formulas distinguishing states which are behaviourally apart? Such results already seem close by if we are able to choose our logic to match the reasoning present in our proof system. Indeed, a proof of behavioural apartness seems close to a proof that a formula is true in one state and its negation is true in the other.

A more recent notion of equivalence based on *codensity lifting* [20,21,31] could also be an approach to more general notions of inequivalence. It has been shown how this can be used to define, e.g., behavioural preorders such as simulation [20] and also quantitative equivalences, which have further been linked to quantitative logics [4,5,10,13,22,23]. We would like to investigate whether the dual notions (which we may call codensity apartness) allow an easier development of expressive logics in these settings. Part of this line of research should be to apply the ideas of Sect. 4 to general liftings beyond that for behavioural apartness (also beyond Set). For instance, can we prove codensity apartness while only looking at states reachable in one step? We hope this would simplify the development of proof systems in the quantitative setting.

References

1. Aczel, P., Mendler, N.: A final coalgebra theorem. In: Pitt, D.H., Rydeheard, D.E., Dybjer, P., Pitts, A.M., Poigné, A. (eds.) Category Theory and Computer Science. LNCS, vol. 389, pp. 357–365. Springer, Heidelberg (1989). https://doi.org/10.1007/BFb0018361
2. Barlocco, S., Kupke, C., Rot, J.: Coalgebra learning via duality. In: Bojańczyk, M., Simpson, A. (eds.) FoSSaCS 2019. LNCS, vol. 11425, pp. 62–79. Springer, Cham (2019). https://doi.org/10.1007/978-3-030-17127-8_4
3. Basold, H.: Mixed inductive-coinductive reasoning types, programs and logic. Ph.D. thesis, Radboud Universiteit Nijmegen (2018)
4. Beohar, H., Gurke, S., König, B., Messing, K.: Hennessy-Milner theorems via Galois connections. In: CSL. LIPIcs, vol. 252, pp. 12:1–12:18. Schloss Dagstuhl - Leibniz-Zentrum für Informatik (2023)
5. Beohar, H., et al.: Expressive quantale-valued logics for coalgebras: an adjunction-based approach. In: STACS. LIPIcs, vol. 289, pp. 10:1–10:19. Schloss Dagstuhl - Leibniz-Zentrum für Informatik (2024)
6. Blok, A.: Interaction, observation and denotation. Master's thesis, Universiteit van Amsterdam (2012)
7. Clouston, R., Bizjak, A., Grathwohl, H.B., Birkedal, L.: Programming and reasoning with guarded recursion for coinductive types. In: Pitts, A. (ed.) FoSSaCS 2015. LNCS, vol. 9034, pp. 407–421. Springer, Heidelberg (2015). https://doi.org/10.1007/978-3-662-46678-0_26
8. Desharnais, J., Edalat, A., Panangaden, P.: Bisimulation for labelled Markov processes. Inf. Comput. **179**(2), 163–193 (2002)
9. Fijalkow, N., Klin, B., Panangaden, P.: Expressiveness of probabilistic modal logics, revisited. In: ICALP. LIPIcs, vol. 80, pp. 105:1–105:12. Schloss Dagstuhl - Leibniz-Zentrum für Informatik (2017)
10. Forster, J., Goncharov, S., Hofmann, D., Nora, P., Schröder, L., Wild, P.: Quantitative Hennessy-Milner theorems via notions of density. In: CSL. LIPIcs, vol. 252, pp. 22:1–22:20. Schloss Dagstuhl - Leibniz-Zentrum für Informatik (2023)
11. Geuvers, H.: Apartness and distinguishing formulas in Hennessy-Milner logic. In: Jansen, N., Stoelinga, M., van den Bos, P. (eds.) A Journey from Process Algebra via Timed Automata to Model Learning. LNCS, vol. 13560, pp. 266–282. Springer, Cham (2022). https://doi.org/10.1007/978-3-031-15629-8_14
12. Geuvers, H., Jacobs, B.: Relating apartness and bisimulation. Log. Methods Comput. Sci. **17**(3) (2021)
13. Goncharov, S., Hofmann, D., Nora, P., Schröder, L., Wild, P.: Kantorovich functors and characteristic logics for behavioural distances. In: Kupferman, O., Sobocinski, P. (eds.) FoSSaCS 2023. LNCS, vol. 13992, pp. 46–67. Springer, Cham (2023). https://doi.org/10.1007/978-3-031-30829-1_3

14. Gumm, H.P.: Elements of the general theory of coalgebras. In: LUATCS 99. Rand Afrikaans University, South Africa (1999)
15. Gumm, H.P.: Copower functors. Theor. Comput. Sci. **410**(12–13), 1129–1142 (2009)
16. Gumm, H.P., Schröder, T.: Monoid-labeled transition systems. In: CMCS. Electronic Notes in Theoretical Computer Science, vol. 44, pp. 185–204. Elsevier (2001)
17. Hasuo, I., Kataoka, T., Cho, K.: Coinductive predicates and final sequences in a fibration. Math. Struct. Comput. Sci. **28**(4), 562–611 (2018)
18. Hermida, C., Jacobs, B.: Structural induction and coinduction in a fibrational setting. Inf. Comput. **145**(2), 107–152 (1998)
19. Jacobs, B.: Introduction to Coalgebra: Towards Mathematics of States and Observation, Cambridge Tracts in Theoretical Computer Science, vol. 59. Cambridge University Press, Cambridge (2016)
20. Katsumata, S., Sato, T., Uustalu, T.: Codensity lifting of monads and its dual. Log. Methods Comput. Sci. **14**(4) (2018)
21. Komorida, Y., et al.: Codensity games for bisimilarity. New Gener. Comput. **40**(2), 403–465 (2022)
22. Komorida, Y., Katsumata, S., Kupke, C., Rot, J., Hasuo, I.: Expressivity of quantitative modal logics: categorical foundations via codensity and approximation. In: LICS, pp. 1–14. IEEE (2021)
23. König, B., Mika-Michalski, C.: (Metric) bisimulation games and real-valued modal logics for coalgebras. In: CONCUR. LIPIcs, vol. 118, pp. 37:1–37:17. Schloss Dagstuhl - Leibniz-Zentrum für Informatik (2018)
24. König, B., Mika-Michalski, C., Schröder, L.: Explaining non-bisimilarity in a coalgebraic approach: games and distinguishing formulas. In: Petrişan, D., Rot, J. (eds.) CMCS 2020. LNCS, vol. 12094, pp. 133–154. Springer, Cham (2020). https://doi.org/10.1007/978-3-030-57201-3_8
25. Larsen, K.G., Skou, A.: Bisimulation through probabilistic testing. Inf. Comput. **94**(1), 1–28 (1991)
26. Lucanu, D., Goriac, E.-I., Caltais, G., Roşu, G.: CIRC: a behavioral verification tool based on circular coinduction. In: Kurz, A., Lenisa, M., Tarlecki, A. (eds.) CALCO 2009. LNCS, vol. 5728, pp. 433–442. Springer, Heidelberg (2009). https://doi.org/10.1007/978-3-642-03741-2_30
27. Moss, L.S.: Coalgebraic logic. Ann. Pure Appl. Log. **96**(1–3), 277–317 (1999)
28. Schröder, L.: Expressivity of coalgebraic modal logic: the limits and beyond. Theor. Comput. Sci. **390**(2–3), 230–247 (2008)
29. Sokolova, A.: Coalgebraic analysis of probabilistic systems. Ph.D. thesis (2005)
30. Sokolova, A.: Probabilistic systems coalgebraically: a survey. Theor. Comput. Sci. **412**(38), 5095–5110 (2011)
31. Sprunger, D., Katsumata, S., Dubut, J., Hasuo, I.: Fibrational bisimulations and quantitative reasoning: extended version. J. Log. Comput. **31**(6), 1526–1559 (2021)
32. Sprunger, D., Moss, L.S.: Precongruences and parametrized coinduction for logics for behavioral equivalence. In: CALCO. LIPIcs, vol. 72, pp. 23:1–23:15. Schloss Dagstuhl - Leibniz-Zentrum für Informatik (2017)
33. Staton, S.: Relating coalgebraic notions of bisimulation. Log. Methods Comput. Sci. **7**(1) (2011)
34. de Vink, E.P., Rutten, J.J.M.M.: Bisimulation for probabilistic transition systems: a coalgebraic approach. Theor. Comput. Sci. **221**(1–2), 271–293 (1999)
35. Wißmann, T., Milius, S., Katsumata, S., Dubut, J.: A coalgebraic view on reachability (2020)

36. Wißmann, T., Milius, S., Schröder, L.: Quasilinear-time computation of generic modal witnesses for behavioural inequivalence. Log. Methods Comput. Sci. **18**(4) (2022)
37. Worrell, J.: On the final sequence of a finitary set functor. Theor. Comput. Sci. **338**(1–3), 184–199 (2005)

A Compositional Framework for Petri Nets

Serge Lechenne[1,2], Clovis Eberhart[2,3(✉)], and Ichiro Hasuo[2,4]

[1] École normale supérieure Paris-Saclay, Gif-sur-Yvette, France
[2] National Institute of Informatics (NII), Tokyo, Japan
eberhart@nii.ac.jp
[3] Japanese-French Laboratory for Informatics, IRL 3527, Tokyo, Japan
[4] SOKENDAI (The Graduate University for Advanced Studies), Hayama, Japan

Abstract. We define a bidirectional compositional framework for Petri nets based on a line of work about compositionally defining games and computation models. This relies on defining structures with open ends that form interfaces they can be composed along. Together with this syntactic construction, we give a graphical language of morphisms in a PROP and a semantic category that describes the evolution of markings in a Petri net. Compared to previous work, the novelty is that computations in a Petri net are stateful, requiring specific care. This framework allows us to solve reachability compositionally.

Keywords: Petri Nets · Category Theory · Compositionality

1 Introduction

Petri nets [13,15,16] are a graph theoretical formalisation of concurrency and parallel programming. Agents are represented by *places* and communication between agents by *transitions*. The main question in this formalism is how information circulates between agents. This information is represented by *tokens* that are contained in places and transmitted to other places by transitions.

A crucial problem on Petri nets is *reachability* [16]: Given an initial *marking* (a number of tokens in each place), is there a sequence of transitions that reaches a given final marking? In this paper, we develop a compositional framework for Petri nets that we call *open Petri nets*. The idea is that we equip Petri nets with *open ends*, which are interfaces along which two Petri nets can be composed. The reachability problem naturally becomes an *open-reachability* problem [19], where the question is to know whether, starting from a given marking on the entry interface of the open Petri net, a given marking on its exit interface is reachable. We then define a graphic language for Petri nets based on PROPs, a categorical framework that has been successfully used to model many different graphical structures [1,5–7]. It abstracts away all the complexity of the definition of open Petri nets by defining a syntax based on generators and equations. Finally, to solve the open-reachability problem, we define a semantic category into which

Published by Springer Nature Switzerland AG 2024
B. König and H. Urbat (Eds.): CMCS 2024, LNCS 14617, pp. 174–193, 2024.
https://doi.org/10.1007/978-3-031-66438-0_9

open Petri nets can be interpreted and where open-reachability can be solved. The base idea is that this semantic category is morally a category of relations between entry markings (markings on the entry interface) and exit markings (markings on the exit interface).

This development follows previous work on different open structures: open parity games [20], open Markov decision processes [21], and open mean-payoff games [22]. However, there are several fundamental differences between previous work and the current article. The first one is that the nature of the structure studied here is different from those of the games studied before. All the games previously studied are sequential in the sense that they are games that consist of passing around a single token. Therefore, when studying a game composed of many sub-games, only one sub-game may play: the game that currently contains the token. In Petri nets, there are many tokens passed around, and the nature of the structure is, therefore, completely concurrent. This means that the semantic composition of Petri nets should be different from those of previous structures.

Another difference is that all the structures we have previously studied are memoryless, in the sense that solving problems on those structures only require the knowledge of which place the token is in. This means that most of the structure can be abstracted away in the semantics. In an open Petri net, a possible exit marking is dependent on the entry marking, but also on the internal state of the Petri net, which is therefore relevant to our approach. In this sense, we may say that the properties studied in the previous papers in this line of work were static, while we study a dynamic property here. For these reasons, the semantic category of open Petri nets differs from those of the previous papers.

One advantage of our approach is that we derive a complex bidirectional framework, expressed in the language of compact closed categories (CompCCs), from a simpler unidirectional framework, expressed in the language of traced symmetric monoidal categories (TSMCs). This is done for free using the well-known Int-construction [10]. On the level of Petri nets, this means that we derive a category **oPN** of open Petri nets from a category **roPN** of *rightward open Petri nets* where all open ends point towards the right. We do the same at the level of semantics, the semantic category $\mathbb{S}_r$ for the unidirectional framework is lifted to $\mathbb{S}$ in the bidirectional framework using the same construction. Moreover, the Int-construction also lifts the interpretation of open Petri nets into the semantic category from the unidirectional level ($\mathcal{S}_r$) to the bidirectional one ($\mathcal{S}$). This is the result of Lemma 6, which is illustrated in Fig. 1.

$$\mathbf{roPN} \xrightarrow{\quad \mathcal{S}_r \quad} \mathbb{S}_r \qquad \mathbf{oPN} = \mathrm{Int}(\mathbf{roPN}) \xrightarrow{\mathcal{S} = \mathrm{Int}(\mathcal{S}_r)} \mathbb{S} = \mathrm{Int}(\mathbb{S}_r)$$

Fig. 1. Lifting from the unidirectional framework to the bidirectional framework

Related Work. As previously mentioned, this work is a continuation of a line of work on compositional structures [20–22], which aims at using compositionality to design faster algorithms to compute properties of structures.

Our graphical language is based on PROPs, which are a specific type of traced symmetric monoidal categories. Many lines of work have used TSMCs as graphical languages. This dates back to [12], see [17] for a survey. For graph-like structures, [8] considers acyclic graphs. In recent years, PROPs have been used to study many graphical structures such as networks [1], signal flow diagrams [5], and quantum graphic calculi [6,7].

Many other works describe compositional approaches to Petri nets. In [2,3], open Petri nets are defined as cospans and composed by pushout. While this is close to our presentation, the fact that open Petri nets are cospans means that they are not exactly graphical objects. In [4], the authors define a graphical language for linear systems. By interpreting this language in the right category, they can encode Petri nets. While their approach is also based on string diagrams, it is a bit farther from graphical intuition than ours, as places, transitions, and their interactions, are encoded as non-trivial string diagrams.

In [18,19], the authors develop a compositional approach to Petri nets, called "Petri nets with boundaries". One advantage is that the categorical structure can be used to derive efficient algorithms. Petri nets with boundaries, in the fashion of a compositional approach, simplify model checking by applying a "divide and conquer" approach, as explained in [14]. We here present a systematic derivation of a compositional Petri net framework, separating the essence of Petri nets with boundaries from specific details. Our framework, called "open Petri nets", is derived systematically, while Petri nets with boundaries were derived in a somewhat ad hoc manner. This approach is the closest to ours, and we offer a more thorough comparison in the article.

Plan. In Sect. 2, we give some reminders on Petri nets and introduce rightward open Petri nets. Then, in Sect. 3, we study the structure of rightward open Petri nets and show that they form a traced symmetric monoidal category (TSMC), and compare them to Petri nets with boundaries. Section 4 is dedicated to defining the semantic category and proving that it is also a TSMC, as well as the interpretation of rightward open Petri nets into the semantic category and prove that it respects the TSMC structure. Finally, we derive the bidirectional framework in Sect. 5 as well as a syntax to inductively compute on open Petri nets, and use this to solve the open-reachability problem compositionally.

Notations and Prerequisites. We assume that the reader is familiar with some basic notions of category theory, especially monoidal categories. For all $k \in \mathbb{N}$, we write $[k]$ for the set $\{1, 2, \ldots, k\}$. Given a proposition P, we write δ_P for the Kronecker symbol, i.e., $\delta_P = 1$ if P holds, and $\delta_P = 0$ otherwise. Given two functions $f\colon X \to Y$ and $f'\colon X' \to Y'$, we name $f + f'\colon X + X' \to Y + Y'$ the function obtained by universal property of coproduct. Given two pairs of natural numbers $n = (a, b)$ and $m = (c, d)$, we write $n + m$ for the pair $(a + c, b + d)$.

2 Rightward Open Petri Nets

After some brief reminders on Petri nets, this section describes our first contribution, *rightward open Petri nets*, which are a compositional approach to Petri nets. For Petri nets, we follow the notations and definitions from [19].

2.1 Petri Nets

Definition 1 (Petri nets). *A* Petri net *is a tuple* $(P, T, {}^\bullet(\_), (\_)^\bullet)$ *where*

- P *(resp.* T*) is a set whose elements are called* places *(resp.* transitions*),*
- ${}^\bullet(\_)$ *and* $(\_)^\bullet$ *are functions* $T \to \mathcal{P}(P)$.

Petri nets admit a graph-theoretical representation [13,15,16] where the places are drawn as circles and the transitions as rectangles, sometimes with their labels written around them. The function ${}^\bullet(\_)$ is represented by arrows from place to transitions, and the function $(\_)^\bullet$ by arrows from transitions to places. For example, the Petri net with $P = \{p_1, p_2, p_3\}$, $T = \{t\}$, ${}^\bullet t = \{p_1\}$, and $t^\bullet = \{p_2, p_3\}$ can be represented as on the left of Fig. 2.

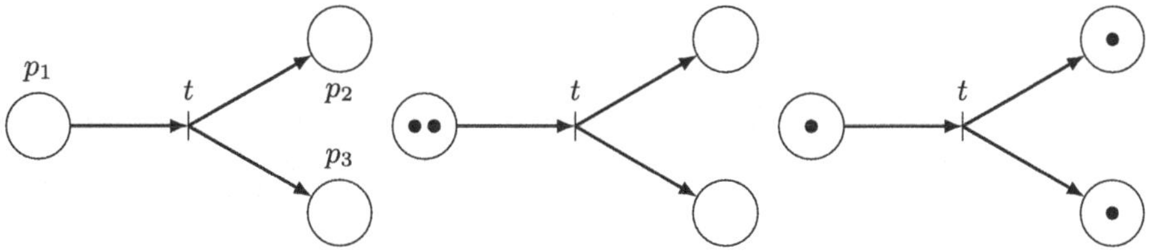

Fig. 2. A Petri net, a marking of it, and the marking reached after firing transition t

The data, named tokens, is exchanged through the transitions from place to place. Intuitively, the data does not represent any physical object but information that agents can freely exchange, create or destroy. This information circulating in a network is modelled in the form of a marking function.

Definition 2 (marking). *A* marking function *(or* marking*) on a Petri net is a function* $\mu : P \to \mathbb{N}$.

Information is transmitted between agents by selecting a transition t, consuming one token from every place leading to t (i.e., every place in ${}^\bullet t$) and placing one token in every place reached by t (i.e., every place in $t^\bullet$).

Definition 3 (enabling, firing). *Given a Petri net, a transition* t*, and a marking* μ*, we say that* t *is* enabled *by* μ *if* $\forall x \in {}^\bullet t, \mu(x) \geq 1$.

A transition t *enabled by a marking* μ *can modify it into a new marking* μ' *defined by* $\mu'(x) = \mu(x) + \delta_{x \in t^\bullet} - \delta_{x \in {}^\bullet t}$. *This operation is called* firing t, *and is denoted* $\mu \xrightarrow{t} \mu'$.

In the definition above, μ' removes one token from μ in all places in $t^\bullet$, and adds one to all places in ${}^\bullet t$.

We extend the notation of firing to sequences $s = t_1, \ldots, t_n \in T^*$ and write $\mu \xrightarrow{s} \mu'$ to mean that each t_i is enabled on the marking reached from μ after firing all the previous t_j's and that μ' is reached after firing t_n. We will also write $t(\mu)$ the marking μ' (or $t_N(\mu)$ when the Petri net N is not explicit).

We now illustrate how the graphical representation accommodates markings and firings of transitions. Tokens are represented as black dots, and a marking μ is represented by placing $\mu(x)$ tokens in each place x. For example, remembering the example Petri net on the left of Fig. 2, the drawing in the middle of Fig. 2 represents the marking μ defined by $\mu(p_1) = 2$ and $\mu(p_2) = \mu(p_3) = 0$. The transition t is enabled by μ in this net, and firing it outputs the marking on the right of Fig. 2.

2.2 Rightward Open Petri Nets

We now define rightward open Petri nets, which have open ends along which they can be composed. We then adapt all the definitions of Petri nets to rightward open Petri nets.

Definition 4 (rightward open Petri net). *A* rightward open Petri net *is a tuple* $N = (m, n, P, T, {}^\circ(\_), (\_)^\circ, {}^\bullet(\_), (\_)^\bullet)$ *where*

- $m = (m_b, m_r)$ *and* $n = (n_b, n_r)$ *are pairs of natural numbers called the (*left *and* right*)* interfaces *of* N*,*
- P *(resp.* T*) is a set whose elements are called* places *(resp.* transitions*),*
- ${}^\circ(\_) : m_r \to T + n_r, m_b \to P + n_b$ *is a function such that* $\forall j \in [n_b] + [n_r], i_1, i_2 \in [m_b] + [m_r], {}^\circ(i_1) = j = {}^\circ(i_2) \Rightarrow i_1 = i_2$*,*
- $(\_)^\circ : P \to \mathcal{P}([n_r]), T \to \mathcal{P}([n_b])$ *is a function such that* $\forall t, t' \in T \cup P, (t)^\circ \cap (t')^\circ \neq \emptyset \Rightarrow t = t'$*,*
- ${}^\bullet(\_)$ *and* $(\_)^\bullet$ *are functions* $T \to \mathcal{P}(P)$*.*

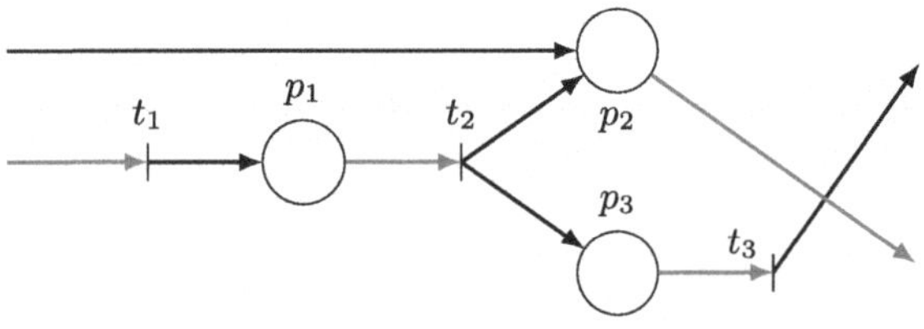

Fig. 3. A rightward open Petri net with interfaces $(1, 1)$ and $(1, 1)$

We now explain in detail what each element of the tuple represents. First of all, P, T, ${}^\bullet(\_)$, and $(\_)^\bullet$ represent the same elements as for regular Petri nets. The left and right interfaces, m and n, respectively, encode the number of open ends on the left-hand and right-hand side of the rightward open Petri net. They are pairs of numbers because open ends that point to places are different from open ends that point to transitions; m_b and n_b represent open ends that point to places, while m_r and n_r represent open ends that point to transitions. The subscripts b and r stand for "black" and "red" respectively, corresponding to the graphical representation we show below. The constraints on $^\circ(\_)$ and $(\_)^\circ$ mean that there is at most one arrow pointing to each open end on the right.

The function $^\circ(\_)$ (which is formally two functions, but for which we use the same symbol) encodes how each open end on the left (in $[m_b]$ or $[m_r]$) is connected to a place, a transition, or an open end on the right (i.e., in $[n_b]$ or $[n_r]$). The type of the function ensures that black open ends only point to either places or black open ends on the right, and red open ends only to either transitions or red open ends on the right.

The function $(\_)^\circ$ (which is again formally two functions) encodes the connection of places and transitions to open ends on the right.

We also adapt the graphical representation of Petri nets to rightward open Petri nets by graphically drawing $^\circ(\_)$ as arrows coming from the left open ends and leading to places, transitions, or to the right open ends. We draw $(\_)^\circ$ similarly from places and transitions to right open ends. Figure 3 shows an example of the graphic representation of a rightward open Petri net. It describes the rightward open Petri net with $m = n = (1, 1)$, $P = \{p_1, p_2, p_3\}$, $T = \{t_1, t_2, t_3\}$, $^\circ 1_b = p_2$, $^\circ 1_r = t_1$, $p_1^\circ = p_2^\circ = t_1^\circ = t_3^\circ = \emptyset$, $p_3^\circ = 1_r$, $t_3^\circ = 1_b$, $^\bullet t_1 = \emptyset$, $^\bullet t_2 = \{p_1\}$, $^\bullet t_3 = \{p_3\}$, $t_1^\bullet = \{p_1\}$, $t_2^\bullet = \{p_2, p_3\}$, and $t_3^\bullet = \emptyset$, where 1_b, 1_r are used to distinguish between black and red open ends (even though it is unambiguous even without this distinction because of the typing).

Open ends are numbered, and verticality in the graphical representation encodes the order between open ends. Note that we always draw black open ends above red ones for interfaces, but that is just a graphical convention: there is no order between open ends of different colours.

We can now define configurations, which are the counterparts of markings for traditional Petri nets. We can then define the way roPN configurations evolve. Like Petri nets, they evolve through firing transitions. The main difference with traditional Petri nets is that they can also evolve by tokens "sliding" from the entry interface to the exit interface.

Definition 5 (Markings, configurations). *Given a rightward open Petri net with interfaces m and n and a set P of places, a* marking *is a function $\mu\colon P \to \mathbb{N}$, an* entry marking *is a function $\mu_\mathrm{i}\colon [m_b] + [m_r] \to \mathbb{N}$, and an* exit marking *is a function $\mu_\mathrm{o}\colon [n_b] + [n_r] \to \mathbb{N}$. A* configuration *is an element $(\mu_\mathrm{i}, \mu, \mu_\mathrm{o})$ of $\mathbb{N}^{m_b+m_r} \times \mathbb{N}^P \times \mathbb{N}^{m_b+m_r}$.*

Definition 6 (firing, sliding). *Given a rightward open Petri net N, a configuration $c = (\mu_\mathrm{i}, \mu, \mu_\mathrm{o})$, and a transition t, we say that t is* enabled *on c if for all places $p \in {}^\bullet t$, $\mu(p) > 0$, and for all entries $i \in [m_r]$ such that $^\circ i = t$, $\mu_\mathrm{i}(i) > 0$. Similarly, an entry $i \in [n_b]$ is enabled on c if $\mu(i) > 0$, and an exit $j \in [m_r]$, for which there exists $p \in P$ with $j \in p^\circ$, is enabled if $\mu(p) > 0$.*

The firing of t *enabled on* c *turns it into* $(\mu_{\mathrm{i}}', \mu', \mu_{\mathrm{o}}')$ *such that* $\mu_{\mathrm{i}}'(i) = \mu_{\mathrm{i}}(i) - \delta_{{}^{\circ}i=t}$, $\mu'(p) = \mu(p) - \delta_{p\in{}^{\bullet}t} + \delta_{p\in t^{\bullet}t}$, *and* $\mu_{\mathrm{o}}'(j) = \mu_{\mathrm{o}}(j) + \delta_{j\in t^{\circ}}$.

The sliding *of entry* i *enabled on* c *turns it into* $(\mu_{\mathrm{i}}', \mu', \mu_{\mathrm{o}}')$ *with* $\mu_{\mathrm{i}}'(i') = \mu_{\mathrm{i}}(i') - \delta_{i'=i}$, $\mu'(p) = \mu(p) + \delta_{{}^{\circ}i=p}$, $\mu_{\mathrm{o}}'(j) = \mu_{\mathrm{o}}(j) + \delta_{{}^{\circ}i=j}$.

The sliding *of exit* j *enabled on* c *turns it* $(\mu_{\mathrm{i}}', \mu', \mu_{\mathrm{o}}')$ *with* $\mu_{\mathrm{i}}' = \mu_{\mathrm{i}}$, $\mu'(p) = \mu(p) - \delta p^{\circ} = j$, *and* $\mu_{\mathrm{o}}'(j') = \mu_{\mathrm{o}}(j') + \delta_{j=j'}$.

In this definition, sliding is rather straightforward: tokens may slide freely between open ends and places (as these correspond to transitions being fired outside of the open Petri net). To fire a transition, all arrows pointing to it must contain at least one token: both from places and from open ends.

We denote by $t(c)$ the configuration obtained by firing transition t from c. Similarly, we denote by $l_i(c)$ (resp. $r_j(c)$) the configuration obtained by sliding the ith entry (resp. jth exit) from c.

3 The Categorical Structure of roPN

We now move on to explicitating the categorical structure of rightward open Petri nets, viewed as arrows between interfaces. We explicitate the categorical structure of rightward open Petri nets, using the language of monoidal categories [9,11]. We will then compare our approach and the work done in [18,19].

3.1 The Category of Rightward Open Petri Nets

We start by establishing their structure as a category by defining the identities and composition. When a Petri net N has interfaces m and n, we write $N\colon m \to n$. Indeed, in this section we define the category **roPN** or rightward open Petri nets whose objects are interfaces are morphisms are roPNs.

Definition 7 (composition). *Given* $N\colon m \to n$ *and* $N'\colon n \to k$*, we name* $N;N'\colon m \to k$ *the rightward open Petri net made from the following data:* $P_{N;N'} = P \cup P'$, $T_{N;N'} = T \cup T'$,

$$
{}^{\bullet}t_{N;N'} = \begin{cases} {}^{\bullet}t_N & \text{if } t \in T \\ {}^{\bullet}t_{N'} \cup \{p \in P \mid \exists j \in [m_r]\,.j \in p_N^{\circ} \wedge {}^{\circ}j_{N'} = t\} & \text{otherwise,} \end{cases}
$$

$$
t^{\bullet}_{N;N'} = \begin{cases} t^{\bullet}_N \cup \{{}^{\circ}j_{N'} \in P' \mid j \in t^{\circ}_N\} & \text{if } t \in T \\ t^{\bullet}_{N'} & \text{otherwise,} \end{cases}
$$

$$
{}^{\circ}i_{N;N'} = \begin{cases} {}^{\circ}({}^{\circ}i_N)_{N'} & \text{if } {}^{\circ}i_N \in [j_b] \\ {}^{\circ}i_N & \text{otherwise,} \end{cases}
$$

$$
x^{\circ}_{N;N'} = \begin{cases} x^{\circ}_{N'} & \text{if } {}^{\circ}(x^{\circ}_N)_{N'} \in P \cup T \\ \{{}^{\circ}j_{N'} \in [k_b] + [k_r] \mid j \in x^{\circ}_N\} & \text{otherwise.} \end{cases}
$$

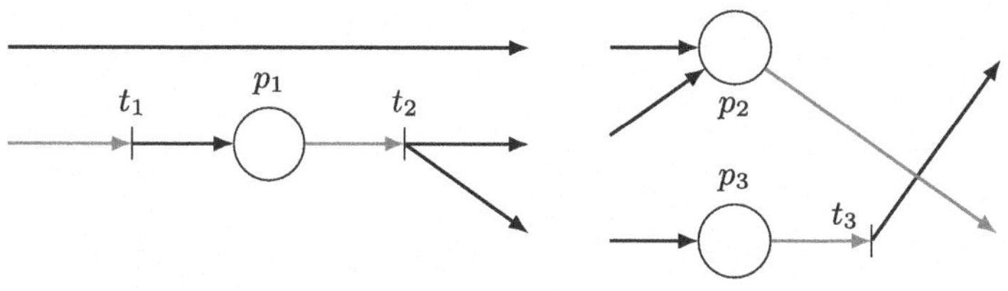

Fig. 4. Two roPNs with interfaces $(1, 1) \to (3, 0)$ and $(3, 0) \to (1, 1)$ respectively

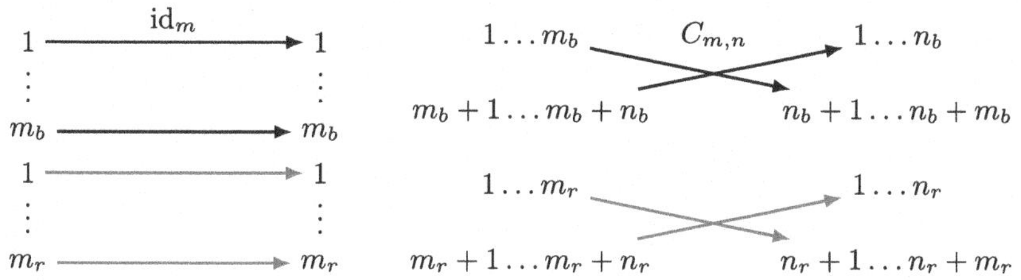

Fig. 5. Identities and swaps

Checking that ${}^\circ(\_)_{N;N}$ and $(\_)^\circ_{N;N}$ satisfy the condition is easy.

The definitions of $(\_)^\circ$, ${}^\circ(\_)$, ${}^\bullet(\_)$, and $(\_)^\bullet$ encode the fact that edges that point to and from the middle interface get merged in the composition $N; N'$. Figure 4 shows two roPNs whose composition is the roPN in Fig. 3.

Definition 8 (identities, swaps). *Given an interface m, the* identity *roPN* $\mathrm{id}_m : m \to m$ *is given by the following data: $P = T = \emptyset$, ${}^\circ i = i$ (the rest of the data being trivial since $P = T = \emptyset$).*

Given two interfaces m and n, the swap $C_{m,n} : m + n \to n + m$ *is given by the following data: $P = T = \emptyset$, ${}^\circ i$ is $i + m_b$ if $i \leq m_b$ and $i - m_b$ otherwise for $i \in [m_b + n_b]$, and ${}^\circ i$ is $i + m_r$ if $i \leq m_r$ and $i - m_r$ otherwise for $i \in [m_r + n_r]$.*

The identities and swaps are illustrated in Fig. 5. We can now organise rightward open Petri nets into a category **roPN**.

Definition 9 (roPN). *We name* **roPN** *the category whose objects are pairs of elements $(m_r, m_b) \in \mathbb{N} \times \mathbb{N}$ and arrows $N : (m_r, m_b) \to (n_r, n_b)$ are rightward open Petri nets with the corresponding numbers of open ends. Composition is defined by ; and the identity of $m \in \mathbb{N} \times \mathbb{N}$ is the rightward open Petri net id_m.*

3.2 Monoidal Categorical Structure

We now further explore the categorical structure of **roPN**, showing that it is a traced symmetric monoidal category.

Definition 10 (tensor product). *Given $N : m \to n$ and $N' : m' \to n'$, we define $N \otimes N' : m + m' \to n + n'$ the rightward open Petri net given by the following data: $P_{N \otimes N'} = P \cup P'$, $T_{N \otimes N'} = T \cup T'$, ${}^\circ(\_)_{N \otimes N'} = {}^\circ(\_)_N + {}^\circ(\_)_{N'}$, $(\_)^\circ_{N \otimes N'} = (\_)^\circ_N + (\_)^\circ_{N'}$, ${}^\bullet(\_)_{N \otimes N'} = {}^\bullet(\_)_N + {}^\bullet(\_)_{N'}$, and $(\_)^\bullet_{N \otimes N'} = (\_)^\bullet_N + (\_)^\bullet_{N'}$.*

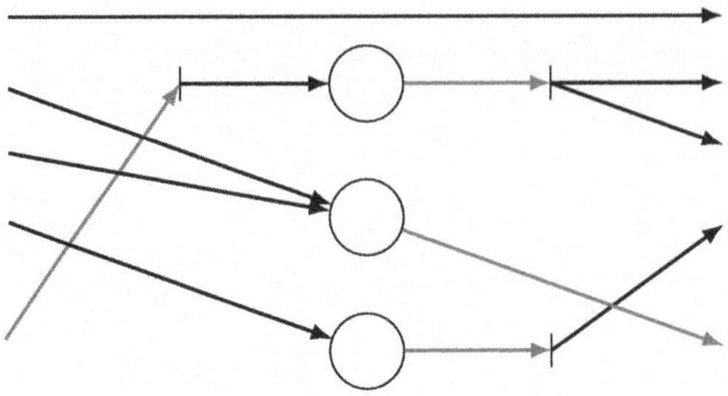

Fig. 6. An roPN with interfaces $(4, 1)$ and $(4, 1)$

Figure 6 shows the tensor product of the two roPNs from Fig. 4. Note that this graphically corresponds to stacking the roPNs (except for the interfaces, where the red ends are drawn below the black ones, but which is only a convention).

Lemma 1. roPN *forms a symmetric monoidal category with* $+$ *as tensor on objects, and* $\otimes$ *on morphisms, and* $C_{m,n}$ *as the symmetries.*

3.3 Trace Operator

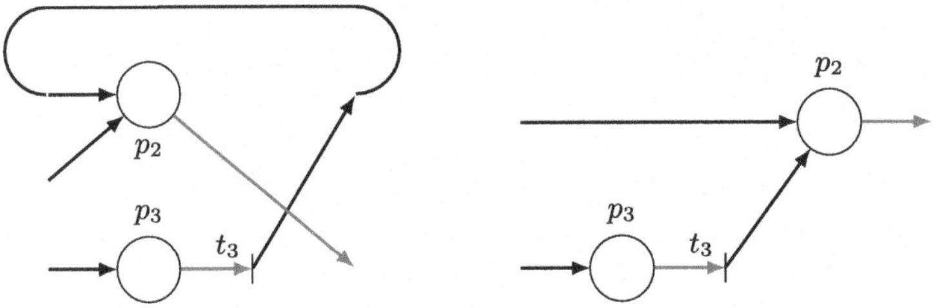

Fig. 7. The trace of the roPN on the right in Fig. 4 along $(1, 0)$ and a simpler representation

Our goal is to derive a bidirectional framework from the unidirectional one. To do this, we need to show that **roPN** is traced. We first introduce some notions, as defining the trace is trickier than for composition or tensor. Graphically, the trace operator "loops" over the first black and red inputs and outputs of the Petri net, as shown in Fig. 7.

This construction has to involve some form of recursive definition to connect inputs and outputs. Figure 8 shows an example that illustrates that if something (a place, a transition, or a left open end) is connected to one of the right open ends affected by the trace, simply looking at the value of the corresponding open end by $(\_)^\circ$ is not enough, since we can "stay" in the loop. In order to define the $^\bullet(-)$ function on the first open end of the traced Petri net, we have to compute

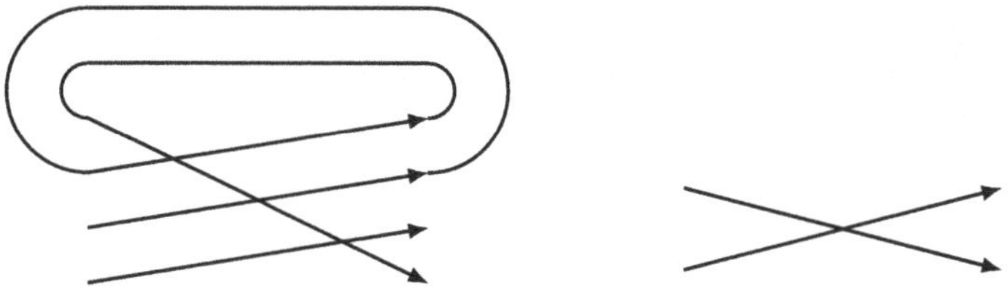

Fig. 8. An roPN with interfaces $(4, 0)$ and $(4, 0)$, traced along $(2, 0)$, and a simpler representation

${}^{\bullet}(-)$ of the original Petri net on three different open ends. The first open end in the trace corresponds to the third open end, which is connected to the second, which is connected to the first, which is finally connected to the fourth. In other words if we denote by ${}^{\bullet}(-)'$ the accessibility function of the traced Petri net and ${}^{\bullet}(-)$ that of the original Petri net, then the formula to compute ${}^{\bullet}(1)'$ is ${}^{\bullet}({}^{\bullet}({}^{\bullet}(1+2)))$ (where 2 is the number of black wires we trace over). In general, this can take an arbitrary number of steps, which justifies the complexity of the definition.

Hence, given a rightward open Petri net $N : (m_b+p, m_r+q) \to (n_b+p, n_r+q)$, we need to define functions ϕ_p (for the p black open ends that are looped) that output the final open end that does not lead back to $[p]$ (in the loop). To ensure such a function ϕ_p can be defined, we must prove that such an exiting open end exists. The following lemma ensures this.

Lemma 2. *Let $j \in [p]$, then there is at most one $i \in [m_b + p]$ such that ${}^{\circ}i_N = j$.*

This is easily proven using the conditions imposed over $(x)^{\circ}_N$. Finally, because $[p]$ is finite, we define $\phi_p(x)$ by induction: $\phi_p(x) = x$ if ${}^{\circ}x \notin [p]$, and $\phi_p(x) = \phi_p({}^{\circ}x)$ otherwise. This definition is well-founded because of Lemma 2 and the fact that $[p]$ is finite. We define ϕ_q equivalently.

Definition 11 (Trace operator). *Given a rightward open Petri net $N : m + p \to n + p$, we define $N' = \mathrm{Tr}^{p}_{m,n}(N) \colon m \to n$ as the rightward open Petri net with the following data: $P' = P$, $T' = T$,*

$$
{}^{\circ}(x)' = \begin{cases} {}^{\circ}\phi_p(x) & \text{if } x \in [m_b] \\ {}^{\circ}\phi_q(x) & \text{if } x \in [m_r], \end{cases}
$$

$$
(x)^{\circ\prime} = \{{}^{\circ}(\phi_p(h) \mid h \in x^{\circ} \cap [p]\} \cup \{{}^{\circ}\phi_q(h) \mid h \in x^{\circ} \cap [q]\} \cup (x^{\circ} \setminus ([p] \cup [q]))
$$

$$
{}^{\bullet}(t)' = \{x \in P \mid \exists h \in [q], h \in x^{\circ} \wedge \phi_q(h) \in {}^{\bullet}t\} \cup {}^{\bullet}t
$$

$$
(t)^{\bullet\prime} = \{x \in P \mid \exists h \in [q]\,, h \in t^{\circ} \wedge {}^{\circ}\phi_p(h) = x\} \cup t^{\bullet}.
$$

Lemma 3 (trace). $\mathrm{Tr}^{p,q}_{m,n}(-)$ *defines a trace over* **roPN**.

Theorem 1. **roPN** *is a traced symmetric monoidal category (TSMC).*

3.4 Comparison with Petri Nets with Boundaries

Now that the categorical structure is spelt out, we move on to pointing out the syntactic and operational differences between **roPN** and **PNB**, the category of Petri net with boundaries defined in [19].

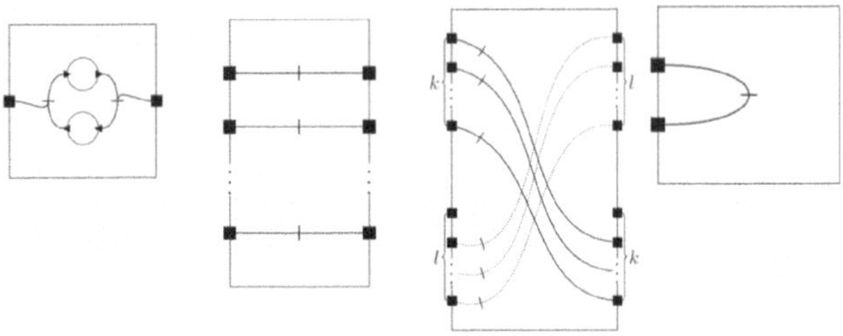

Fig. 9. Some examples of Petri nets with boundaries

We first give out a quick summary of Petri net with boundaries. A Petri net with boundaries is thought of as a Petri net with extra structure, namely a net $(P, T, {}^\bullet(\_), (\_)^\bullet)$ as defined in our work to which is added two finite ordinals that act as the interface, and two accessibility functions, called "in" and "out" (those are not the original notations introduced in [19]), encoding how the interface is linked to transitions. Petri nets with boundaries admit a graphical representation, as shown in Fig. 9 (taken from [19]), where the boundaries are represented as black squares located on the edges of a box enclosing the Petri net. The two accessibility functions are represented as regular edges, connecting the black squares and the transition.

The first main difference, and the most important one, is that the boundaries representing the interface are connected to transitions *only*, meaning they are not leading in or out of the Petri net by default. Here, it is the function in (resp. out) that makes one boundary in the interface leading to or coming from transitions.

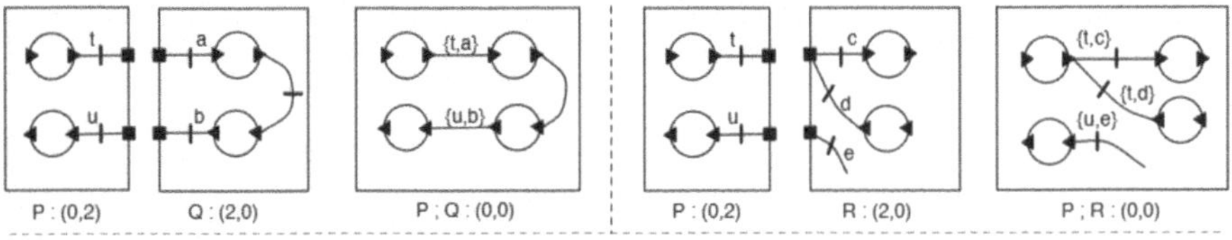

Fig. 10. composing Petri net with boundaries

The fact that boundaries are only connected to transitions also affects the composition of Petri nets with boundaries: we have chosen our definition of a rightward open Petri net to have an "implicit" interface, which allows us to follow graphical intuition of fusing edges. In the Petri net with boundary framework, the fusing is intuitively done on the boundaries, as demonstrated in the two examples shown in Fig. 10 (taken from [18]). However, when fusing the edges, the transitions these edges are connected to also need to be fused (this is called a minimal synchronisation in [19]). This fusing process is destructive in regard to the structure of the Petri net, as transitions can be either created or deleted, like in the rightmost net in Fig. 10. For example, the middle two Petri nets with boundaries in Fig. 9 are the categorical identities and crossing of the category **PNB**, and they only satisfy the required naturality equations *up to isomorphism*, making the categorical structure lax monoidal, contrary to **roPN**.

Another consequence of this fact is that by connecting two boundaries of the same interface via a transition, one can create the counit of **PNB** (depicted as the rightmost net in Fig. 9). As the snake equation is satisfied, this means that **PNB** is a compact closed (lax) category, where the trace is defined by the canonical construction described in [10]. In comparison, in the case of open Petri nets, the compact closed structure is derived from the more primitive traced symmetric monoidal structure that describes the unidirectional case.

4 The Semantic Structure

In this section, we define the semantic category $\mathbb{S}_r$, prove that it has a traced symmetric monoidal structure, define a semantic functor $\mathcal{S}_r\colon \mathbf{roPN} \to \mathbb{S}_r$ and prove that it respects that traced symmetric monoidal structure.

4.1 The Semantic Category $\mathbb{S}_r$

The main idea is that what can be observed of an open Petri net is how many tokens enter and exit it through its interfaces. However, that is not enough to characterise the behaviour of an open Petri net. Indeed, some token can be left in the net, therefore modifying the behaviour of said net. The marking acts like an internal state to the Petri net, and that is how we represent it in the semantics.

Definition 12 (The semantic category $\mathbb{S}_r$). *We denote by $\mathbb{S}_r$ the* semantic category. *Its objects are pairs (n_b, n_r) of natural numbers. The morphisms from (m_b, m_r) to (n_b, n_r) are pairs of a set Q (called the set of* internal states*) and a function from $\mathbb{N}^{m_b+m_r} \times Q \times \mathbb{N}^{n_b+n_r}$ to $\mathcal{P}(\mathbb{N}^{m_b+m_r} \times Q \times \mathbb{N}^{n_b+n_r})$ that satisfies the following conditions*

- *(reflexivity)* $(\mu_\mathrm{i}, q, \mu_\mathrm{o}) \in f(\mu_\mathrm{i}, q, \mu_\mathrm{o})$,
- *(transitivity) if* $(\mu'_\mathrm{i}, q', \mu'_\mathrm{o}) \in f(\mu_\mathrm{i}, q, \mu_\mathrm{o})$ *and* $(\mu''_\mathrm{i}, q'', \mu''_\mathrm{o}) \in f(\mu'_\mathrm{i}, q', \mu'_\mathrm{o})$, *then* $(\mu''_\mathrm{i}, q'', \mu''_\mathrm{o}) \in f(\mu_\mathrm{i}, q, \mu_\mathrm{o})$,
- *(additivity)* $(\mu'_\mathrm{i}, q, \mu'_\mathrm{o}) \in f(\mu_\mathrm{i}, q, \mu_\mathrm{o})$ *iff* $(\mu'_\mathrm{i} + \tilde{\mu_\mathrm{i}}, q, \mu'_\mathrm{o} + \tilde{\mu_\mathrm{o}}) \in f(\mu_\mathrm{i} + \tilde{\mu_\mathrm{i}}, q, \mu_\mathrm{o} + \tilde{\mu_\mathrm{o}})$
- *(monotonicity) if* $(\mu'_\mathrm{i}, q, \mu'_\mathrm{o}) \in f(\mu_\mathrm{i}, q, \mu_\mathrm{o})$, *then* $\mu'_\mathrm{i} \leq \mu_\mathrm{i}$ *and* $\mu_\mathrm{o} \leq \mu'_\mathrm{o}$.

For simplicity, we sometimes just write $\mathbb{N}^m$ for $\mathbb{N}^{m_b+m_r}$. We write $(\mu_\mathrm{i}, q, \mu_\mathrm{o})$ for an element in $\mathbb{N}^m \times Q \times \mathbb{N}^n$, where μ_i represents the entry marking, q the internal state, and μ_o the exit marking. Note that the function is equivalent to a relation on $\mathbb{N}^m \times Q \times \mathbb{N}^n$, and we write $(\mu_\mathrm{i}, q, \mu_\mathrm{o}) \to_f (\mu'_\mathrm{i}, q', \mu'_\mathrm{o})$ to denote that these two elements are related by f (i.e., that $(\mu'_\mathrm{i}, q', \mu'_\mathrm{o}) \in f((\mu_\mathrm{i}, q, \mu_\mathrm{o}))$).

The idea is that two elements $(\mu_\mathrm{i}, q, \mu_\mathrm{o})$ and $(\mu'_\mathrm{i}, q', \mu'_\mathrm{o})$ are related by the interpretation of a Petri net N if, starting with an entry marking μ_i in internal state q, and assuming that N has already output μ_o, then there exists a sequence of firings that consumes tokens from the entry marking until exactly μ'_i is left, modifying the internal state to q', and outputting tokens to reach exactly μ'_o.

There are two differences between the semantic category here and those of previous work. The first one is insignificant: in previous work, the objects were natural numbers rather than pairs, as there was a single type of edges. More importantly, in previous work, the semantic category was always defined via a monad, but this is not the case here. The current definition may seem too complicated, but we argue that such complexity is necessary. We go through several simpler ideas and explain why they fail.

As mentioned above, the internal state of the Petri net dynamically modifies its behaviour, hence if we considered a morphism from m to n to only be a function $\mathbb{N}^m \to \mathcal{P}(\mathbb{N}^n)$ (which would fall under the monad construction), it would be impossible to interpret roPNs into this semantic category.

A second idea would then be to define morphisms as functions $f: \mathbb{N}^m \times Q \to \mathcal{P}(Q \times \mathbb{N}^n)$, which solves the problem of taking the internal state of the Petri net into account, and it would be possible to interpret rightward-open Petri nets into this semantic category. The problem lies with what it means for (q', μ_o) to be in $f(\mu_\mathrm{i}, q)$. It would mean that μ_o is reachable by consuming all the tokens from μ_i. However, it would then be impossible to define a trace operator that is compatible with that of roPNs, since some tokens should remain in the entry marking between two loops in order to reach some exit markings. Hence, this idea does not lead to a trace that is compatible with that of **roPN**.

This explains why we need to remember the entry marking even in the output of the function. Technically, it is possible to define the semantics as a category of functions of type $\mathbb{N}^m \times Q \to \mathcal{P}(\mathbb{N}^m \times Q \times \mathbb{N}^n)$, but we chose to add the exit marking to the argument of the function, making it symmetric and therefore, equivalent to a category of relations.

To define $\mathbb{S}_r$, we first need to define the identities and composition.

Definition 13 (Identities, composition of $\mathbb{S}_r$). *The identity on n is* $\mathrm{id}_n: n \to n$ *defined as the set* $\{*\}$ *of internal states and the relation* $(\mu_\mathrm{i}, *, \mu_\mathrm{o}) \to_{\mathrm{id}_n} (\mu'_\mathrm{i}, *, \mu'_\mathrm{o})$ *if and only if* $\mu'_\mathrm{i} \leq \mu_\mathrm{i}$ *and* $\mu_\mathrm{i} + \mu_\mathrm{o} = \mu'_\mathrm{i} + \mu'_\mathrm{o}$.

Let $f: n \to m$ *and* $g: m \to p$ *be two morphisms in* $\mathbb{S}_r$*, we define their composition* $f;g$ *as having* $Q_f \times Q_g$ *as its set of internal states, and* $(\mu_\mathrm{i}, (q_f, q_g), \mu_\mathrm{o}) \to_{f;g} (\mu'_\mathrm{i}, (q'_f, q'_g), \mu'_\mathrm{o})$ *if and only if there exists* μ^* *such that* $(\mu_\mathrm{i}, q_f, 0^m) \to_f (\mu'_\mathrm{i}, q'_f, \mu^*)$ *and* $(\mu^*, q_g, \mu_\mathrm{o}) \to_g (0^m, q'_g, \mu'_\mathrm{o})$*. With this data,* $\mathbb{S}_r$ *forms a category.*

The idea of identities is that they connect each entry i to its corresponding exit, and each wire can slide any number of tokens from the entry marking to the exit one. Note that additivity and monotonicity are necessary to prove that $\mathbb{S}_r$ forms a category.

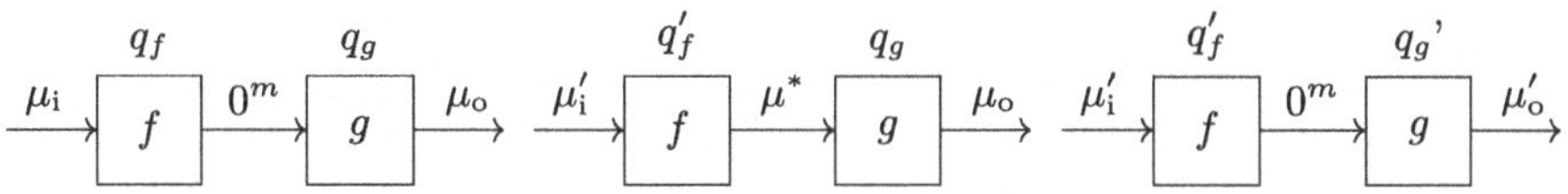

Fig. 11. Composition in the semantic category $\mathbb{S}_r$

We depict how composition intuitively works in Fig. 11. In the configuration on the left, before the Petri net starts evolving, there should be no "floating" tokens on the interface between f and g. The execution starts from f, which gives a new initial marking and internal states, as well as tokens on the interface between f and g, as depicted in the middle of Fig. 11. Then g is executed, but its execution should consume all the tokens on the interface, so that there are no tokens floating on the interface after execution, as depicted on the right of Fig. 11.

4.2 The Monoidal Categorical Structure

We now show that $\mathbb{S}_r$ has a monoidal structure as a first step towards showing that it is a TSMC.

Definition 14 (tensor product of $\mathbb{S}_r$). *Given $n, m \in \mathbb{N}$, we define $\otimes$ as the addition. Given $f_1 \colon m_1 \to n_1$ and $f_2 \colon m_2 \to n_2$, we define $f_1 \otimes f_2 \colon m_1 + m_2 \to n_1 + n_2$ as the morphism with the internal state $Q_1 \times Q_2$ and $(\mu_{\mathrm{i},1} \otimes \mu_{\mathrm{i},2}, (q_1, q_2), \mu_{\mathrm{o},1} \otimes \mu_{\mathrm{o},2}) \to_{f_1 \otimes f_2} (\mu'_{\mathrm{i},1} \otimes \mu'_{\mathrm{i},2}, (q'_1, q'_2), \mu'_{\mathrm{o},1} \otimes \mu'_{\mathrm{o},2})$ if and only if $(\mu_{\mathrm{i},i}, q_i \mu_{\mathrm{o},i}) \to_{f_i} (\mu'_{\mathrm{i},i}, q'_i, \mu'_{\mathrm{o},i})$ for $i \in \{1, 2\}$.*

Definition 15 (Symmetries of $\mathbb{S}_r$). *Given two pairs of natural numbers m and n, we define the symmetry $C_{m,n} : m+n \to n+m$ as the morphism whose set of internal states is $\{*\}$ and such that $(\mu_\mathrm{i} \otimes \nu_\mathrm{i}, *, \nu_\mathrm{o} \otimes \mu_\mathrm{o}) \to_{C_{n,m}} (\mu'_\mathrm{i} \otimes \nu'_\mathrm{i}, *, \nu'_\mathrm{o} \otimes \mu'_\mathrm{o})$ if and only if for all $i \in [m_b] + [m_r]$, there exists k such that $\mu'_\mathrm{i}(i) = \mu_\mathrm{i}(i) - k$ and $\mu'_\mathrm{o}(i) = \mu_\mathrm{o}(i) + k$, and similarly for all $i \in [n_b] + [n_r]$, there exists k such that $\nu'_\mathrm{i}(i) = \nu_\mathrm{i}(i) - k$ and $\nu'_\mathrm{o}(i) = \nu_\mathrm{o}(i) + k$.*

Finally, $\mathbb{S}_r$ also possesses a trace operator, as defined in

Definition 16 (trace of $\mathbb{S}_r$). *Given three pairs of natural numbers m, n, p and a morphism $f \colon p + m \to p + n$, the* trace $\mathrm{Tr}^p_{m,n}(f)$ *of f along p is defined as having Q as its set of internal states, and $(\mu_\mathrm{i}, q, \mu_\mathrm{o}) \to_{\mathrm{Tr}^p_{m,n}(f)} (\mu'_\mathrm{i}, q', \mu'_\mathrm{o})$ if and only if there exist $(\mu^0_\mathrm{i}, q^0, \mu^0_\mathrm{o}), \ldots, (\mu^n_\mathrm{i}, q^n, \mu^n_\mathrm{o})$ such that*

- $\mu^0_\mathrm{i} = 0^p \otimes \mu_\mathrm{i}$, $q^0 = q$, *and* $\mu^0_\mathrm{o} = 0^p \otimes \mu_\mathrm{o}$,

- $\mu_{\rm i}^n = 0^p \otimes \mu_{\rm i}'$, $q^n = q'$, *and* $\mu_{\rm o}^n = 0^p \otimes \mu_{\rm o}'$, *and*
- *for all* $j < n$, $(\mu_{\rm i}^{\leftarrow,j}, q^j, \mu_{\rm o}^{\rightarrow,j}) \rightarrow_f (\mu_{\rm i}^{j+1}, q^{j+1}, \mu_{\rm o}^{j+1})$, *where* $\mu_{\rm o}^{\rightarrow,j} = 0^p \otimes \mu^j_{{\rm o}|>p}$, $\mu_{\rm i}^{\leftarrow,j} = (\mu_{\rm i}^j + \mu_{\rm o}^j)_{|\leq p} \otimes \mu^j_{{\rm i}\,|>p}$, *and* $(\mu_i)_{i<n|P} = (\mu_i)_{i<n \wedge P(i)}$ *is the restriction of* μ *to those indices that satisfy* P.

This definition is somewhat similar to that of trace on **Rel** [10], but taking the internal state into account. It must also take into account the fact that the tokens that have reached a looping exit must be made available in the corresponding entry, which is encoded into $\mu_{\rm o}^{\rightarrow,j}$ and $\mu_{\rm i}^{\leftarrow,j}$. They can be viewed as "reloading" the Petri net by moving all tokens that have reached a looping exit to the corresponding looping entry: $\mu_{\rm o}^{\rightarrow,j}$ has no tokens on the first p exits and is equal to $\mu_{\rm o}^j$ on the other ones, while $\mu_{\rm i}^{\leftarrow,j}$ is equal to $\mu_{\rm i}^j + \mu_{\rm o}^j$ on the looping entries (gaining tokens from $\mu_{\rm o}^j$) and to $\mu_{\rm i}^j$ on the other ones.

Lemma 4. *$\mathbb{S}_r$ is a traced symmetric monoidal category.*

We need reflexivity and transitivity in Definition 12 to prove that $\mathrm{Tr}^p_{m,n}(-)$ is a trace operator.

4.3 The Semantic Functor $\mathcal{S}_r$

We now define the semantic function $\mathcal{S}_r$: **roPN** $\rightarrow \mathbb{S}_r$ that interprets rightward open Petri nets in the semantic category.

Definition 17 (the functor $\mathcal{S}_r$). *We define the* semantic functor $\mathcal{S}_r$: **roPN** $\rightarrow$ $\mathbb{S}_r$ *on objects by* $\mathcal{S}_r(n) = n$ *and on arrows* $N\colon m \rightarrow n$ *as* $\mathcal{S}_r(N)\colon m \rightarrow n$ *whose set of internal states is* $Q = \mathbb{N}^P$ *the set of all markings of* N, *and* $(\mu_{\rm i}, \mu, \mu_{\rm o}) \rightarrow_{\mathcal{S}_r(N)} (\mu_{\rm i}', \mu', \mu_{\rm o}')$ *if and only if* $(\mu_{\rm i}', \mu', \mu_{\rm o}')$ *is reachable from* $(\mu_{\rm i}, \mu, \mu_{\rm o})$ *in* N *by a sequence of firings and slidings.*

We aim to prove that $\mathcal{S}$ is a traced symmetric monoidal functor. The difficult points to prove are $\mathcal{S}_r(N; N') = \mathcal{S}_r(N); \mathcal{S}_r(N')$ and $\mathcal{S}_r(\mathrm{Tr}^p_{m,n}(N)) = \mathrm{Tr}^p_{m,n}(\mathcal{S}_r(N))$. We only give informal proof of these two identities.

We start by proving $\mathcal{S}_r(N; N') = \mathcal{S}_r(N); \mathcal{S}_r(N')$, which amounts to proving $\mathcal{S}_r(N; N')(c) = (\mathcal{S}_r(N); \mathcal{S}_r(N'))(c)$ for all $c = (\mu_{\rm i}, \mu, \mu_{\rm o})$. First, given two sequences $(\mu_{\rm i}, \mu_N, 0^n) \rightarrow \ldots \rightarrow (\mu_{\rm i}', \mu_N', \mu^*)$ in N and $(\mu^*, \mu_M, \mu_{\rm o}) \rightarrow \ldots \rightarrow (0^n, \mu_M', \mu_{\rm o}')$, it is simple to build a sequence $c \rightarrow \ldots \rightarrow (\mu_{\rm i}', \mu', \mu_{\rm o}')$ in $N; N'$ by concatenating the two sequences and turning pairs of transitions/slidings that touch the middle interface to a single transition/sliding in $N; N'$.

It is slightly more difficult to reconstruct μ^* and the sequences on N and N' from the sequence on $N; N'$, for which the following lemma is convenient.

Lemma 5 (Priority). *Given Petri nets* $N\colon m \rightarrow n$ *and* $N' : n \rightarrow p$, *a configuration* c *for* $N; N'$, $k \in T \cup (l_i)_{i \in [m_b]^+}$, $k' \in T' \cup (r_j)_{j \in [p_r]}$ *if* k' *is enabled on* c *and* k *is enabled on* $k'(c)$, *then* k *is enabled on* c, k' *on* $k(c)$, *and* $k'(k(c)) = k(k'(c))$.

This lemma is crucial, as given a sequence of transitions and slidings in $N; N'$, it guarantees that all the sliding and transitions in N can be prioritised over the ones in N'. This means we can always assume that all transitions and sliding of N are fired before those of N'. Given a sequence of transitions and slidings on $N; N'$, we can permute all entry slidings and transitions in N first. We can then rebuild μ^* by counting how many tokens are output by the input slidings and transitions in N, then reconstruct the sequence on N (resp. N') by copying those slidings in the sequence on $N; N'$ that are in N (resp. N'). This proves that $\mathcal{S}_r(N; N') = \mathcal{S}_r(N); \mathcal{S}_r(N')$.

Finally, we move on to proving $\mathcal{S}_r(\mathrm{Tr}^p_{m,n}(N)) = \mathrm{Tr}^p_{m,n}(\mathcal{S}_r(N))$. Here again, the idea is to turn sequences $c \to \ldots \to c'$ in $\mathrm{Tr}^p_{m,n}(N)$ to sequences in N for the (semantic) trace and vice-versa. The only difficulty is managing tokens on the looping part of the interface. Given a sequence $c \to \ldots \to c'$ in $\mathrm{Tr}^p_{m,n}(N)$ whose general term is $(\mu^j_{\mathrm{i}}, \mu^j, \mu^j_{\mathrm{o}})$, we can recreate a sequence for the trace by copying all transitions/slidings and "reloading" the net (see Definition 16) after each move $\max\{\phi_p(i) \mid i \in [p]\}$ times, so that reloading moves no tokens anymore. Its general term is $(\tilde{\mu}^j_{\mathrm{i}} \otimes \mu^j_{\mathrm{i}}, \mu^j, \tilde{\mu}^j_{\mathrm{o}} \otimes \mu^j_{\mathrm{o}})$, and is a witness that $c \to \ldots \to c'$ in the trace. The other direction is simpler. Given a sequence of configurations such that $(\mu^{\to,j}_{\mathrm{i}}, \mu^j, \mu^{\to,j}_{\mathrm{o}}) \to (\mu^{j+1}_{\mathrm{i}}, \mu^{j+1}, \mu^{j+1}_{\mathrm{o}})$ in N, we can create a sequence for $\mathrm{Tr}^p_{m,n}(N)$ by simply copying all firings/slidings.

5 The Semantic Interpretation Functor $[\![.]\!]$

5.1 Moving to the Bidirectional Case

One idea of this line of work [20–22] is that the bidirectional framework can be derived from the unidirectional one automatically using the Int-construction, instead of being an ad hoc construction. Here, we define the category **oPN** of open Petri nets from that of rightward open Petri nets. This simplifies the development, readability, and conciseness of the ideas, in addition to giving a compact closed structure "for free" to **oPN** (and the semantic domain $\mathbb{S}$).

The basic idea of the Int-construction [10] is that objects of $\mathrm{Int}(\mathbb{C})$ are pairs (X, Y) of objects of $\mathbb{C}$ where X represents objects going "forward", while Y represents objects going "backward". Morphisms from (X, Y) to (X', Y') in $\mathrm{Int}(\mathbb{C})$ are morphisms $X \otimes Y' \to Y \otimes X'$ (notice that the Y's are reversed). In our case, this allows us to model open ends going leftward, allowing for simpler modelling especially for looping structures, which do not need to use trace explicitly.

Definition 18. *We define* **oPN** $= \mathrm{Int}(\mathbf{roPN})$ *and* $\mathbb{S} := \mathrm{Int}(\mathbb{S}_r)$ *the category of open Petri nets and semantic category, respectively. Moreover, because* $\mathcal{S}_r$ *respects the traced symmetric monoidal structure, the* Int*-construction applies to it too, and we define* $\mathcal{S} = \mathrm{Int}(\mathcal{S}_r)\colon$ **oPN** $\to \mathbb{S}$.

From the property of the Int-construction, it follows that :

Lemma 6. **oPN** *and* $\mathbb{S}$ *are compact closed categories and* $\mathcal{S}$ *is a compact closed functor.*

This lemma gives a formal meaning to the claim we have been making that the bidirectional framework is derived canonically from the unidirectional one, and gives meaning to Fig. 1.

5.2 The Coloured Graphical PROP

Until now, we have considered **oPN** (or **roPN**) to be our syntax and $\mathbb{S}$ to be our semantics. However, the definitions of the different elements of **oPN** such as composition and trace are hard to read and to work with, and they obscure their graphical essences. In this section, we give a simpler syntax for open Petri nets based on free coloured PROPs.

PROPs are a categorical concept widely used in many works to describe various graphical structures and fit well to define a categorical syntax in the form of string diagrams. Precisely, a PROP is a symmetric monoidal category whose objects are the elements of the free monoid on a single object. Concretely, the elements are built as the powers of a single base object; hence, every object admits a unique decomposition. In particular, this allows us to draw string diagrams where wires do not carry specific information.

This formalism is expanded upon by considering the free monoid on a finite set C (called the set of *colours*) in an object called a coloured PROP [7]. Concretely, its objects are lists of elements of C. Here, the usage of coloured PROPs is doubly beneficial: because of the nature of Petri nets, we need to consider two colours (as explained before) for the arrows, and we have to double the number of colours (going from 2 to 4) to encompass the bidirectional setting. We define our syntax category as a free PROP on a set of generators Σ^C_{PN} and equations E^C_{PN}. A similar approach is presented in [20–22], from which we will adapt the results without fully detailing them.

Let $C = \{\bullet_L, \bullet_R, \bullet_L, \bullet_R\}$ be the set of colours. The colour $\bullet_L$ (resp. $\bullet_R$) correspond to red arrows going leftward (resp. red arrows going rightward), and analogously for $\bullet_L$ and $\bullet_L$. We now move on to defining the signature Σ^C_{PN}, containing the generators. They are illustrated in Fig. 12.

Definition 19 (Σ^C_{PN}). *We pose $\Sigma^C_{PN} : C^* \times C^* \to$ **Set** the functor defined by $\Sigma^C_{PN}(\epsilon, \bullet_R\bullet_L) = \{\eta_b, t_{0,2}\}$, $\Sigma^C_{PN}(\bullet_L\bullet_R, \epsilon) = \{\epsilon_b, p_{2,0}\}$, $\Sigma^C_{PN}(\epsilon, \bullet_R\bullet_L) = \{\eta_r, p_{0,2}\}$, $\Sigma^C_{PN}(\bullet_L\bullet_R, \epsilon) = \{\epsilon_r, t_{2,0}\}$, $\Sigma^C_{PN}(u, v) = \{p_{|u|,|v|}\}$ if $u \in \bullet^*_L$ and $v \in \bullet^*_L$, $\Sigma^C_{PN}(u, v) = \{t_{|u|,|v|}\}$ if $u \in \bullet^*_L$ and $v \in \bullet^*_L$, and $\Sigma^C_{PN}(u, v) = \emptyset$ otherwise.*

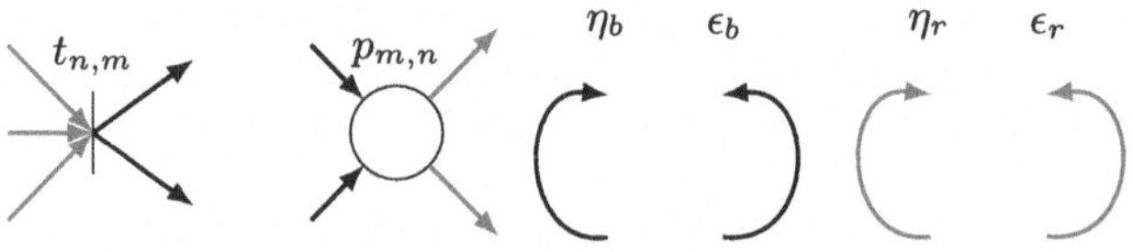

Fig. 12. The generators in Σ^C_{PN}

We define E^C_{PN} to be the set containing the equations $(\epsilon_b \otimes \mathrm{id}_{\bullet_L}) \circ (\mathrm{id}_{\bullet_L} \otimes \eta_b) = \mathrm{id}_{\bullet_L}$, $(\mathrm{id}_{\bullet_R} \otimes \epsilon_b) \circ (\eta_b \otimes \mathrm{id}_{\bullet_R}) = \mathrm{id}_{\bullet_R}$, $(\epsilon_r \otimes \mathrm{id}_{\bullet_L}) \circ (\mathrm{id}_{\bullet_L} \otimes \eta_r) = \mathrm{id}_{\bullet_L}$, $(\mathrm{id}_{\bullet_R} \otimes \epsilon_r) \circ (\eta_r \otimes \mathrm{id}_{\bullet_R}) = \mathrm{id}_{\bullet_R}$, i.e. the black and red snake equations for ϵ_b, ϵ_r, η_b, η_r. Our syntactic PROP is $\mathbb{F}(\Sigma^C_{PN}, E^C_{PN})$, where $\mathbb{F}$ is defined as the free construction from signatures (pairs of generators and equations) to PROPs detailed in [7]. The free property of $\mathbb{F}(\Sigma^C_{PN}, E^C_{PN})$ is that any valuation V of Σ^C_{PN} to a PROP $\mathbb{C}$ (i.e., a pair of a set function from colours to objects of $\mathbb{C}$ and a set function from Σ^C_{PN} to morphisms of $\mathbb{C}$ of the corresponding type) uniquely extends to a PROP morphism $\mathbb{F}(\Sigma^C_{PN}, E^C_{PN}) \to \mathbb{C}$.

This freeness allows us to define the realisation functor $\mathcal{R}\colon \mathbb{F}(\Sigma^C_{PN}, E^C_{PN}) \to$ **oPN** canonically, by considering the valuation $V\colon \Sigma^C_{PN} \to \mathrm{Int}(\mathbf{roPN})$ defined by $V(\epsilon_b) = V(\eta_b) = \mathrm{id}_{(1,0)}$, $V(\epsilon_r) = V(\eta_r) = \mathrm{id}_{(0,1)}$, $V(p_{m,n})$ the open Petri net $m \to n$ with a single place with n inputs and m outputs and no transitions, and $V(t_{m,n})$ the open Petri net $m \to n$ with no places and a single transition with n inputs and m outputs.

Not only is $\mathbb{F}(\Sigma^C_{PN}, E^C_{PN})$ the free PROP over $(\Sigma^C_{PN}, E^C_{PN})$ by definition, it also possesses a more interesting property here.

Lemma 7. *There is a signature Σ such that $\mathbb{F}(\Sigma^C_{PN}, E^C_{PN}) \cong \mathrm{Int}(\mathbb{F}_{tr}(\Sigma))$ the free traced symmetric monoidal category over Σ.*

The nature of the signature Σ is not useful on its own for this work but is developed in [20]. This theorem allows us to consider the second universal property of $\mathbb{F}(\Sigma^C_{PN}, E^C_{PN})$, as it is also a free compact closed category. First, we take the valuation V defined earlier and name $\mathcal{R}$ the unique compact closed functor that makes the leftmost triangle in Fig. 13 commute. Second, we take the valuation $\mathcal{S} \circ V : \Sigma^C_{PN} \to \mathbb{S}$, and we name $[\![.]\!] : \mathbb{F}(\Sigma^C_{PN}, E^C_{PN}) \to \mathbb{S}$ the unique functor that makes the larger triangle in Fig. 13 commute.

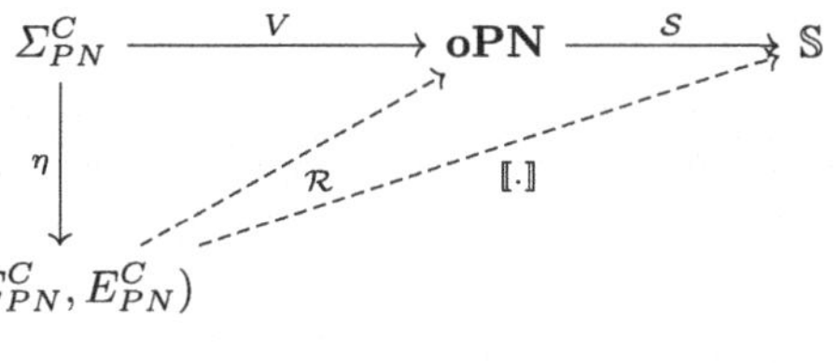

Fig. 13. The construction of $\mathcal{R}$ and $[\![.]\!]$

This means that both $\mathcal{S} \circ \mathcal{R}$ and $[\![.]\!]$ satisfies the lifting property of $\mathbb{F}(\Sigma^C_{PN}, E^C_{PN})$ as the free compact closed category, hence the following result.

Theorem 2. $[\![.]\!] = \mathcal{S} \circ \mathcal{R}$.

We could have defined $[\![.]\!]$ as $\mathcal{S} \circ \mathcal{R}$, but this would have hidden the fact that $[\![.]\!]$ is given by a freeness property and can thus be computed inductively. This property is important to solve the open reachability problem inductively.

6 Conclusion and Future Work

In this work, we have designed a compositional approach to Petri nets to solve open reachability. To achieve such a result, we have defined a category of open Petri nets, as well as a graphical language of string diagram and a semantic domain in which open Petri nets can be interpreted.

Directions for future work include the bounded case, in which there is a finite limit to the number of tokens present on places/open ends. However, semantic composition is more complex in this case. Moreover, we could implement an algorithm in the bounded case. Finally, a general work on compositionality, following [20,21] and the current work, would prove useful to design frameworks for open structures without starting from scratch every time.

Acknowledgments. The authors are supported by CREST ZT-IoT Project (No. JPMJCR21M3), ERATO HASUO Metamathematics for Systems Design Project (No. JPMJER1603), and ASPIRE Grant No. JPMJAP2301, JST.

References

1. Baez, J.C., Coya, B., Rebro, F.: Props in network theory. Theory Appl. Categ. **33**(25), 727–783 (2018)
2. Baez, J.C., Maste, J.: Open Petri nets. Math. Struct. Comput. Sci. **30**(3), 314–341 (2020)
3. Baez, J.C., Pollard, B.S.: A compositional framework for reaction networks. Rev. Math. Phys. **29**(09), 1750028 (2017)
4. Bonchi, F., Holland, J., Piedeleu, R., Sobociński, P., Zanasi, F.: Diagrammatic algebra: from linear to concurrent systems. Proc. ACM Program. Lang. **3**(POPL), 1–28 (2019)
5. Bonchi, F., Sobociński, P., Zanasi, F.: Interacting Hopf algebras. J. Pure Appl. Algebra **221**(1), 144–184 (2017)
6. Carette, T., Horsman, D., Perdrix, S.: SZX-calculus: scalable graphical quantum reasoning. In: 44th International Symposium on Mathematical Foundations of Computer Science, MFCS 2019, 26–30 August 2019, Aachen, Germany, vol. 138, pp. 55:1–55:15 (2019)
7. Carette, T., Perdrix, S.: Colored props for large scale graphical reasoning (2020)
8. Fiore, M., Devesas Campos, M.: The algebra of directed acyclic graphs. In: Coecke, B., Ong, L., Panangaden, P. (eds.) Computation, Logic, Games, and Quantum Foundations. The Many Facets of Samson Abramsky. LNCS, vol. 7860, pp. 37–51. Springer, Heidelberg (2013). https://doi.org/10.1007/978-3-642-38164-5_4
9. Joyal, A., Street, R.: Braided tensor categories. Adv. Math. **102**(1), 20–78 (1993)
10. Joyal, A., Street, R., Verity, D.: Traced monoidal categories. In: Mathematical Proceedings of the Cambridge Philosophical Society, vol. 119, pp. 447–468. Cambridge University Press (1996)
11. Mac Lane, S.: Categories for the Working Mathematician, vol. 5. Springer, New York (2013). https://doi.org/10.1007/978-1-4757-4721-8
12. Penrose, R.: Applications of negative dimensional tensors. Comb. Math. Appl. **1**, 221–244 (1971)
13. Peterson, J.L.: Petri nets. ACM Comput. Surv. (CSUR) **9**(3), 223–252 (1977)

14. Rathke, J., Sobociński, P., Stephens, O.: Compositional reachability in petri nets. In: Ouaknine, J., Potapov, I., Worrell, J. (eds.) RP 2014. LNCS, vol. 8762, pp. 230–243. Springer, Cham (2014). https://doi.org/10.1007/978-3-319-11439-2_18
15. Reisig, W.: Petri Nets: An Introduction, vol. 4. Springer, Heidelberg (2012). https://doi.org/10.1007/978-3-642-69968-9
16. Reisig, W.: Understanding Petri Nets. Springer, Heidelberg (2016). https://doi.org/10.1007/978-3-642-33278-4
17. Selinger, P.: A survey of graphical languages for monoidal categories. In: Coecke, B. (ed.) New Structures for Physics, pp. 289–355. Springer, Heidelberg (2010). https://doi.org/10.1007/978-3-642-12821-9_4
18. Sobociński, P.: Compositional model checking of concurrent systems, with petri nets. arXiv preprint arXiv:1603.00976 (2016)
19. Stephens, O.: Compositional specification and reachability checking of net systems. Ph.D. thesis, University of Southampton (2015)
20. Watanabe, K., Eberhart, C., Asada, K., Hasuo, I.: A compositional approach to parity games. arXiv preprint arXiv:2112.14058 (2021)
21. Watanabe, K., Eberhart, C., Asada, K., Hasuo, I.: Compositional probabilistic model checking with string diagrams of MDPs. In: 35th International Conference on Computer Aided Verification (CAV 2023) (2023)
22. Watanabe, K., Eberhart, C., Asada, K., Hasuo, I.: Compositional solution of mean payoff games by string diagrams. arXiv preprint arXiv:2307.08034 (2023)

Correspondence Between Composite Theories and Distributive Laws

Aloïs Rosset[1(✉)], Maaike Zwart[2], Helle Hvid Hansen[3], and Jörg Endrullis[1]

[1] Vrije Universiteit Amsterdam, Amsterdam, Netherlands
{a.rosset,j.endrullis}@vu.nl
[2] IT University of Copenhagen, Copenhagen, Denmark
[3] University of Groningen, Groningen, Netherlands
h.h.hansen@rug.nl

Abstract. Composite theories are the algebraic equivalent of distributive laws. In this paper, we delve into the details of this correspondence and concretely show how to construct a composite theory from a distributive law and vice versa. Using term rewriting methods, we also describe when a minimal set of equations axiomatises the composite theory.

Keywords: monad · distributive law · algebraic theory · composite theory · term rewriting

1 Introduction

Monads are categorical structures [4,20] with many applications in (co)algebraic approaches to program semantics, notably to model effects such as nondeterminism, probabilities and exceptions [6,17,24,27]. Monads that occur in the specification of programs and are used in reasoning about programs are often finitary and Set-based, and hence can be presented as algebraic theories [1,7,21].

The algebraic view on monads has been especially useful when studying monad compositions [8,14,25,26,36]. Composing monads is a way to combine multiple computational effects, and is usually done categorically via a distributive law [5,22]. However, the required distributive laws do not always exist, and the use of algebraic theories was instrumental in proving so-called no-go theorems, which tell us when two finitary monads cannot be composed via a distributive law [36].

Central to these results is the correspondence between composites of algebraic theories, and distributive laws between the corresponding monads. Briefly stated, a composite of two algebraic theories $\mathbb{S}$ and $\mathbb{T}$ is a theory $\mathbb{U}$ that contains all the function symbols and equations of $\mathbb{S}$ and $\mathbb{T}$ as well as a set of distribution axioms that specify how equality of mixed terms can be reduced to equality in $\mathbb{S}$ and $\mathbb{T}$. Composite theories were originally studied by Cheng [8] on the abstract level of Lawvere theories. Piróg & Staton [26] formulated them in the more concrete setting of algebraic theories.

Published by Springer Nature Switzerland AG 2024
B. König and H. Urbat (Eds.): CMCS 2024, LNCS 14617, pp. 194–215, 2024.
https://doi.org/10.1007/978-3-031-66438-0_10

While Piróg & Staton state the correspondence between composite theories and distributive laws, they do not provide a proof, referring instead to Cheng. In her thesis, Zwart [35] gives a constructive version of this correspondence for the category Set, but she does not prove directly that the algebraic theory she constructs from a distributive law is indeed a composite theory.

Furthermore, the theory Zwart constructs is given via a set E_λ that contains all possible equations with interaction between the theories $\mathbb{S}$ and $\mathbb{T}$. While this axiomatisation does the job, it is neither elegant nor practical to work with. Composite theories can often be described in terms of a few simple distribution axioms. A classic example is the theory of rings, which is a composite of the theories of monoids and Abelian groups via the two 'times over plus' distribution axioms. A systematic approach to identify such a minimal set of distribution axioms for a composite theory would be far more practical than the set E_λ.

In this paper, we present a full and self-contained proof of the correspondence between composite theories $\mathbb{U}$ (of $\mathbb{T}$ after $\mathbb{S}$) and distributive laws $\lambda\colon ST \to TS$, where $\mathbb{S}$ and $\mathbb{T}$ are algebraic theories and S, T are their corresponding finitary Set-monads. Section 4 shows how to get a distributive law from a composite theory, and Sect. 5 shows how to construct a composite theory from a distributive law. The proof of the latter uses term rewriting techniques. In particular, we introduce *functor rewriting systems* in order to reason about strings of functors, and to obtain a separation of $\mathbb{U}$-terms.

In addition, in Sect. 6 we give criteria that ensure that a certain minimal set of distribution axioms $E' \subseteq E_\lambda$ suffices to axiomatise $\mathbb{U}$. The natural candidate for E' consists of equations in which the left-hand side is a term consisting of exactly one $\mathbb{S}$-operation symbol, which has exactly one $\mathbb{T}$-operation symbol among its arguments. We prove that if a term rewriting system corresponding to E' is terminating, then $E_\mathbb{S} \cup E_\mathbb{T} \cup E'$ axiomatises $\mathbb{U}$. To illustrate that this criterion is not trivially satisfied, we give an example in which E' does not terminate and indeed does not axiomatise $\mathbb{U}$. Finally, we show that we have termination if the right-hand sides of the equations in E' are of a certain form, and apply our results to establish presentations of some composite monads/theories.

2 Preliminaries

We assume that the reader is familiar with basic notions of category theory [3,20,28]. This section recalls basic definitions and results concerning monads, algebraic theories, and term rewriting systems, and fixes notation for the concepts we use in this paper.

2.1 Monads

Definition 1. *A* **monad** (M, η, μ) *on a category* C *is a triple consisting of an endofunctor* $M : \mathsf{C} \to \mathsf{C}$*, and two natural transformations, the* **unit** $\eta : \mathrm{id} \Rightarrow M$ *and the* **multiplication** $\mu : M^2 \Rightarrow M$ *that make* (1) *and* (2) *commute. For convenience, we often refer to a monad* (M, η, μ) *by its functor part* M*.*

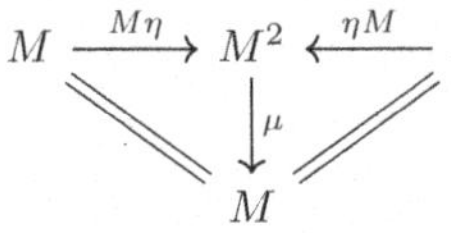

(1)

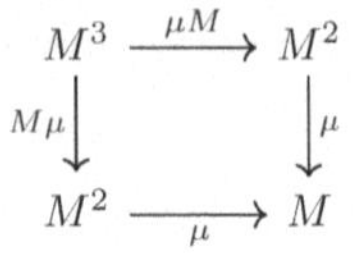

(2)

Example 2. Here are some examples of Set-monads, where we always mean the finitary versions. For more details on these monads, see e.g. [13, Sect. 1.2.1].

- The *list* and *non-empty list* monads L and L^+, with $\eta^L_X(x) = \eta^{L^+}_X(x) = [x]$, and $\mu^L = \mu^{L^+}$ being concatenation.
- The *multiset* monad $\mathcal{M}$, with $\eta^{\mathcal{M}}(x) = \wr x \wr$ and $\mu^{\mathcal{M}}$ taking the union, adding multiplicities. Taking multiplicities in $\mathbb{Z}$ gives the *Abelian group* monad $\mathcal{A}$.
- The *distribution* monad $\mathcal{D}$, with $\eta^{\mathcal{D}}(x) = 1x$ and a weighted average of $\mu^{\mathcal{D}}$.
- The *reader* monad $R_A(X) = X^A$, where A is a finite set, with η^R the constant function and μ^R reading the same element twice.

Definition 3. *Given two monads* (M, η^M, μ^M) *and* (T, η^T, μ^T) *on a category* C, *a* **monad morphism** *from* M *to* T *is a natural transformation* $\theta : M \Rightarrow T$ *that makes* (3) *and* (4) *commute, where* $\theta\theta := \theta_T \cdot M\theta = T\theta \cdot \theta_M$ *(called* horizontal composition*). If each component of* θ *is an isomorphism, we say that the two monads are* **isomorphic**.

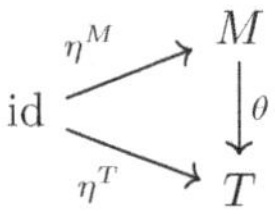

(3)

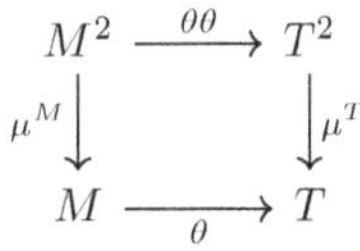

(4)

Definition 4. *Let* (M, η, μ) *be a monad on category* C. *An (Eilenberg-Moore)* M-**algebra** *is a* C*-morphism* $\alpha : MX \to X$ *for some* $X \in$ C, *denoted* (X, α) *for short, such that* (5) *and* (6) *commute. An* M*-algebra* **homomorphism** $f : (X, \alpha) \to (Y, \beta)$ *between two* M*-algebras is a function* $f : X \to Y$ *such that* (7) *commutes. The category of* M*-algebras and* M*-algebra homomorphisms is denoted* **EM**(M) *and called the* **Eilenberg-Moore category** *of* M.

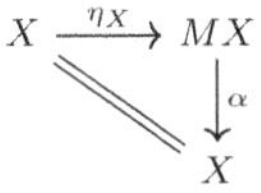

(5)

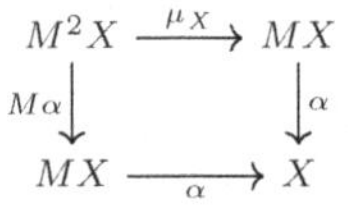

(6)

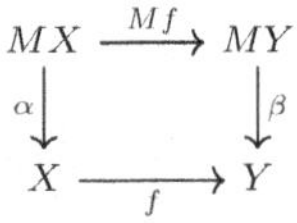

(7)

Definition 5. *Let* S, T *be monads. A* **distributive law** $\lambda : ST \Rightarrow TS$ *between monads is a natural transformation satisfying* (8)–(11). *A* **weak distributive law** $\lambda : ST \Rightarrow TS$ *is a natural transformation satisfying* (9)–(11).

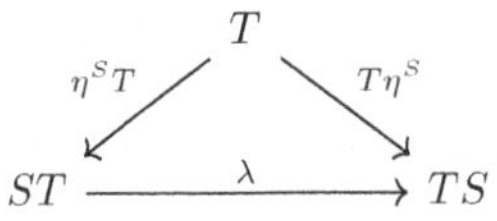

(8)

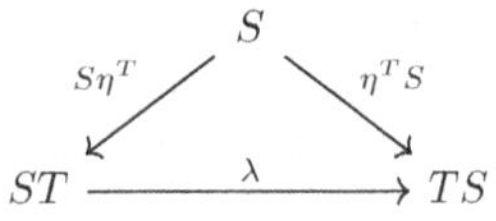

(9)

$$\begin{array}{ccccc} SST & \xrightarrow{S\lambda} & STS & \xrightarrow{\lambda S} & TSS \\ \downarrow{\scriptstyle \mu^S T} & & & & \downarrow{\scriptstyle T\mu^S} \\ ST & & \xrightarrow{\lambda} & & TS \end{array} \tag{10}$$

$$\begin{array}{ccccc} STT & \xrightarrow{\lambda T} & TST & \xrightarrow{T\lambda} & TTS \\ \downarrow{\scriptstyle S\mu^T} & & & & \downarrow{\scriptstyle \mu^T S} \\ ST & & \xrightarrow{\lambda} & & TS \end{array} \tag{11}$$

A distributive law $\lambda : ST \to TS$ induces a monad structure on the functor TS as follows [5, Sect. 1]:

$$\Big(TS,\ \eta^{TS} := \big(\mathrm{id} \xrightarrow{\eta^T\eta^S} TS\big),\ \mu^{TS} := \big(TSTS \xrightarrow{T\lambda S} TTSS \xrightarrow{\mu^T\mu^S} TS\big)\Big) \tag{12}$$

The algebras for this composite monad are algebras that are simultaneously S-algebras and T-algebras. This is visible through the isomorphism $\mathbf{EM}(TS) \cong \mathbf{Alg}(\lambda)$ [5, Sect. 2], where the category $\mathbf{Alg}(\lambda)$ of λ-algebras is defined as follows:

Definition 6. *Given monads S, T and distributive law $\lambda : ST \to TS$, then the objects of the category $\mathbf{Alg}(\lambda)$ are triples (X, σ, τ), such that (X, σ) is an S-algebra and (X, τ) is a T-algebra, and the diagram on the right commutes. The morphisms of $\mathbf{Alg}(\lambda)$ are C-morphisms that are both S- and T-algebra homomorphisms.*

$$\begin{array}{ccccc} STX & & \xrightarrow{\lambda} & & TSX \\ {\scriptstyle S\tau}\downarrow & & & & \downarrow{\scriptstyle T\sigma} \\ SX & & & & TX \\ & {\scriptstyle \sigma}\searrow & & \swarrow{\scriptstyle \tau} & \\ & & X & & \end{array}$$

2.2 Algebraic Theories

Definition 7. *An* **algebraic theory** *is a pair (Σ, E) consisting of an algebraic signature Σ and set of equations E over Σ defined as follows.*

- *An* **algebraic signature** *Σ is a set of operation symbols. Each $\mathsf{op}^{(n)} \in \Sigma$ has an arity $n \in \mathbb{N}$.*
- *The set $\mathcal{T}(\Sigma, X)$, also denoted $\Sigma^* X$, of* **Σ-terms** *over a set X is defined inductively: elements in X are terms, and given terms $t_1, \ldots, t_n$ and $\mathsf{op}^{(n)} \in \Sigma$, then $\mathsf{op}(t_1, \ldots, t_n)$ is a term.*
- *An* **equation** *over a signature Σ is a pair (s, t) of Σ-terms.*

For the rest of this paper, we fix a set $\mathcal{V} = \{v_1, v_2, v_3, \ldots\}$ of variables. The subset of $\mathcal{V}$ appearing in a term t is denoted as $\mathsf{var}(t)$. Functions of the form $v : \mathcal{V} \to Y$ are called **variable assignments**.

Notation 8. In this paper we make heavy use of substitutions. For readability, we pick from the following notations for substitutions, depending on context. Given terms $t(x_1, .., x_n)$ and $s_1, \ldots, s_n$, and variable assignment $h : \mathcal{V} \to \mathcal{T}(\Sigma, \mathcal{V})$ defined as $x_1 \mapsto s_{x_1}, \ldots, x_n \mapsto s_{x_n}$ and identity elsewhere, we denote the term t where each x_i is substituted with s_i (for $i = 1, \ldots, n$) by either $t[h]$, $t[s_1, ..., s_n]$, or $t[s_x/x]$ (or even $t[s_x]$) for short, where x ranges over all variables in t. Moreover, given a family of terms $(t_x[s_{x,y}/y])_{x \in X}$, we will simply write each term $t_x[s_y]$, as we can assume that each t_x has distinct variables by choosing the (say m) variables of t_{x_1} to be $y_1, \ldots, y_m$, the variables of t_{x_2} to start at y_{m+1}, and so on.

Definition 9. *The category* $\mathbf{Alg}(\Sigma, E)$ *consists of* (Σ, E)*-algebras and homomorphisms between them.*

- *A* Σ-**algebra** *is a pair* $(X, [\![\cdot]\!])$ *consisting of a set* X *and a collection of interpretations: for each* $\mathsf{op}^{(n)} \in \Sigma$, *we have* $[\![\mathsf{op}]\!] : X^n \to X$. *Any function* $f : X \to Y$ *extends to a unique homomorphism,* $[\![\cdot]\!]_f : \mathcal{T}(\Sigma, X) \to Y$, *as given by Eqs.* (13) *and* (14) *below. When* $f = \mathrm{id}_X$, *we omit the subscript.*

$$[\![x]\!]_f := f(x), \text{ and} \tag{13}$$

$$[\![\mathsf{op}(t_1, \ldots, t_n)]\!]_f := [\![\mathsf{op}]\!]([\![t_1]\!]_f, \ldots, [\![t_n]\!]_f). \tag{14}$$

- *A* (Σ, E)-**algebra** $(X, [\![\cdot]\!])$ *is a* Σ*-algebra whose* $[\![\cdot]\!]$ *satisfies all equations in* E, *i.e., for each* $(s, t) \in E$ *and all variable assignments* v, $[\![s]\!]_v = [\![t]\!]_v$.
- *A* (Σ, E)*-algebra* **homomorphism** $f : (X, [\![\cdot]\!]) \to (X', [\![\cdot]\!]')$ *is a function* $f : X \to X'$ *such that* $f[\![\mathsf{op}]\!] = [\![\mathsf{op}]\!]' f^n$, *for all* $\mathsf{op}^{(n)} \in \Sigma$.

Given an algebraic theory $\mathbb{T} = (\Sigma_{\mathbb{T}}, E_{\mathbb{T}})$ and $\Sigma_{\mathbb{T}}$-terms s and t, we write $s =_{\mathbb{T}} t$ to denote that the equality $s = t$ is derivable from the axioms $E_{\mathbb{T}}$ in equational logic. The inference rules of equational logic are in [30, Sect. 8.1].

Definition 10. *There is a free-forgetful adjunction* $F\colon \mathsf{Set} \rightleftarrows \mathbf{Alg}(\Sigma, E) : U$.

- *The* **free**(Σ, E)**–algebra** *on set* X *is the* (Σ, E)*-algebra* $(\mathcal{T}(\Sigma,X)/{=_{(\Sigma,E)}}, [\![\cdot]\!])$ *with carrier* $\mathcal{T}(\Sigma, X)$ *modulo* $=_{(\Sigma,E)}$. *The equivalence class of a term* t *is denoted* $\overline{t}^{(\Sigma,E)}$ *or* $\overline{t}$ *if the theory is clear from context. The interpretation of* $\mathsf{op}^{(n)} \in \Sigma_{\mathbb{T}}$ *is* $[\![\mathsf{op}]\!](\overline{t_1}, \ldots, \overline{t_n}) := \overline{\mathsf{op}(t_1, \ldots, t_n)}$.
- *The* **free functor** $F : \mathsf{Set} \to \mathbf{Alg}(\Sigma, E)$ *sends* X *to its free* (Σ, E)*-algebra, and any function* $f : X \to Y$ *to* $Ff : FX \to FY$ *defined by* $Ff(\overline{t}) := \overline{t[f]}$.

The fact that F *is a well-defined functor is well-known and an account of it is provided in the extended version of the paper [30]. Composing the adjoint functors gives a monad* $(T := UF, \eta, \mu)$, *called the* **free algebra monad** *[20, VI.1]. The unit is* $\eta : x \mapsto \overline{x}$ *and the multiplication is* $\mu : \overline{t[\overline{t_i}/v_i]} \mapsto \overline{t[t_i/v_i]}$.

Definition 11 ([29, Def. 5, Lem. 8])**.** *An algebraic theory* (Σ, E) *is an* **algebraic presentation** *of a* Set*-monad* (M, η^M, μ^M) *if we have an isomorphism of monads* $(T, \eta^T, \mu^T) \cong (M, \eta^M, \mu^M)$, *where* T *is the free algebra monad of* (Σ, E). *An equivalent formulation is that both categories of algebras are concretely isomorphic*[1]*:* $\mathbf{EM}(M) \cong_{\mathsf{conc}} \mathbf{Alg}(\Sigma, E)$. *The former isomorphism relates the monads on a syntactic level, whereas the latter relates them semantically.*

Note that a monad can have multiple presentations.

Example 12. Here are algebraic presentations of the monads from Example 2.

[1] "concrete" means that both functors of this isomorphism commute with the forgetful functors $\mathbf{EM}(M) \to \mathsf{Set}$ and $\mathbf{Alg}(\Sigma, E) \to \mathsf{Set}$. In other words it sends an M-algebra $(X, x : MX \to X)$ to a (Σ, E)-algebra with same carrier $(X, [\![\cdot]\!])$ and vice-versa.

- The *list* monad L is presented by the theory of *monoids*.
- The *non-empty list* monad L^+ is presented by the theory of *semigroups*.
- The *multiset* monad $\mathcal{M}$ is presented by the theory of *commutative monoids*.
- The *Abelian group* monad $\mathcal{A}$ is presented by the theory of *Abelian groups*.
- The *distribution* monad $\mathcal{D}$ is presented by the theory of *convex algebras* [15].
- The *reader* monad R_A is presented by the theory of *local states* [27] consisting of a single $|A|$-ary operation symbol, satisfying idempotence and diagonal equations (e.g. in the case $|A| = 2$: $a * a = a$ and $(a * b) * (c * d) = (a * d)$).

2.3 Term Rewriting Systems

We only briefly explain the basic concepts and results of term rewriting systems (TRS) that we need in our proofs. For more background, we recommend the book "Term Rewriting Systems" by Terese [32].

Definition 13. *Given a signature Σ, a* **rewrite rule** *$(l \to r)$ is a pair of Σ-terms (l, r) such that l is not a variable, and all variables in the right occur also in the left:* $\mathsf{var}(l) \supseteq \mathsf{var}(r)$. *A* **term rewriting system** *$\mathcal{R} = (\Sigma, R)$ consists of a signature Σ and a set of rewrite rules R. The rewrite relation $\to_{\mathcal{R}}$ is the smallest relation on $\mathcal{T}(\Sigma, X)$ that contains $\mathcal{R}$ and is closed under substitution and under context.*[2] *We simply write $\to$ when $\mathcal{R}$ is clear from the context. The transitive and reflexive closure of $\to$ is written as $\twoheadrightarrow$. When all operation symbols in Σ have arity* 1, *then $\mathcal{R} = (\Sigma, R)$ is called a* **string rewriting system**.

Example 14. Let $\Sigma := \{0^{(0)}, s^{(1)}, +^{(2)}\}$ and $\mathcal{R} = \{x + 0 \to x,\ x + s(y) \to s(x + y)\}$. A rewrite sequence is for instance

$$s(s(0)) + s(0) \quad \to \quad s(s(s(0)) + 0) \quad \to \quad s(s(s(0))).$$

Definition 15. *Let $\mathcal{R} := (\Sigma, R)$ be a TRS.*

- $\mathcal{R}$ *is* **terminating** *or* **strongly normalising** (SN) *if every rewriting sequence is finite $t_0 \to t_1 \to \ldots \to t_n \not\to$.*

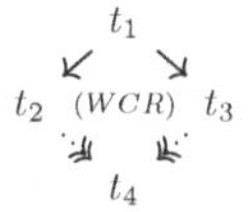

- $\mathcal{R}$ *is* **locally confluent** *or* **weak Church-Rosser** (WCR) *if for all terms t_1, t_2, t_3 with $t_2 \leftarrow t_1 \to t_3$, there exists a term t_4 with $t_2 \twoheadrightarrow t_4 \twoheadleftarrow t_3$.*
- $\mathcal{R}$ *is* **confluent** *or* **Church – Rosser** (CR) *if for all terms t_1, t_2, t_3 with $t_2 \twoheadleftarrow t_1 \twoheadrightarrow t_3$, there exists a term t_4 with $t_2 \twoheadrightarrow t_4 \twoheadleftarrow t_3$.*

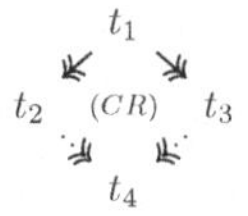

A term is called a *normal form*, if it cannot be rewritten any further. If a TRS is terminating (SN) and confluent (CR), then each term can be rewritten to a unique normal form.

A well-known result says that in the presence of termination, local confluence is enough to entail confluence.

[2] For the definition of context, see [32, Sect. 2.1.1].

Lemma 16 (Newman's Lemma). *If a TRS is terminating* (SN) *and locally confluent* (WCR), *then it is also confluent* (CR).

Two common techniques to prove termination are the *polynomial interpretation* over $\mathbb{N}$ [32, Sect. 6.2.2] and the *multiset path order* [31]. The idea of polynomial interpretation over $\mathbb{N}$ is to choose a Σ-algebra $(\mathbb{N}, [\![\cdot]\!])$ where every interpretation $[\![\mathsf{op}]\!]$ is a monotone polynomial on $\mathbb{N}$. If each rule (l, r) of a system is strictly decreasing, $[\![l]\!] > [\![r]\!]$, then termination follows by well-foundedness of $\mathbb{N}$.

Example 17. The TRS in Example 14 is terminating. To see this, take as polynomial interpretation for example $[\![0]\!] = 1$, $[\![s(x)]\!] = x+1$, and $[\![x+y]\!] = x+2y+1$. These polynomials are monotone and every rule is strictly decreasing:

$$[\![x+0]\!] = x + 2\cdot 1 + 1 = x + 3 > x = [\![x]\!],$$
$$[\![x+s(y)]\!] = x + 2y + 3 > x + 2y + 2 = [\![s(x+y)]\!].$$

The multiset path order method uses a decreasing sequence of multisets to show termination. We explain this briefly in the arXiv version of the paper [30].

A common technique for proving local confluence is to prove convergence of *critical pairs* [32, Sect. 2.7]. Informally, a critical pair is formed when two rewrite rules can be applied to the same term while overlapping on one or more function symbols, creating two different terms. A critical pair *converges* if the two mentioned terms can be rewritten to the same term.

Lemma 18 (Critical pair lemma). *A TRS is locally confluent* (WCR) *if and only if all its critical pairs converge.*

3 Composite Theories

We introduce the concept of *composite theories*. Our definition is slightly different from, but equivalent to, the original definition by Piróg & Staton [26, Def. 3] and equivalent formulations in Zwart's thesis [35, Def. 3.2, Prop. 3.4].

Definition 19. *Let* $\mathbb{U}, \mathbb{S}, \mathbb{T}$ *be algebraic theories. Suppose* $\mathbb{U}$ *contains* $\mathbb{S}$ *and* $\mathbb{T}$, *meaning* $\Sigma_{\mathbb{S}}, \Sigma_{\mathbb{T}} \subseteq \Sigma_{\mathbb{U}}$ *and* $E_{\mathbb{S}}, E_{\mathbb{T}} \subseteq E_{\mathbb{U}}$.

- *A* $\mathbb{U}$*-term is* **separated** *if it is of the form* $t[s_x/x]$, *where* t *is a* $\mathbb{T}$*-term and* $\{s_x \mid x \in \mathsf{var}(t)\}$ *is a family of* $\mathbb{S}$*-terms.*
- *Two separated terms* $t[s_x]$ *and* $t'[s'_y]$ *are* **equal** *modulo* $(\mathbb{S}, \mathbb{T})$ *if their TS-equivalence classes are equal in TSV:* $\overline{t[\overline{s_x}^{\mathbb{S}}]}^{\mathbb{T}} = \overline{t'[\overline{s'_y}^{\mathbb{S}}]}^{\mathbb{T}}$.
- $\mathbb{U}$ *is a* **composite theory** *of* $\mathbb{T}$ *after* $\mathbb{S}$ *if every* $\mathbb{U}$*-term* u *is equal to a separated term* $u =_{\mathbb{U}} t[s_x/x]$, *that we call a* **separation** *of* u, *and for any two separated terms* v, v', *if* $v =_{\mathbb{U}} v'$ *then* v *and* v' *must be equal modulo* $(\mathbb{S}, \mathbb{T})$.

Lemma 20. *For any two separated terms* $t[s_x/x]$ *and* $t'[s_y/y]$ *in a composite theory, the following are equivalent:*

1. $t[s_x/x]$ *and* $t'[s'_y/y]$ *are equal modulo* $(\mathbb{S}, \mathbb{T})$ *in the sense of Definition 19.*
2. $t[s_x/x]$ *and* $t'[s'_y/y]$ *are equal modulo* $(\mathbb{S}, \mathbb{T})$ *in the sense of [35, Definition. 3.2].* ■[3]

Example 21. Two $\mathbb{S}$-terms s and s' are equal modulo $(\mathbb{S}, \mathbb{T})$ if and only if $s =_{\mathbb{S}} s'$, and similarly for $\mathbb{T}$-terms.

Example 22. The prime example of a composite theory is the theory of rings $\mathbb{U} := \mathsf{Ring}$. It contains the theories $\mathbb{S} := \mathsf{Mon}$ of monoids and $\mathbb{T} := \mathsf{AbGrp}$ of Abelian groups. We recall their signatures to fix notation: $\Sigma_{\mathsf{Mon}} := \{\cdot^{(2)}, 1^{(0)}\}$ and $\Sigma_{\mathsf{AbGrp}} := \{0^{(0)}, +^{(2)}, -^{(1)}\}$. We sometimes omit the "multiplication" symbol $\cdot$ for simplicity. The signature of rings is given by $\Sigma_{\mathsf{Ring}} := \Sigma_{\mathsf{Mon}} \uplus \Sigma_{\mathsf{AbGrp}}$. The equations of rings are given by the equations of monoids, Abelian groups, and two distributivity axioms:

$$E_{\mathsf{Ring}} := E_{\mathsf{Mon}} \cup E_{\mathsf{AbGrp}} \cup \left\{ \begin{array}{l} x(y+z) = (xy) + (xz), \\ (y+z)x = (yx) + (zx) \end{array} \right\}.$$

A separated term $t[s_x/x]$ in Ring is an Abelian group term t, with monoid terms $\{s_x\}$ substituted for its variables. We give some examples of non-separated terms, of possible separations for them, and of equality modulo $(\mathsf{Mon}, \mathsf{AbGrp})$ between the separations.

The term $x(y+z)$ is non-separated. Possible separations are e.g. $xy + xz$ and $(xy + xz) + 0$. Both are equal modulo $(\mathsf{Mon}, \mathsf{AbGrp})$, as their monoid parts are identical and their Abelian group parts $t = (x_1 + x_2) + 0$ and $t' = x_1 + x_2$ are equal in the theory of Abelian groups.

The term $x \cdot 0$ is also non-separated. It is equal in Ring to the separated terms 0 and $(1 \cdot x) + (-(x \cdot 1))$. To see that these separations are equal modulo $(\mathsf{Mon}, \mathsf{AbGrp})$, notice that $1 \cdot x =_{\mathsf{Mon}} x \cdot 1$, and that the terms 0 and $x_1 + (-x_2)$ are equal in Abelian groups when $x_1 = x_2$. Thus: $\overline{0}^{\mathsf{AbGrp}} = \overline{(\overline{1 \cdot x}^{\mathsf{Mon}}) + (-(\overline{x \cdot 1}^{\mathsf{Mon}}))}^{\mathsf{AbGrp}}$.

We now show that distributive laws between monads correspond one-to-one to composite theories.

4 From Composite Theory to Distributive Law

We first show how to construct a distributive law from a given composite theory.

Theorem 23 ([35, Theorem 3.8]). *Let* $\mathbb{S}, \mathbb{T}$ *be algebraic theories with free algebra monads* S, T *respectively. Let* $\mathbb{U}$ *be a composite theory of* $\mathbb{T}$ *after* $\mathbb{S}$, *with free algebra monad* U. *Then the following defines a distributive law* $\lambda : ST \Rightarrow TS$ *such that* $\mathbb{U}$ *is an algebraic presentation of the resulting monad* TS, *where* $t'[s'_x]$ *is a separation of* $s[t_x]$:

$$\lambda_{\mathcal{V}} : ST\mathcal{V} \to TS\mathcal{V} : \quad \overline{s[\overline{t_x}^{\mathbb{T}}/x]}^{\mathbb{S}} \mapsto \overline{t'[\overline{s'_x}^{\mathbb{S}}/x]}^{\mathbb{T}}$$

[3] The symbol ■ denotes that the proof is in the extended version on arXiv [30].

Proof. Instead of directly checking the axioms for a distributive law, we prove an equivalent characterisation given by Beck [5, p. 122]. That is, we claim that there exist a natural transformation $\mu^{TS} : TSTS \Rightarrow TS$ such that:

(i) $(TS, \eta^{TS} := \eta^T \eta^S, \mu^{TS})$ is a monad.
(ii) The natural transformations $\eta^T S$ and $T\eta^S$ are monad morphisms.
(iii) The middle unitary law holds: $\mu^{TS} \cdot T\eta^S \eta^T S = \mathrm{id}\, TS$.

It follows then that the monad $(TS, \eta^T \eta^S, \mu^{TS})$ does indeed come from a distributive law, which is given by: $\lambda = \mu^{TS} \cdot \eta^T ST\eta^S$. A simple but tedious calculation shows that indeed $\lambda(\overline{s[\overline{t_x}^{\mathbb{T}}/x]}^{\mathbb{S}}) = \overline{t'[\overline{s'_x}^{\mathbb{S}}/x]}^{\mathbb{T}}$. The details of this calculation are in the extended version [30].

To define μ^{TS}, we use the fact that the *functors* U and TS are isomorphic. Indeed, since $\mathbb{U}$ is a composite theory, every $\mathbb{U}$-term u has a separation $u =_{\mathbb{U}} t[s_x/x]$. Hence $\phi : U \Rightarrow TS$ and $\psi : TS \Rightarrow U$ given below are inverse natural transformations. Using ϕ, ψ, and the multiplication μ^U, we can then define μ^{TS}.

$$\phi(u) := \overline{t[\overline{s_x}^{\mathbb{S}}/x]}^{\mathbb{T}} \tag{15}$$

$$\psi(\overline{t[\overline{s_x}^{\mathbb{S}}/x]}^{\mathbb{T}}) := \overline{t[s_x/x]}^{\mathbb{U}} \tag{16}$$

$$\mu^{TS} := (TSTS \xrightarrow{\psi\psi} UU \xrightarrow{\mu^U} U \xrightarrow{\phi} TS). \tag{17}$$

Notice that ϕ is well-defined, as the choice of the separation $t[s_x/x]$ does not matter by equality modulo $(\mathbb{S}, \mathbb{T})$. To see that ψ is also well-defined, take $\overline{t[\overline{s_x}^{\mathbb{S}}/x]}^{\mathbb{T}} = \overline{t'[\overline{s'_x}^{\mathbb{S}}/x]}^{\mathbb{T}}$. The $\mathbb{T}$- and $\mathbb{S}$-proofs of that equality are also $\mathbb{U}$-proofs by definition of $\mathbb{U}$, implying that $\overline{t[\overline{s_x}^{\mathbb{U}}/x]}^{\mathbb{U}} = \overline{t'[\overline{s'_x}^{\mathbb{U}}/x]}^{\mathbb{U}}$ and hence that $\overline{t[s_x/x]}^{\mathbb{U}} = \overline{t'[s'_x/x]}^{\mathbb{U}}$ by applying $\mu^{\mathbb{U}}$ on both sides. The proofs of (i)-(iii) are in [30]. □

5 From Distributive Law to Composite Theory

We now show how to construct a composite theory from a given distributive law.

Theorem 24. *Let S, T be two monads algebraically presented by two algebraic theories $\mathbb{S}$ and $\mathbb{T}$, respectively. Let $\lambda : ST \Rightarrow TS$ be a distributive law. We define a set E_λ of equations and a theory $\mathbb{U}^\lambda$ as follows [35, Definition 3.8].*

$$E_\lambda := \left\{ (s[t_x/x], t[s_y/y]) \mid \lambda_{\mathcal{V}}(\overline{s[\overline{t_x}^{\mathbb{T}}/x]}^{\mathbb{S}}) = \overline{t[\overline{s_y}^{\mathbb{S}}/y]}^{\mathbb{T}} \right\}.$$

$$\Sigma_{\mathbb{U}^\lambda} := \Sigma_{\mathbb{S}} \uplus \Sigma_{\mathbb{T}},$$

$$E_{\mathbb{U}^\lambda} := E_{\mathbb{S}} \cup E_{\mathbb{T}} \cup E_\lambda.$$

Then, $\mathbb{U}^\lambda$ is a composite theory of $\mathbb{T}$ after $\mathbb{S}$.

To prove Theorem 24, we observe that every $\mathbb{U}^\lambda$-term u can be assigned a regular set $\mathsf{type}(u)$ in $\{S, T\}^* \mathcal{V}$, expressing how u nests $\mathbb{S}$ and $\mathbb{T}$ operation symbols. We give an example below in Example 27. We obtain a TS-separated term by first mapping u to the equivalence class $\overline{u}$ in $\mathsf{type}(u)$, now viewed as a set. We then apply λ, μ^S and μ^T to $\overline{u}$ until we reach an equivalence class $\overline{t[\overline{s_x}^{\mathbb{S}}]}^{\mathbb{T}} \in TS\mathcal{V}$, where we use the axiom of choice to choose a representative $t[s_x]$. The axioms of the three natural transformations ensure that $\overline{t[\overline{s_x}^{\mathbb{S}}]}^{\mathbb{T}}$ does not depend on the order in which they were applied.

The termination of the procedure of applying λ, μ^S and μ^T and the uniqueness of $\overline{t[\overline{s_x}^{\mathbb{S}}]}^{\mathbb{T}}$ are intuitively clear, yet showing it formally is not trivial. In the following definitions we formalise the separation procedure that we described here. We then give a proof of termination using rewriting techniques. We denote string concatenation with "::".

Definition 25. *We define a function* $\mathsf{type} : \Sigma^*_{\mathbb{U}\lambda}\mathcal{V} \to \{S, T\}^*\mathcal{V}$ *recursively:*

- *For* $v \in \mathcal{V}$, *then* $\mathsf{type}(v) := \mathcal{V}$.
- *For* $s[u_1, \ldots, u_n]$, *where* $s \in \mathcal{T}(\Sigma_{\mathbb{S}}, \mathcal{V})$, *and* $u_1, \ldots, u_n \in \Sigma^*_{\mathbb{U}\lambda}\mathcal{V}$ *do not have an* $\mathbb{S}$*-symbol as root, let* w *be longest word in the set* $\{\mathsf{type}(u_1), \ldots, \mathsf{type}(u_n)\}$, *then* $\mathsf{type}(s[u_1, \ldots, u_n]) := S{::}w$.
- *The* $t[u_1, \ldots, u_n]$ *case, where* $u_1, \ldots, u_n$ *do not start with a* $\mathbb{T}$*-symbol, is dual.*

Informally, $\mathsf{type}(u)$ is the shortest string $w\mathcal{V}$ such that u belongs to an equivalence class in the set $w\mathcal{V}$. We will formally define this equivalence class in Definition 26 below. Furthermore, it can be seen that $\mathsf{type}(u)$ does not contain successive occurrences of S, similarly for T.

Definition 26. *For* $u \in \Sigma^*_{\mathbb{U}\lambda}\mathcal{V}$ *and* $w \in \{S, T\}^*$ *such that* $\mathsf{type}(u)$ *is a substring*[4] *of* $w\mathcal{V}$, *we recursively define* $\overline{u}^w \in w\mathcal{V}$:

- *For* $v \in \mathcal{V}$, $\overline{v}^{\varepsilon} := v$, $\overline{v}^{S::w'} := \overline{\overline{v}^{w'}}^{\mathbb{S}}$, *and* $\overline{v}^{T::w'} := \overline{\overline{v}^{w'}}^{\mathbb{T}}$.
- *For* $s[u_1, \ldots, u_n]$ *where* $s \in \Sigma^*_{\mathbb{S}}\mathcal{V}$, *and* $u_1, \ldots, u_n \in \Sigma^*_{\mathbb{U}\lambda}\mathcal{V}$ *that are either variables or have root symbols in* $\Sigma_{\mathbb{T}}$,

$$\overline{s[u_1, \ldots, u_n]}^{S::w'} := \overline{s[\overline{u_1}^{w'}, \ldots, \overline{u_n}^{w'}]}^{\mathbb{S}}.$$

- *The* $t[u_1, \ldots, u_n]$ *case, where* $u_1, \ldots, u_n$ *do not start with a* $\mathbb{T}$*-symbol, is dual.*

If $\mathsf{type}(u)$ *is not a substring of* $w\mathcal{V}$, *then* $\overline{u}^w$ *is undefined.*

Example 27. Take $f^{(2)} \in \Sigma_{\mathbb{S}}$, and $g^{(1)} \in \Sigma_{\mathbb{T}}$.
For $u := f(f(x, g(x)), g(f(x, x)))$, we have

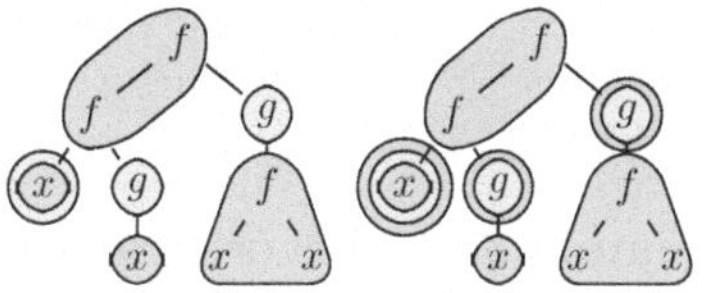

$$\mathsf{type}(u) = ST S\mathcal{V}.$$

$$\overline{u}^{STS} = \overline{f(f(\overline{\overline{x}^{\mathbb{S}}}^{\mathbb{T}}, \overline{g(\overline{x}^{\mathbb{S}})}^{\mathbb{T}}), g(\overline{\overline{f(x,x)}^{\mathbb{S}}}^{\mathbb{T}}))}^{\mathbb{S}}.$$

$$\overline{u}^{SSTS} = \overline{f(f(\overline{\overline{\overline{x}^{\mathbb{S}}}^{\mathbb{T}}}^{\mathbb{S}}, \overline{\overline{g(\overline{x}^{\mathbb{S}})}^{\mathbb{T}}}^{\mathbb{S}}), g(\overline{\overline{\overline{f(x,x)}^{\mathbb{S}}}^{\mathbb{T}}}^{\mathbb{S}}))}^{\mathbb{S}}.$$

[4] We say that w is a substring of w' if w can be obtained by deleting zero or more letters from w'.

Before we formalise the remainder of the separation procedure, we interpret functors and natural transformations as a term rewriting system.

Definition 28. *Let $\Sigma := \{F_i \mid i \in I\}$ be a finite set of (names of) functors, and $\mathcal{R} := \{\alpha_j : w_j \to w'_j \mid w_j, w'_j \in \Sigma^*, j \in J\}$ be a finite set of (names of) natural transformations. We call $(\Sigma, \mathcal{R})$ a* **functor rewriting system** (**FRS**)*.*

The name "functor rewriting system" is motivated by seeing each natural transformation $(\alpha : w \to w') \in \mathcal{R}$ as a rewrite rule on strings of functors in Σ^*. For all functor strings $w_0, w_1 \in \Sigma^*$, the natural transformation $w_0\alpha w_1 : w_0ww_1 \to w_0w'w_1$ (sometimes called a *whiskering*) is seen as a rewrite step, with w_0 as left-context and w_1 as right-context. Note that the only valid rewrite steps are those resulting from natural transformations in $\mathcal{R}$. If the functors in Σ satisfy (semantic) identities like $FG = H$ that are not represented by some $\alpha \in \mathcal{R}$, then we do not allow rewrite steps that use this identity.

Remark 29. Kozen [19] introduced *rewrite categories* for applying rewriting concepts to categorical reasoning, including reasoning about monad compositions. A functor rewrite system $(\Sigma, \mathcal{R})$ is the rewrite category $(\Sigma^*, \mathcal{R})$. For Set-monads (S, μ^S, η^T) and (T, μ^T, η^T) and distributive law $\lambda\colon ST \to TS$, the FRS $\mathcal{R}^{sep}$ defined (below) in Definition 33 is the rewrite category $(\{S,T\}^*, \{\mu^S, \mu^T, \lambda\})$ viewed as a subcategory of the 2-category presented by $(\mathcal{O}, \mathcal{F}, \mathcal{R}, \mathcal{E})$ where $\mathcal{O} = \{\mathsf{Set}\}$, $\mathcal{F} = \{S, T\}$, $\mathcal{R} = \{\mu^S, \mu^T, \lambda\}$ and $\mathcal{E}$ consists of the Eq. (2) for μ^S and μ^T, and the distributive law axioms (10) and (11) involving λ and μ^S, μ^T. See also Sect. 7.1 for further discussion.

The functors and natural transformations in an FRS carry categorical structure in the form of commuting diagrams, allowing a variation of (local) confluence [19, Sect. 3.1].

Definition 30. *A functor rewriting system is (read $\circlearrowleft$ as "commuting")*

- WCR$\circlearrowleft$ *if for all $w_0 \xleftarrow{\alpha} w \xrightarrow{\beta} w_1$ there exists $T_0 \overset{\gamma}{\twoheadrightarrow} w' \overset{\delta}{\twoheadleftarrow} T_1$ s.t. $\gamma\alpha = \delta\beta$.*
- CR$\circlearrowleft$ *if for all $w_0 \overset{\alpha}{\twoheadleftarrow} w \overset{\beta}{\twoheadrightarrow} w_1$ there exists $w_0 \overset{\gamma}{\twoheadrightarrow} w' \overset{\delta}{\twoheadleftarrow} w_1$ s.t. $\gamma\alpha = \delta\beta$.*

There are equivalents to Newman's Lemma (Lemma 16) and the Critical Pair Lemma (Lemma 18). The proofs are in the extended version [30].

Lemma 31 (FRS Newman's lemma). *If a functor rewriting system is terminating* (SN) *and locally confluent-commuting* (WCR$\circlearrowleft$)*, then it is confluent-commuting* (CR$\circlearrowleft$)*.* ■

Lemma 32 (FRS critical pair lemma). *A functor rewriting system is locally confluent-commuting* (WCR$\circlearrowleft$) *if and only if all critical pairs converge with a commuting diagram.* ■

We use the following FRS for our separation procedure.

Definition 33. *We define a functor rewriting system* $\mathcal{R}^{sep} = (\Sigma, R)$*, where* $\Sigma := \{S, T\}$ *and* $R := \{\lambda : ST \to TS,\ \mu^S : SS \to S,\ \mu^T : TT \to T\}$.

Lemma 34. $\mathcal{R}^{sep}$ *is terminating* (SN) *and confluent-commuting* (CR↻). *Hence each functor string has a unique normal form in* $\mathcal{R}^{sep}$.

Proof. We show termination (SN) of $\mathcal{R}^{sep}$ using polynomial interpretation over $\mathbb{N}$. Let $[\![S]\!](x) := 2x + 1$ and $[\![T]\!](x) := x + 1$, which are indeed monotone in x. The three rewrite rules are strictly decreasing with respect to that order:

$$\begin{aligned} [\![ST]\!](x) &= 2x + 3 > 2x + 2 = [\![TS]\!](x), \\ [\![SS]\!](x) &= 4x + 3 > 2x + 1 = [\![S]\!](x), \\ [\![TT]\!](x) &= x + 2 > x + 1 = [\![T]\!](x). \end{aligned}$$

We now prove that $\mathcal{R}^{sep}$ is CR↻. Since we have termination (SN) it suffices to prove WCR↻ by Lemma 31. To invoke Lemma 32, we check that all critical pairs converge. Because we consider the objects purely syntactically as strings/words, we can enumerate all possible overlaps of left-hand sides of rules, giving rise to exactly 4 critical pairs, that indeed all converge:

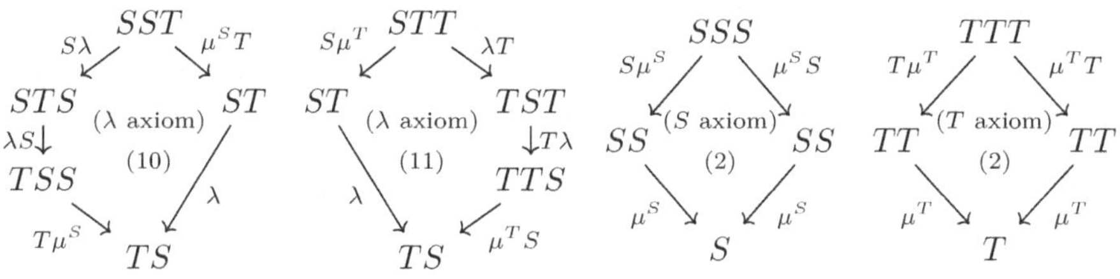

□

We now have the required tools to formalise the separation procedure and show that every term in $\mathbb{U}^\lambda$ can be separated. The first step is to define a function sep that maps a $\mathbb{U}^\lambda$-term u to a separated term $\mathsf{sep}(u)$.

Definition 35. *For* $u \in \Sigma^*_{\mathbb{U}^\lambda}$*, we define* $\mathsf{sep}(u)$ *as follows. Let* $w \in \{S, T\}^*$ *be such that* $\mathsf{type}(u) = w\mathcal{V}$*. Let* $\alpha\colon w \twoheadrightarrow w'$ *be a* $\mathcal{R}^{sep}$*-rewrite sequence to the unique normal form* w' *of* w *in* $\mathcal{R}^{sep}$*. By the axiom of choice, there is a choice function* $\rho_{w'\mathcal{V}}$ *that selects a term representative* $\rho_{w'\mathcal{V}}(c)$ *for each equivalence class* $c \in w'\mathcal{V}$*. We define* $\mathsf{sep}(u) := \rho_{w'\mathcal{V}}(\alpha_{\mathcal{V}}(\overline{u}^w))$.

Remark 36. In general, we need the Axiom of Choice to obtain sep. However, if the theory $\mathbb{S}$ and $\mathbb{T}$ can be oriented[5] to give terminating and confluent TRSs, then we can make ρ select the unique normal form making sep constructive.

Lemma 37. *For all* $u \in \Sigma^*_{\mathbb{U}^\lambda}$, $\mathsf{sep}(u)$ *is a well-defined, separated* $\mathbb{U}^\lambda$*-term and* $u =_{\mathbb{U}^\lambda} \mathsf{sep}(u)$.

[5] By *orientation*, we mean turning an equation $l = r$ into a rewrite rule, either from left to right $l \to r$ or right to left $l \leftarrow r$.

Proof. To see that $\mathsf{sep}(u)$ is well defined, note that if α and β are rewrite sequences $w \twoheadrightarrow w'$ from $w = \mathsf{type}(u)$ to its normal form w', then by CR$\circlearrowleft$, we have $\alpha = \beta$.

To see that $\mathsf{sep}(u)$ is separated, note that the normal form w' is equal to TS, T or S, since any other string will contain a reducible expression (redex). Hence $\alpha_{\mathcal{V}}(\overline{u}^{w}) \in TS\mathcal{V}, T\mathcal{V}$ or $S\mathcal{V}$, so any representative selected by $\rho_{w'\mathcal{V}}$ is separated.

To see that $u =_{\mathbb{U}^\lambda} \mathsf{sep}(u)$, recall that $\alpha\colon w \twoheadrightarrow w'$ is composed of λ, μ^S and μ^T, possibly applied within a context. By substitution and congruence rules, it suffices to prove that for all terms u, u' (of compatible type), if $\overline{u'}^{TS} = \lambda(\overline{u}^{ST})$ then $u =_{\mathbb{U}^\lambda} u'$, and similarly for μ^S and μ^T. That is, representatives of the input are $\mathbb{U}^\lambda$-equal to representatives of the output. For λ, this holds by definition of E_λ. For μ^S, if $u \in \overline{s[\overline{s_x}^{\mathbb{S}}]}^{\mathbb{S}}$ then $u =_{\mathbb{U}^\lambda} s[s_x]$, and if $u' \in \overline{s[s_x]}^{\mathbb{S}}$ then $u' =_{\mathbb{U}^\lambda} s[s_x]$. Hence by transitivity, $u =_{\mathbb{U}^\lambda} u'$. Similarly for μ^T. □

Lemma 38. *For all* $\mathbb{S}$*-terms* s, $\mathsf{sep}(s) =_{\mathbb{S}} s$, *for all* $\mathbb{T}$*-terms* t, $\mathsf{sep}(t) =_{\mathbb{T}} t$, *and for any separated term* $t[s_x/x]$, $\mathsf{sep}(t[s_x/x])$ *is equal to* $t[s_x/x]$ *modulo* $(\mathbb{S}, \mathbb{T})$.

Proof. For an $\mathbb{S}$-term s, we have $\mathsf{type}(s) = S\mathcal{V}$ and $\overline{s}^{S} = \overline{s}^{\mathbb{S}}$. By definition $\mathsf{sep}(s) = \rho_{S\mathcal{V}}(\overline{s}^{\mathbb{S}})$ is a representative of $\overline{s}^{\mathbb{S}}$, hence $\mathsf{sep}(s) =_{\mathbb{S}} s$. The arguments for $\mathbb{T}$-terms and for separated terms $t[s_x/x]$ are similar. □

We now apply Lemma 37 to show that any two separated terms that are equal in $\mathbb{U}^\lambda$, are equal modulo $(\mathbb{S}, \mathbb{T})$.

Lemma 39. *Any two separated terms equal in* $\mathbb{U}^\lambda$ *are equal modulo* $(\mathbb{S}, \mathbb{T})$.

Proof. Suppose two separated terms $t_0[s_x/x]$ and $t'_0[s'_y/y]$ are equal in $\mathbb{U}^\lambda$. Let $\mathfrak{T}$ be a $\mathbb{U}^\lambda$-derivation tree of this equality $t_0[s_x/x] =_{\mathbb{U}^\lambda} t'_0[s_y/y]$ in equational logic. By an induction on the structure of $\mathfrak{T}$, we prove that for each equation $u = u'$ in $\mathfrak{T}$, $\mathsf{sep}(u)$ and $\mathsf{sep}(u')$ are equal modulo $(\mathbb{S}, \mathbb{T})$. By Lemma 38 and transitivity of equality modulo $(\mathbb{S}, \mathbb{T})$, we then conclude that $t_0[s_x/x]$ and $t'_0[s'_y/y]$ are equal modulo $(\mathbb{S}, \mathbb{T})$.

The base cases are the Axiom and Reflexivity rules. The induction steps are the Symmetry, Transitivity, Congruence, and Substitution rules. We show only the cases of Congruence and Substitution here, as these are the only interesting cases. The full proof is in the extended version [30].

– Congruence: Given $\mathsf{op}^{(n)} \in \Sigma_{\mathbb{U}^\lambda}$, consider $\dfrac{u_1 = u'_1 \qquad \dots \qquad u_n = u'_n}{\mathsf{op}(u_1, \dots, u_n) = \mathsf{op}(u'_1, \dots, u'_n)}$
  Let $t_i[s_i] := \mathsf{sep}(u_i)$ and $t'_i[s'_i] := \mathsf{sep}(u'_i)$ for $i = 1, \dots, n$. The IH is that $\overline{t_i[\overline{s_i}^{\mathbb{S}}]}^{\mathbb{T}} = \overline{t'_i[\overline{s'_i}^{\mathbb{S}}]}^{\mathbb{T}}$. We consider the cases in which op is a $\mathbb{T}$-symbol or an $\mathbb{S}$-symbol separately.

- Suppose $\mathsf{op} \in \Sigma_{\mathbb{T}}$. Here is a sketch of the reasoning:

$$\begin{aligned}
\overline{\mathsf{sep}(\mathsf{op}(u_1, \ldots, u_n))}^{TS} &= \mu^T_{S\mathcal{V}}\left(\overline{\mathsf{op}(\overline{t_1[\overline{s_1}^{\mathbb{S}}]}^{\mathbb{T}}, \ldots, \overline{t_n[\overline{s_n}^{\mathbb{S}}]}^{\mathbb{T}})}^{\mathbb{T}}\right) \\
&= \mu^T_{S\mathcal{V}}\left(\overline{\mathsf{op}(\overline{t'_1[\overline{s'_1}^{\mathbb{S}}]}^{\mathbb{T}}, \ldots, \overline{t'_n[\overline{s'_n}^{\mathbb{S}}]}^{\mathbb{T}})}^{\mathbb{T}}\right) && \text{by IH} \\
&= \overline{\mathsf{sep}(\mathsf{op}(u'_1, \ldots, u'_n))}^{TS}.
\end{aligned}$$

  The first and third equalities are intuitively clear. The details can be found in the extended version [30].
- Suppose $\mathsf{op} \in \Sigma_{\mathbb{S}}$. Here is a sketch of the reasoning:

$$\begin{aligned}
\overline{\mathsf{sep}(\mathsf{op}(u_1, \ldots, u_n))}^{TS} &= T\mu^S_{\mathcal{V}} \cdot \lambda_{\mathcal{V}}\left(\overline{\mathsf{op}(\overline{t_1[\overline{s_1}^{\mathbb{S}}]}^{\mathbb{T}}, \ldots, \overline{t_n[\overline{s_n}^{\mathbb{S}}]}^{\mathbb{T}})}^{\mathbb{S}}\right) \\
&= T\mu^S_{\mathcal{V}} \cdot \lambda_{\mathcal{V}}\left(\overline{\mathsf{op}(\overline{t'_1[\overline{s'_1}^{\mathbb{S}}]}^{\mathbb{T}}, \ldots, \overline{t'_n[\overline{s'_n}^{\mathbb{S}}]}^{\mathbb{T}})}^{\mathbb{S}}\right) && \text{by IH} \\
&= \overline{\mathsf{sep}(\mathsf{op}(u'_1, \ldots, u'_n))}^{TS}.
\end{aligned}$$

  The first and third equalities are intuitively clear. The details can be found in the extended version [30].

– Substitution: Given a substitution f, consider $\dfrac{u = u'}{u[f] = u'[f]}$.
Let $t[s_x] := \mathsf{sep}(u)$ and $t'[s'_x] := \mathsf{sep}(u')$. The IH is that $\overline{t[\overline{s_x}^{\mathbb{S}}]}^{\mathbb{T}} = \overline{t'[\overline{s'_x}^{\mathbb{S}}]}^{\mathbb{T}}$. We start by separating all terms in the image of f. This gives another substitution $g := \mathsf{sep} \cdot f$. We denote $t_y[s_z] := g(y)$ for all $y \in \mathsf{var}(u_1) \cup \mathsf{var}(u_2)$. Here is a sketch of the reasoning:

$$\begin{aligned}
\overline{\mathsf{sep}(u[f])}^{TS} &= \mu^{TS}(\overline{t[s_x[t_y[s_z]]]}^{TSTS}) \\
&= \mu^{TS}(\overline{t'[s'_x[t_y[s_z]]]}^{TSTS}) && \text{by IH} \\
&= \overline{\mathsf{sep}(u'[f])}^{TS}.
\end{aligned}$$

The first and third equalities are intuitively clear. The details can be found in the extended version [30].

□

The proof of Theorem 24 now follows from Lemmas 37 and 39.

The next theorem was given in Zwart's thesis [35, Theorem 3.9] but not published elsewhere. We have updated the reasoning and obtained a much shorter proof using the shortcut $\mathbf{EM}(TS) \cong_{\mathsf{conc}} \mathbf{Alg}(\lambda)$.

Theorem 40. *Let S and T be the free algebra monads of algebraic theories $\mathbb{S}$ and $\mathbb{T}$. If there is a distributive law $\lambda : ST \Rightarrow TS$, then the monad $(TS, \eta^T\eta^S, \mu^T\mu^S \cdot T\lambda S)$ is presented algebraically by $\mathbb{U}^\lambda$.* ■

6 Axiomatisations of Composite Theories

In Theorem 24, we showed how to obtain an algebraic presentation $\mathbb{U}^\lambda$ of the composite monad arising from a distributive law $\lambda : ST \to TS$. However, the set of equations E_λ accounting for the interactions between $\mathbb{S}$- and $\mathbb{T}$-terms is maximal in the sense that it contains all possible equations that consist of representatives of some pair $(u, \lambda(u))$ in the graph of λ. In practice, we would like to have a minimal description of E_λ, such as the one for Ring in Example 22, which only adds two distribution axioms to the theories of monoids and Abelian groups.

In this section, we identify criteria on the shape of axioms that allow us to prove that certain minimal subsets of E_λ suffice to generate the whole of E_λ. We apply term rewriting methods for proving the necessary claims.

The shape of axioms will be described in terms of *layers*.

Definition 41. *Let $\mathbb{S}$ and $\mathbb{T}$ be two algebraic theories. Given a term $s[t_x/x] \in \Sigma_{\mathbb{S}}^* \Sigma_{\mathbb{T}}^* V$, its $ST-$**layers** are described by the pair (m, n) of natural numbers where $m := \mathsf{depth}(s)$ and $n := \max\{\mathsf{depth}(t_x) \mid x \in \mathsf{var}(s)\}$, where depth denotes the maximal number of nested (possibly nullary) operation symbols. This corresponds to the inductively defined notion of depth of term trees where constants have depth 1, and variables depth 0. TS-layers are defined similarly for terms in $\Sigma_{\mathbb{T}}^* \Sigma_{\mathbb{S}}^* V$.*

Example 42. We illustrate ST-layers in Ring (where $\mathbb{S} = \mathsf{Mon}, \mathbb{T} = \mathsf{AbGrp}$).

ST-Layers	$(0,0)$	$(0,1)$	$(1,0)$	$(1,1)$	$(0,2)$	$(2,0)$
Examples	x	0	1	$x \cdot 0$	$x + 0$	$x \cdot 1$
	y	$x + y$	$x \cdot y$	$(x + y) \cdot (y + z)$	$(x + y) + z$	$x \cdot (y \cdot z)$

For the remainder of this section, we assume that $\mathbb{S}$, $\mathbb{T}$, λ, $E_{\mathbb{S}}$, $E_{\mathbb{T}}$, E_λ, and $\mathbb{U}^\lambda$ are as in Theorem 24.

Lemma 43. *For all $E' \subseteq E_\lambda$ such that for each $f^{(n)} \in \mathbb{S}$, $g^{(m)} \in \mathbb{T}$ and each $i \in \{1, \ldots, n\}$, E' contains one equation of the form $l = r$, where $l = f(x_1, \ldots, x_{i-1}, g(\vec{y}), x_{i+1}, \ldots, x_n)$ and $r \in \lambda_V(\overline{\overline{l}^{\mathbb{T}}}^{\mathbb{S}})$, if the TRS $(\Sigma_{\mathbb{U}^\lambda} = \Sigma_{\mathbb{S}} \uplus \Sigma_{\mathbb{T}}, E')$ is terminating, then $E_{\mathbb{S}} \cup E_{\mathbb{T}} \cup E'$ generates the same congruence on $\mathbb{U}^\lambda$-terms as $E_{\mathbb{S}} \cup E_{\mathbb{T}} \cup E_\lambda$.* ■

Proof. Let us show why $(\Sigma_{\mathbb{U}^\lambda} = \Sigma_{\mathbb{S}} \uplus \Sigma_{\mathbb{T}}, E')$ is a TRS. First, no left-hand side is a variable by definition of E'. Second, $A := \mathsf{var}(s[t_x]) \supseteq \mathsf{var}(t[s_y])$ holds for all $(s[t_x], t[s_y]) \in E'$. This is the case since $\lambda_A : STA \to TSA : \overline{s[\overline{t_x}^{\mathbb{T}}]}^{\mathbb{S}} \mapsto \overline{t[\overline{s_y}^{\mathbb{S}}]}^{\mathbb{T}}$ forces the equivalence class of $t[s_y]$ to be in TSA and therefore to only use the variables in A.

Now let us argue why the congruence relation is left unchanged. Take an equation $(u, u') \in E_\lambda \cup E_{\mathbb{S}} \cup E_T$. The goal is to obtain this equation using only $E_{\mathbb{S}} \cup E_{\mathbb{T}} \cup E'$.

- First, using only equations in E', the $\mathbb{U}^\lambda$-terms u and u' can be separated. Indeed, we assume that the TRS $(\Sigma_{\mathbb{U}^\lambda}, E')$ is terminating, thus both u and u' can be rewritten to normal forms. The equations E' are exhaustive in the following sense: every term containing a $\Sigma_{\mathbb{T}}$-symbol below an $\Sigma_{\mathbb{S}}$-symbol is reducible (not in normal form). Thus the normal forms of u and u' must be in $\Sigma_{\mathbb{T}}^*\Sigma_S^*\mathcal{V}$, i.e., separated. Let us denote them $t[s_x/x]$ and $t'[s'_y/y]$.
- Since $\mathbb{U}^\lambda$ is a composite theory (proven in Theorem 24), and the separated normal forms $t[s_x/x]$ and $t'[s'_y/y]$ are $\mathbb{U}^\lambda$-equal, they must also be equal modulo $(\mathbb{S}, \mathbb{T})$. By equality modulo $(\mathbb{S}, \mathbb{T})$, we have a proof of $t[s_x/x] = t'[s'_y/y]$ using only equations from $E_{\mathbb{S}}$ and $E_{\mathbb{T}}$ (explicitly so when using the equivalent formulation (4) of equality modulo $(\mathbb{S}, \mathbb{T})$ in [35, Prop. 3.4]). □

In order to obtain an E' for Lemma 43, one can take equations of the form $l = \mathsf{sep}(l)$, but Lemma 43 also applies to other choices of r. As mentioned in Remark 36, if the theories $\mathbb{S}$ and $\mathbb{T}$ can be oriented to obtain a confluent and terminating TRS, then $\mathsf{sep}(l)$ can be chosen to be a normal form. For example, in [26], the theory of left-zero monoids and the theory with a unary idempotent operation were both oriented, allowing for a practical presentation of the composite theory that the authors called CUT.

Example 44. Let us retrieve the axiomatisation of Ring as given in Example 22, but starting from its corresponding distributive law $\lambda : L\mathcal{A} \to \mathcal{A}L$ [5, §4]. The set E will only contain equations whose left-hand side is among $(x+y)z$, $x(y+z)$, $0 \cdot x$, $x \cdot 0$, $(-x)y$, and $x(-y)$. For each of those, there are infinitely many choices for the right-hand side. For instance $(x \cdot 0, 0)$, $(x \cdot 0, 0+0)$, etc. Thankfully, there is an easy choice for the right-hand side r, because the theory Mon can be oriented, $(xy)z \to x(yz)$, $1 \cdot x \to x$, and $x \cdot 1 \to x$, as can the theory AbGrp without the commutativity axiom. Not taking the commutativity axiom into account simply means that we have to choose one equation between $((x+y)z,\ xz+yz)$ and $((x+y)z,\ yz+xz)$. We end up with 6 equations:

$$
\begin{aligned}
(x+y)z &= xz+yz, & x \cdot 0 &= 0, & (-x)y &= -(xy),\\
z(x+y) &= zx+zy, & 0 \cdot x &= 0, & x(-y) &= -(xy).
\end{aligned}
$$

Reducing from 6 to only the 2 equations of left and right distributivity can be done using automated tools. In our case, we used Prover9 [23] and obtained the result instantaneously [30, §8.7].

Note that if $E' \subseteq E_\lambda$ is not terminating, then the conclusion is not guaranteed to hold. The example below exhibits a situation where the set E' of equations as defined in Lemma 43 is not enough to generate all of the E_λ equations.

Example 45. We show that the subset of equations of E_λ where all left-hand sides have layers $(1, 1)$ is not always sufficient (together with $E_{\mathbb{S}}$ and $E_{\mathbb{T}}$) to generate all E_λ equations obtained from a distributive law λ. This example is an extension of the well-known non-terminating TRS $ab \to bbaa$ [32, Ex.2.3.9].

Consider the theories $\mathbb{S}$ and $\mathbb{T}$, with signatures $\Sigma_{\mathbb{S}} := \{a^{(1)}\}$ and $\Sigma_{\mathbb{T}} := \{b^{(1)}\}$, and equations $E_{\mathbb{S}} := \{aaa = aa\}$ and $E_{\mathbb{T}} := \{bbb = bb\}$. We use some string rewriting notations, such as aax or a^2x as shorthand for $a(a(x))$, etc. The set of equivalence classes of $\mathbb{S}$ is $S\mathcal{V} = \{\overline{a^2x}^{\mathbb{S}}, \overline{ax}^{\mathbb{S}}, \overline{x}^{\mathbb{S}} \mid x \in \mathcal{V}\}$. Similarly, $T\mathcal{V} = \{\overline{b^2x}^{\mathbb{T}}, \overline{bx}^{\mathbb{T}}, \overline{x}^{\mathbb{T}} \mid x \in \mathcal{V}\}$. We define a mapping

$$\begin{aligned}
&\lambda\colon ST\mathcal{V} \to TS\mathcal{V} \\
&\overline{a^n\overline{b^m x}^{\mathbb{T}}}^{\mathbb{S}} \mapsto \overline{b^2\overline{a^2x}^{\mathbb{S}}}^{\mathbb{T}}, \quad \text{for } n, m \in \{1, 2\} \\
&\overline{a^n\overline{x}^{\mathbb{T}}}^{\mathbb{S}} \mapsto \overline{\overline{a^n x}^{\mathbb{S}}}^{\mathbb{T}}, \quad \text{for } n \in \{1, 2\} \\
&\overline{\overline{b^n x}^{\mathbb{T}}}^{\mathbb{S}} \mapsto \overline{b^n\overline{x}^{\mathbb{S}}}^{\mathbb{T}}, \quad \text{for } n \in \{1, 2\} \\
&\overline{\overline{x}^{\mathbb{T}}}^{\mathbb{S}} \mapsto \overline{\overline{x}^{\mathbb{S}}}^{\mathbb{T}}
\end{aligned}$$

We show that λ is a distributive law:

- Unit law (8): $\lambda_{\mathcal{V}}(S\eta^T_{\mathcal{V}}(\overline{a^n x}^{\mathbb{S}})) = \lambda_{\mathcal{V}}(\overline{a^n\overline{x}^{\mathbb{T}}}^{\mathbb{S}}) = \overline{\overline{a^n x}^{\mathbb{S}}}^{\mathbb{T}} = \eta^T_{S\mathcal{V}}(\overline{a^n x}^{\mathbb{S}})$.
- Unit law (9): $\lambda_{\mathcal{V}}(\eta^S_{T\mathcal{V}}(\overline{b^n x}^{\mathbb{T}})) = \lambda_{\mathcal{V}}(\overline{\overline{b^n x}^{\mathbb{T}}}^{\mathbb{S}}) = \overline{b^n\overline{x}^{\mathbb{S}}}^{\mathbb{T}} = T\eta^S_{\mathcal{V}}(\overline{b^n x}^{\mathbb{T}})$.
- Multiplication law (10): We only show the case for $n, m, k \geqslant 1$. Other cases can be easily verified in a similar manner.

$$\begin{array}{ccccc}
\overline{a^n\overline{a^m\overline{b^k x}^{\mathbb{T}}}^{\mathbb{S}}}^{\mathbb{S}} \in SST & \xrightarrow{S\lambda} & \overline{a^n\overline{b^2\overline{a^2x}^{\mathbb{S}}}^{\mathbb{T}}}^{\mathbb{S}} \in STS & \xrightarrow{\lambda S} & \overline{b^2\overline{a^2\overline{a^2x}^{\mathbb{S}}}^{\mathbb{S}}}^{\mathbb{T}} \in TSS \\
\downarrow{\scriptstyle \mu^S T} & & & & \downarrow{\scriptstyle T\mu^S} \\
\overline{a^{n+m}\overline{b^k x}^{\mathbb{T}}}^{\mathbb{S}} \in ST & & \xrightarrow[\lambda]{\qquad\qquad} & & \overline{b^2\overline{a^2x}^{\mathbb{S}}}^{\mathbb{T}} = \overline{b^2\overline{a^4x}^{\mathbb{S}}}^{\mathbb{T}} \in TS
\end{array}$$

- Multiplication law (11): Analogous to the previous point.

From Theorem 24, defining the set E_λ of distributivity equations as below ensures that $E_{\mathbb{S}} \cup E_{\mathbb{T}} \cup E_\lambda$ is an axiomatization of the composite theory $\mathbb{U}^\lambda$.

$$\begin{aligned}
E_\lambda = \{&a^n b^m x = b^2a^2x \mid m, n \geqslant 1, x \in \mathcal{V}\} \cup \\
\{&a^n x = a^n x, b^n x = b^n x \mid n \in \{0, 1, 2\}, x \in \mathcal{V}\}
\end{aligned}$$

The subset of equations of E_λ that have left-hand side with ST-layers $(1, 1)$ is $E' = \{ab = b^2a^2\}$. However, we claim that $E_{\mathbb{S}} \cup E_{\mathbb{T}} \cup E'$ cannot derive all equations in E_λ. Indeed, we observe that the distributivity equation $aab =_{E_\lambda} bbaa$ cannot be derived. Trying to do so leaves us stuck in a loop: (we underline the part where an equation is applied)

$$\begin{aligned}
a\underline{ab} &=_{E'} \underline{ab}baa =_{E'} bba\underline{ab}aa =_{E'} bbabb\underline{aaaa} =_{E_{\mathbb{S}}} bb\underline{ab}baa \\
&=_{E'} \underline{bbbb}aabaa =_{E_{\mathbb{T}}} bbaabaa =_{E'} \ldots \text{ (loop)}
\end{aligned}$$

It is not hard to see that there are no other ways of proving $aab = bbaa$ in $E_{\mathbb{S}} \cup E_{\mathbb{T}} \cup E'$. Hence $E_{\mathbb{S}} \cup E_{\mathbb{T}} \cup E'$ does not generate the same congruence as $E_{\mathbb{S}} \cup E_{\mathbb{T}} \cup E_\lambda$. In line with Theorem 43, the above indeed also shows that E', when viewed as a TRS, is not terminating. Note that Theorem 43 only says that termination is a sufficient condition for a (1,1)-axiomatisation. It does not exclude that in some composite theories, the set of equations $E_{\mathbb{S}} \cup E_{\mathbb{T}} \cup E'$ might axiomatise $\mathbb{U}^\lambda$ even in presence of non-termination.

The next lemma identifies a class of equations where termination of the TRS $(\Sigma_{\mathbb{U}} = \Sigma_{\mathbb{S}} \uplus \Sigma_{\mathbb{T}}, E')$ is guaranteed. These are equations in which the right-hand sides have layers $(n, 1)$, which is inspired from similar results for string rewriting obtained by Zantema & Geser [34].

Lemma 46. *Let $\mathbb{S}$ and $\mathbb{T}$ be two algebraic theories. Let R be a set rules of the form $s[t_x/x] \to t[s_y/y]$. Let $Z = \{t_x \mid t_x$ is a variable$\}$, i.e., all $z \in Z$ occur directly below an $\mathbb{S}$-operation in $s[t_x/x]$. If each $s[t_x/x]$ has ST-layers $(1,1)$, each $t[s_y/y]$ has TS-layers $(n,1)$ for some n not fixed, and each s_y is linear*[6] *in Z, then R is terminating.* ■

Example 47. We give some axiomatisations of composite theories resulting from distributive laws in the literature:

1. Let $R(X) = X^A$ be the reader monad, with $A = \{a_1, \ldots, a_n\}$. There is a distributive law of the finite distribution monad $\mathcal{D}$ over R, $\lambda : \mathcal{D}R \to R\mathcal{D}$, that sends $p_1 h_1 + \ldots + p_n h_n$ to $(a \mapsto p_1 h_1(a) + \ldots + p_n h_n(a))$ [13, Example 1.34]. Recall that R is presented algebraically by a single operation $f^{(n)}$ with two equations Example 12, and $\mathcal{D}$ is presented by convex algebras. The distribution axioms as described in Lemma 43 are in our case, for each $p \in [0,1]$
$$f(x_1, \ldots, x_n) \oplus_p y = f(x_1 \oplus_p y, \ldots, x_n \oplus_p y).$$
$$x \oplus_p f(y_1, \ldots, y_n) = f(x \oplus_p y_1, \ldots, x \oplus_p y_n).$$
We see that the right-hand sides of these equations have layers $(1,1)$ and both equations satisfy the linearity requirement of Lemma 46, thus ensuring termination. Hence by Theorem 24 and Lemma 43, the above equations together with the equations for f and for convex algebras present the composite monad on $R\mathcal{D}$ induced by λ. Furthermore, we notice that each of the above equations can be derived from the other one using the axioms of convex algebras. Therefore, we only need to include one of them for each p.
2. There is a distributive law of multisets over distributions $\lambda\colon \mathcal{MD} \to \mathcal{DM}$ called the *parallel multinomial law* in [16], see also [9,11] and [13, Ex. 1.37]. It sends e.g. $\wr p x_1 + (1-p) x_2, y \wr$ to $p \wr x_1, y \wr + (1-p) \wr x_2, y \wr$, which can be expressed in the syntax of convex algebras and commutative monoids as
$$(x_1 \oplus_p x_2) \cdot y = (x_1 \cdot y) \oplus_p (x_2 \cdot y).$$
By Theorem 24, Lemma 43 and Lemma 46 these equations (one for each $p \in [0,1]$), together with the axioms of convex algebras and commutative monoids, present the composite monad on $\mathcal{DM}$ induced by λ.
3. There is a distributive law $\lambda\colon L^+L^+ \to L^+L^+$ for the non-empty list monad over itself [22]. It sends a list of lists to the singleton list containing the list of all heads: $[[a,b],[c],[d,e,f]] \mapsto [[a,c,d]]$. We get the following distributivity axioms for the composite theory:

[6] *Linear* in a TRS sense, i.e. variables appearing *at most* once.

$$a * (b \star c) = a * b$$
$$(a \star b) * c = a * c.$$

Again, the equations satisfy the conditions for Lemma 46, and our results imply that the above equations together with the semigroup axioms for $*$ and $\star$ present the composite monad on L^+L^+ induced by λ.

7 Conclusion

In this paper, we proved the correspondence between composite theories of $\mathbb{T}$ after $\mathbb{S}$ and distributive laws $\lambda\colon ST \to TS$. Furthermore, we gave sufficient criteria for when a minimal set $E' \subseteq E_\lambda$ of distribution equations, along with $E_\mathbb{S}$ and $E_\mathbb{T}$, axiomatises the composite theory.

The set E' itself is unlikely to turn many heads, as distributive laws are often informally described in the literature in terms of such simple distribution axioms. The surprise, however, comes from the fact that E' is not always enough (see Example 45). This is a possible pitfall similar to the 'simplicity' of the various false distributive laws of the powerset monad over itself [18].

7.1 Related Work

In Kozen's work on rewrite categories, he proves that distributive laws yield composite monads in [19, Sect. 4.2], by showing that crucial properties correspond to TS being a terminal object in the rewrite category with $\mu^S, \mu^T, \eta^S, \eta^T, \lambda$. However, we cannot apply these results to prove Theorem 24 since they do not involve composite theories. Another difference with Kozen's approach is that we do not include the monad units in $\mathcal{R}^{sep}$ (Definition 33). By omitting the units, we obtain unique normal forms in $\mathcal{R}^{sep}$ in the classic rewriting sense, but no terminal object in the corresponding rewrite category. This allows our reasoning to follow classic rewrite arguments more closely.

A result akin to Theorem 40 appears in the literature on polygraphs [2, 3.3.6 Theorem]. Polygraphs are generalisations of graphs that can serve as presentations of categories. The notion of distributive law between categories presented by polygraphs seems related to the notion of distributive law between Lawvere Theories as described by Cheng [8], but the precise connection is not explained in [2] and remains to be explored.

7.2 Future Work

There are several directions for future work. We showed that termination of E' (as TRS) is sufficient for $E_\mathbb{S} \cup E_\mathbb{T} \cup E'$ to axiomatise the composite theory (Lemma 43), and that taking equations in E' to have layers $(1, 1) \to (n, 1)$ ensures termination (Lemma 46). We would like to identify other criteria for termination, and make more use of term rewriting techniques. We speculate that one could allow layers $(1, 1) \to (2, 2)$ in which some symbol in the left-hand side is absent from the right-hand side in order to avoid problems such as in Example 45 with $ab \to bbaa$.

In light of negative results concerning monad compositions [10,18,33,36], there has been much interest in understanding the limits of monad composition. Positive results using algebraic methods were given in [9]. Another approach has been to generalise to so-called weak distributive laws [12,13]. Presentations of monads arising from the composition of monads via a weak distributive law, in particular monads for nondeterminism and probabilities, have been given in [6,14]. These presentations are obtained by adding a simple distribution axiom to the two underlying theories, similar to our results in Sect. 6, but the resulting theory is no longer a composite theory as the essential uniqueness modulo $(\mathbb{S}, \mathbb{T})$ is not guaranteed to hold. Another future line of work would be to extend the current correspondence to weak distributive laws [12,13] thereby giving a definition of *weak composite theories*. Such a correspondence would allow for a more thorough study of weak distributive laws on the algebraic level, and could perhaps lead to no-go theorems for weak distributive laws.

Alternatively, the current correspondence could also be extended to account for multi-sorted algebraic theories, and by such means defining *multi-sorted distributive law*.

Acknowledgments. We thank all anonymous reviewers for their valuable feedback and suggestions. Aloïs Rosset and Jörg Endrullis received funding from the Netherlands Organization for Scientific Research (NWO) under the Innovational Research Incentives Scheme Vidi (project. No. VI.Vidi.192.004).

References

1. Aczel, P., Adámek, J., Milius, S., Velebil, J.: Infinite trees and completely iterative theories: a coalgebraic view. Theoret. Comput. Sci. **300**(1–3), 1–45 (2003). https://doi.org/10.1016/S0304-3975(02)00728-4
2. Ara, D., Burroni, A., Guiraud, Y., Malbos, P., Métayer, F., Mimram, S.: Polygraphs: from rewriting to higher categories. CoRR abs/2312.00429 (2023). https://doi.org/10.48550/ARXIV.2312.00429
3. Awodey, S.: Category Theory. Oxford Logic Guides. Ebsco Publishing, Ipswich (2006)
4. Barr, M., Wells, C.: Toposes, Triples and Theories. Comprehensive Studies in Mathematics, Springer, New York (1985)
5. Beck, J.: Distributive laws. In: Eckmann, B. (ed.) Seminar on Triples and Categorical Homology Theory. LNM, vol. 80, pp. 119–140. Springer, Heidelberg (1969). https://doi.org/10.1007/BFb0083084
6. Bonchi, F., Sokolova, A., Vignudelli, V.: The theory of traces for systems with nondeterminism and probability. In: 34th Annual ACM/IEEE Symposium on Logic in Computer Science. LICS 2019, pp. 1–14. IEEE (2019). https://doi.org/10.1109/LICS.2019.8785673
7. Borceux, F.: Handbook of Categorical Algebra. 2, Encyclopedia of Mathematics and its Applications, vol. 51. Cambridge University Press, Cambridge (1994), categories and structures

8. Cheng, E.: Distributive laws for Lawvere theories. Compositionality (2020). https://doi.org/10.32408/compositionality-2-1
9. Dahlqvist, F., Parlant, L., Silva, A.: Layer by layer – combining monads. In: Fischer, B., Uustalu, T. (eds.) ICTAC 2018. LNCS, vol. 11187, pp. 153–172. Springer, Cham (2018). https://doi.org/10.1007/978-3-030-02508-3_9
10. Dahlqvist, F., Neves, R.: Compositional semantics for new paradigms: probabilistic, hybrid and beyond (2018). https://doi.org/10.48550/ARXIV.1804.04145
11. Dash, S., Staton, S.: A monad for probabilistic point processes. In: Spivak, D.I., Vicary, J. (eds.) Proceedings of the 3rd Annual International Applied Category Theory Conference 2020, Cambridge, USA, 6–10 July 2020. Electronic Proceedings in Theoretical Computer Science, vol. 333, pp. 19–32. Open Publishing Association (2021). https://doi.org/10.4204/EPTCS.333.2
12. Garner, R.: The Vietoris monad and weak distributive laws. Appl. Categ. Struct. **28**(2), 339–354 (2020). https://doi.org/10.1007/s10485-019-09582-w
13. Goy, A.: On the compositionality of monads via weak distributive laws. (Compositionnalité des monades par lois de distributivité faibles). Ph.D. thesis, University of Paris-Saclay, France (2021). https://tel.archives-ouvertes.fr/tel-03426949
14. Goy, A., Petrisan, D.: Combining probabilistic and non-deterministic choice via weak distributive laws. In: Hermanns, H., Zhang, L., Kobayashi, N., Miller, D. (eds.) LICS '20: 35th Annual ACM/IEEE Symposium on Logic in Computer Science, pp. 454–464. ACM (2020). https://doi.org/10.1145/3373718.3394795
15. Jacobs, B.: Convexity, duality and effects. In: Calude, C.S., Sassone, V. (eds.) TCS 2010. IAICT, vol. 323, pp. 1–19. Springer, Heidelberg (2010). https://doi.org/10.1007/978-3-642-15240-5_1
16. Jacobs, B.: From multisets over distributions to distributions over multisets. In: Proceedings of the 36th Annual ACM/IEEE Symposium on Logic in Computer Science. LICS '21. Association for Computing Machinery, New York, NY, USA (2021). https://doi.org/10.1109/LICS52264.2021.9470678
17. Jacobs, B., Silva, A., Sokolova, A.: Trace semantics via determinization. J. Comput. Syst. Sci. **81**(5), 859–879 (2015). https://doi.org/10.1016/j.jcss.2014.12.005
18. Klin, B., Salamanca, J.: Iterated covariant powerset is not a monad. In: Staton, S. (ed.) Proceedings of the Thirty-Fourth Conference on the Mathematical Foundations of Programming Semantics. MFPS 2018. Electronic Notes in Theoretical Computer Science, vol. 341, pp. 261–276. Elsevier (2018). https://doi.org/10.1016/j.entcs.2018.11.013
19. Kozen, D.: Natural transformations as rewrite rules and monad composition. Log. Methods Comput. Sci. **15**(1) (2019). https://doi.org/10.23638/LMCS-15(1:1)2019
20. MacLane, S.: Categories for the Working Mathematician. Graduate Texts in Mathematics, vol. 5. Springer, New York (1971). https://doi.org/10.1007/978-1-4757-4721-8
21. Manes, E.: Algebraic Theories. Graduate Texts in Mathematics, vol. 26. Springer, New York (1976). https://doi.org/10.1007/978-1-4612-9860-1
22. Manes, E., Mulry, P.: Monad compositions. i: general constructions and recursive distributive laws. Theory Appl. Categ. **18**, 172–208 (2007)
23. McCune, W.: Release of prover9. In: Mile High Conference on Quasigroups, Loops and Nonassociative Systems, Denver, Colorado (2005)
24. Moggi, E.: Notions of computation and monads. Inf. Comput. **93**(1), 55–92 (1991). https://doi.org/10.1016/0890-5401(91)90052-4, selections from 1989 IEEE Symposium on Logic in Computer Science

25. Parlant, L.: Monad composition via preservation of algebras. Ph.D. thesis, University College London, UK (2020). https://ethos.bl.uk/OrderDetails.do?uin=uk.bl.ethos.819930
26. Piróg, M., Staton, S.: Backtracking with cut via a distributive law and left-zero monoids. J. Funct. Program. **27**, e17 (2017). https://doi.org/10.1017/S0956796817000077
27. Plotkin, G., Power, J.: Notions of computation determine monads. In: Nielsen, M., Engberg, U. (eds.) FoSSaCS 2002. LNCS, vol. 2303, pp. 342–356. Springer, Heidelberg (2002). https://doi.org/10.1007/3-540-45931-6_24
28. Riehl, E.: Category Theory in Context. Dover Modern Math Originals, Dover Publications, Aurora (2017)
29. Rosset, A., Hansen, H.H., Endrullis, J.: Algebraic presentation of semifree monads. In: Hansen, H.H., Zanasi, F. (eds.) CMCS 2022. LNCS, vol. 13225, pp. 110–132. Springer, Cham (2022). https://doi.org/10.1007/978-3-031-10736-8_6
30. Rosset, A., Zwart, M., Hansen, H.H., Endrullis, J.: Correspondence between composite theories and distributive laws (2024). https://doi.org/10.48550/ARXIV.2404.00581
31. Schneider-Kamp, P., Thiemann, R., Annov, E., Codish, M., Giesl, J.: Proving termination using recursive path orders and SAT solving. In: Konev, B., Wolter, F. (eds.) FroCoS 2007. LNCS (LNAI), vol. 4720, pp. 267–282. Springer, Heidelberg (2007). https://doi.org/10.1007/978-3-540-74621-8_18
32. Terese: Term Rewriting Systems, Cambridge Tracts in Theoretical Computer Science, vol. 55. Cambridge University Press, Cambridge (2003)
33. Varacca, D., Winskel, G.: Distributing probabililty over nondeterminism. Math. Struct. Comput. Sci. **16**, 87–113 (2006). https://doi.org/10.1017/S0960129505005074
34. Zantema, H., Geser, A.: A complete characterization of termination of $0^p1^q -> 1^r0^s$. Appl. Algebra Eng. Commun. Comput. **11**(1), 1–25 (2000). https://doi.org/10.1007/S002009900019
35. Zwart, M.: On the non-compositionality of monads via distributive laws. Ph.D. thesis, Department of Computer Science, University of Oxford (2020)
36. Zwart, M., Marsden, D.: No-go theorems for distributive laws. Log. Methods Comput. Sci. **18** (2022). https://doi.org/10.46298/lmcs-18(1:13)2022

Author Index

Published by Springer Nature Switzerland AG 2024
B. König and H. Urbat (Eds.): CMCS 2024, LNCS 14617, p. 217, 2024.
https://doi.org/10.1007/978-3-031-66438-0

SPRINGER NATURE

GPSR Compliance

The European Union's (EU) General Product Safety Regulation (GPSR) is a set of rules that requires consumer products to be safe and our obligations to ensure this.

If you have any concerns about our products, you can contact us on ProductSafety@springernature.com

In case Publisher is established outside the EU, the EU authorized representative is:

Springer Nature Customer Service Center GmbH
Europaplatz 3
69115 Heidelberg, Germany

The manufacturer's authorised representative in the EU is Springer Nature Customer Service Centre GmbH, Europaplatz 3, 69115 Heidelberg, Germany. If you have any concerns regarding our products, please contact ProductSafety@springernature.com

Printed and bound by CPI Group (UK) Ltd, Croydon, CR0 4YY
26/03/2026
02078975-0002